SCIENCE 1

Teacher Edition

Fourth Edition

bju press®

Greenville, South Carolina

Note: The fact that materials produced by other publishers may be referred to in this volume does not constitute an endorsement of the content or theological position of materials produced by such publishers. Any references and ancillary materials are listed as an aid to the student or the teacher and in an attempt to maintain the accepted academic standards of the publishing industry.

SCIENCE 1 Teacher Edition
Fourth Edition

Coordinating Writer
Leah R. Thomas, MEd

Writers
Rosanna Capellan, MS
Debra Harrold White

Biblical Worldview
Bryan Smith, PhD
Tyler Trometer, MDiv

Academic Oversight
Jeff Heath, EdD

Project Editors
Kristin Kowalk, MA
Joanna Lynch, MA

Consultant
L. Michelle Rosier

Concept Designers
Josh Frederick
Emily Heinz

Designer
Emily Heinz

Cover Designer
Emily Heinz

Cover Illustrator
Dana Thompson

Illustrators
Emily Heinz
Jack Pullan
Dana Thompson

Page Layout
Faith Bupe Mazunda

Project Coordinators
Tony Every
Jennifer Ferguson

Permissions
Sharon Belknap
Sylvia Gass
Carrie Hanna
Ashleigh Schieber
Kathleen Thompson

All trademarks are registered and unregistered marks of their respective owners. BJU Press is in no way affiliated with these companies. No rights are granted by BJU Press to use such marks, whether by implication, estoppel, or otherwise.

Teacher Edition photo credits: **viiit, xiiitr** Lambros Kazan/Shutterstock.com; **viiib, xiiibr** Apollo 14 Shepard/Nasa/Wikimedia Commons/Public Domain; **ix** BJU Press; **xiit** images72/Shutterstock.com; **xiib** "George Washington Carver c1910"/Wikimedia Commons/Public Domain; **xiiil** Evgeny Atamanenko/Shutterstock.com

Student Text photo credits begin on p. A84.

ISBN: 978-1-62856-316-0

15 14 13 12 11 10 9 8 7 6 5 4

Contents

Biblical Worldview Shaping `BWS`

Everyone has a worldview. Simply put, everyone has a way of viewing the world. It is a set of assumptions and beliefs growing out of a big story of the world used to interpret everything that happens. Some base their worldview on God's Word, while others base theirs on mankind's own words and philosophies. Science is one of the most hotly contested areas of human culture. Many people today do not believe in God's existence and try to explain everything from a naturalistic point of view.

In order to think about first-grade science correctly, a student must understand where our world came from, why things work together the way they do, who we are and why we are here, and why science is important. These are essential topics for first-grade students to understand and internalize as they make application to real-world situations.

Where did our world come from? (Origins)

God created a perfect world. He made the earth, sun, moon, and stars. He made plants, animals, and the first people. Some books teach that all things came into being without God—that the universe formed as a result of a big bang and that chaos over time led to order. The Bible tells us that God made all things in the heavens and on the earth in six days only a few thousand years ago. The Bible also tells us that we are to praise Him and thank Him for making the world and for making us.

Why do things work together the way they do? (Design in nature)

The world we study in science is very well designed. God designed each part to fit together with other parts, like gears in a machine. God is a marvelous Engineer and Artist.

Some books state that there is no evidence that God designed a perfect world. They claim that chaos became order. They deny that the pain, suffering, and death in this world are the result of mankind's sin and the Fall.

Those who believe the Bible know that, because of sin, everything in this world does not work perfectly together. The Bible teaches that God will one day restore this world. When that time comes, everything will work together as God designed it to. God's perfect design will be everywhere, and there will be no pain, suffering, or death. Jesus, God's Son, will rule on earth, and everyone who has turned away from sin and trusted in Him will live with God forever.

Who are we? Why are we here? (Human element)

Humans are unique in this world. While some books teach that people are just like animals, the Bible states that God made humans, the most important part of His creation, in His own image or likeness. God has called people to have dominion, or rule, over this world. We are to study this world and figure out ways to use it for God's glory and for the good of others.

The Bible teaches that people are to love God more than anything or anyone else because He made us and takes care of us. We are to love others because they are made in God's image. For this reason, we are to treat each other with love and respect. In all of life, we are to try to be like Jesus, God's Son, who died on the cross to save us from our sin. We should rule over and care for the earth, because it is God's gift to us.

Why is science important? (Purpose of science)

God's first command to humans was for them to have dominion over, or to manage, the earth. Science is a useful tool. We study science to learn how nature works so we can make changes in nature to help, and not hurt, people. Science is a tool to help people rule over and care for this world.

Some books teach that we can use science to answer all of life's important questions. They teach that science is the only way to know something for sure. These books teach that science proves the Bible is wrong and that there is no God.

While science can be a powerful tool to help us answer questions, it is only a tool. It is good at answering some questions, such as What do plants need to grow?, or What causes seasons? But there are other important questions that science cannot answer. Questions like, Who is God?, or Where did this world come from? These are some of the most important questions in life, and science cannot answer them. But these questions can be answered in the Bible. Whatever the Bible says is true, and whenever science disagrees with the Bible, it is science, not the Bible, that is wrong.

Goals

Develop a knowledge of God

▶ Inculcate an understanding of God as the Creator and Sustainer of the universe (Colossians 1:16–17)

▶ Identify the precision of God's creation (Ecclesiastes 3:1–8)

▶ Inspire curiosity, wonder, and appreciation of God's creation (Psalm 19:1)

Encourage Christian growth

▶ Promote good stewardship of God's creation (Genesis 2:15)

▶ Promote Christ-honoring character through cooperative group activities (2 Peter 1:5–8)

▶ Challenge students to use science to show the love of Christ to others (Matthew 22:39–40)

Promote scientific literacy

▶ Establish foundational science facts and skills for further science learning

▶ Balance presentation of information with active participation

▶ Teach the processes involved in the scientific method

▶ Promote disciplined and orderly approaches to problem solving

▶ Show the integration of science into the everyday lives of the students

▶ Effect positive attitudes toward science through active participation and relevant discussions

Instructional Materials

1 Student Edition

The *Science 1* Student Edition provides grade-appropriate information through text, diagrams, and annotated photographs and illustrations. The Student Edition also includes a Big Question at the start of each chapter, interest boxes, and Quick Check questions, as well as background information for each Investigation, Exploration, and STEM Activity.

2 Activities

The *Science 1* Activities provides a variety of pages to aid understanding. Study Guide pages provide a systematic review of concepts. Investigation, Exploration, and STEM Activity lessons apply the scientific method and reinforce the basic science process skills. Other pages help reinforce and expand on the concepts taught in the lessons.

Answers can be found in the *Science 1* Activities Answer Key.

3 Teacher Edition

The *Science 1* Teacher Edition contains 90 lessons. Lessons include additional background information, helps, looking ahead information, and additional science activities for enrichment.

The Teacher Edition also includes useful information about science process skills, the management of activities, and grading. Information about sharing the gospel with children and a Worldview Scope and Sequence are located in the Teacher Resources in the back of the Teacher Edition.

4 Assessments

The *Science 1* Assessments includes eleven chapter tests. Tests include objective items, the use and labeling of diagrams, and application questions. Assessments also includes a rubric tailored for each Investigation, Exploration, and STEM Activity.

An answer key for the tests is available separately.

New to This Edition

If you have used previous editions of BJU Press *Science 1*, you may notice some changes made in the 4th Edition.

- The Student Edition is divided into five units: Let's Learn About Science, Let's Learn About Living Things, Let's Learn About Our Bodies, Let's Learn About Earth and Space, and Let's Learn About Energy. Because humans are different and distinct from other living things, the human body is discussed in its own unit. Lesson numbers appear on the first page of each new lesson. Each chapter begins with a Big Question, to focus attention on a main concept. There are pages to engage the students before each Investigation, Exploration, and STEM lesson. The Investigation and Exploration lessons identify the process skills that are emphasized. STEM activities and STEM careers are designed to build twenty-first century skills and to develop interest in science, technology, engineering, and math. Interest boxes provide additional information about a person or topic. The Glossary and Index have been expanded to include more than the bold-faced vocabulary terms. The Glossary includes illustrations of some terms. Some chapters have been added, while others have been revised or removed to better meet the course objectives.

- The Activities book has additional graphic organizers and activities. Each chapter begins with the Bible verses used in the teaching of the chapter. STEM activities and new graphic organizers have been added. The six foundational process skills are emphasized in the Investigation and Exploration lessons to better prepare the students for the upper grades, while the other six process skills are integrated into these lessons. The Investigation and Exploration pages include Materials and Procedure check boxes to aid in organization. Each chapter includes Study Guides to review concepts that are important to the chapter Assessment. Higher-level thinking skills are developed through application questions (Write About It) on the Study Guides. Many pages include worldview application questions.

- The Teacher Edition includes plans for 90 lessons, including separate Review and Assessment lessons. Four chapters include lessons that allot two days to cover the lesson content. A two-day lesson allows time to include additional activities, demonstrations, or study skills. The front of the book has a new section that introduces the teacher to a biblical worldview of *Science 1*. The Biblical Worldview Shaping objectives are identified by an icon. In addition, a Biblical Worldview Scope and Sequence has been added to the back of the Teacher Edition. Some of the Teacher Edition helps that were previously found in the front of the book have been moved to the "Getting Started" section in the Teacher Resources in the back. The Teacher Edition Index is separate and expanded from the Student Edition Index.

- TeacherToolsOnline.com includes a chapter-by-chapter Materials Checklist. Instructional Aids and Visuals are located at TeacherToolsOnline and in the back of the Teacher Edition.

- The Assessments include chapter tests and rubrics. The eleven chapter tests have been revised to align with the updated material on the Study Guide pages in the Activities book. Each Investigation, Exploration, and STEM Activity has a tailored rubric.

Student Edition Features

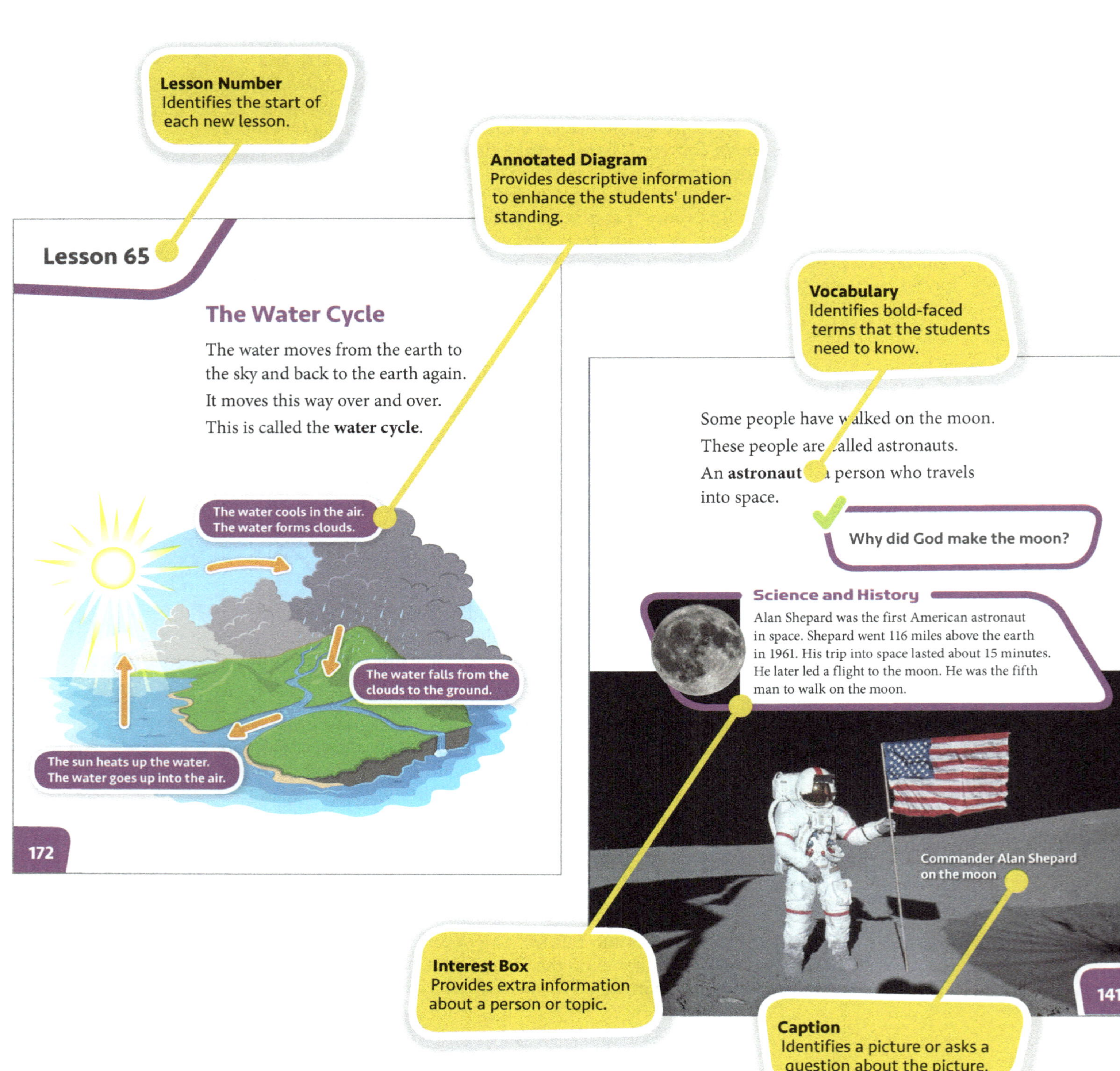

INVESTIGATION

Lesson 19

Plant Needs

Process Skills
- Predict
- Observe

All living things have needs.

What do plants need to survive and grow?

In this Investigation you will predict what will happen to a plant that does not get water.

You will make a hypothesis.

You will observe to check your hypothesis.

Let's investigate!

60

Your Stomach

You need food to survive and grow.

Your teeth chew food.

Then you swallow the broken-up food.

The food travels down a tube to your stomach.

God designed your stomach.

The **stomach** breaks down the s

Your body uses the food from y

survive and grow.

Your body gets energy from the food.

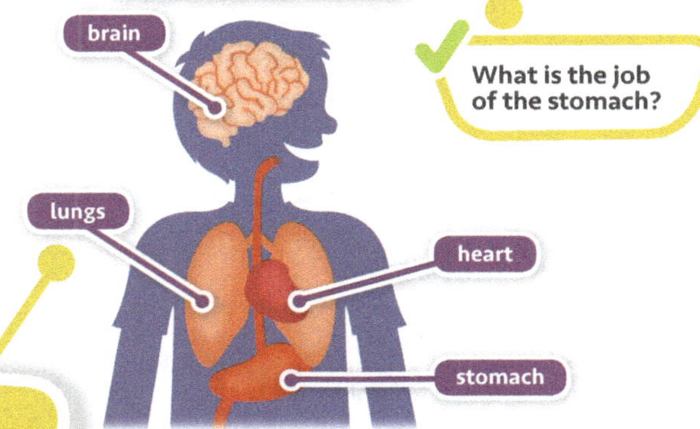

Activities Features

Bible Verses
Provides key Bible verses in written format for each chapter.

Graphic Organizer
Helps to aid the students' understanding. Helps them to see the organization of facts and details.

Study Guide
Reviews chapter terms and concepts. "Review" tab identifies Study Guide page. Material for the chapter assessment will come from the Study Guides.

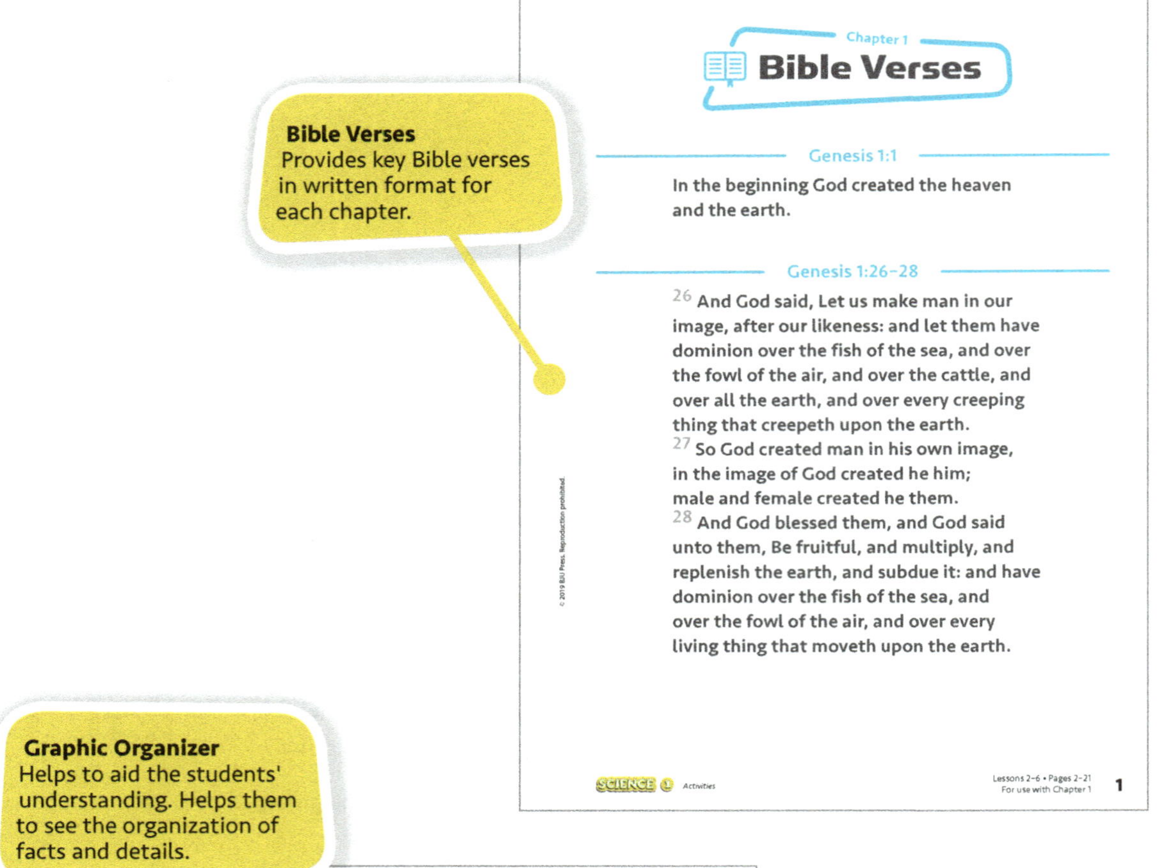

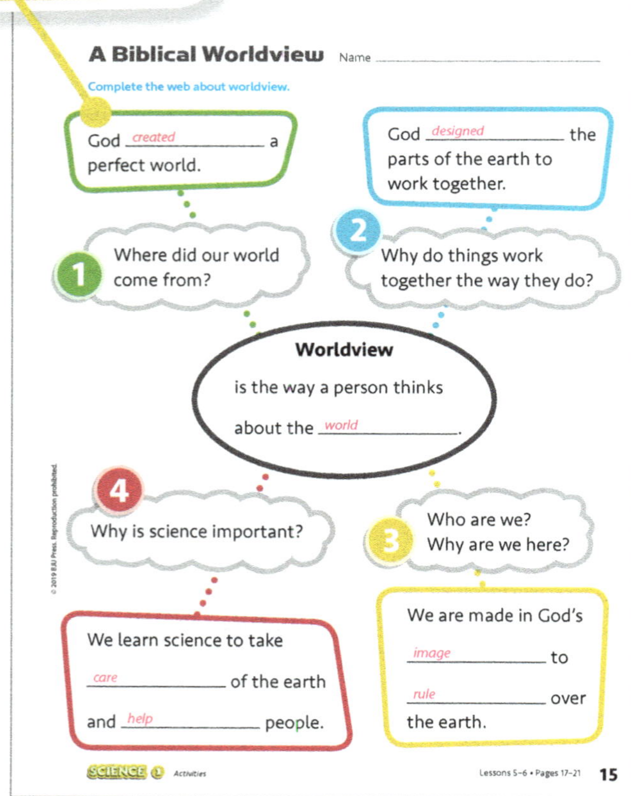

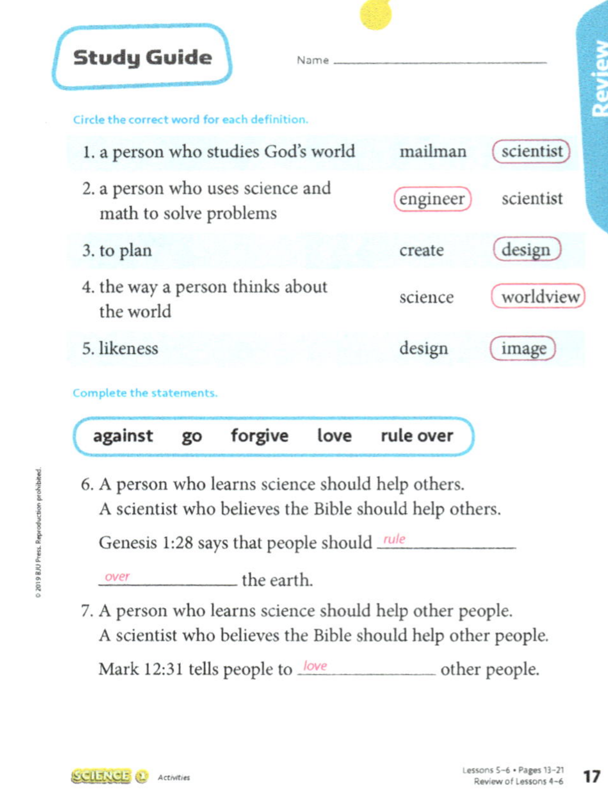

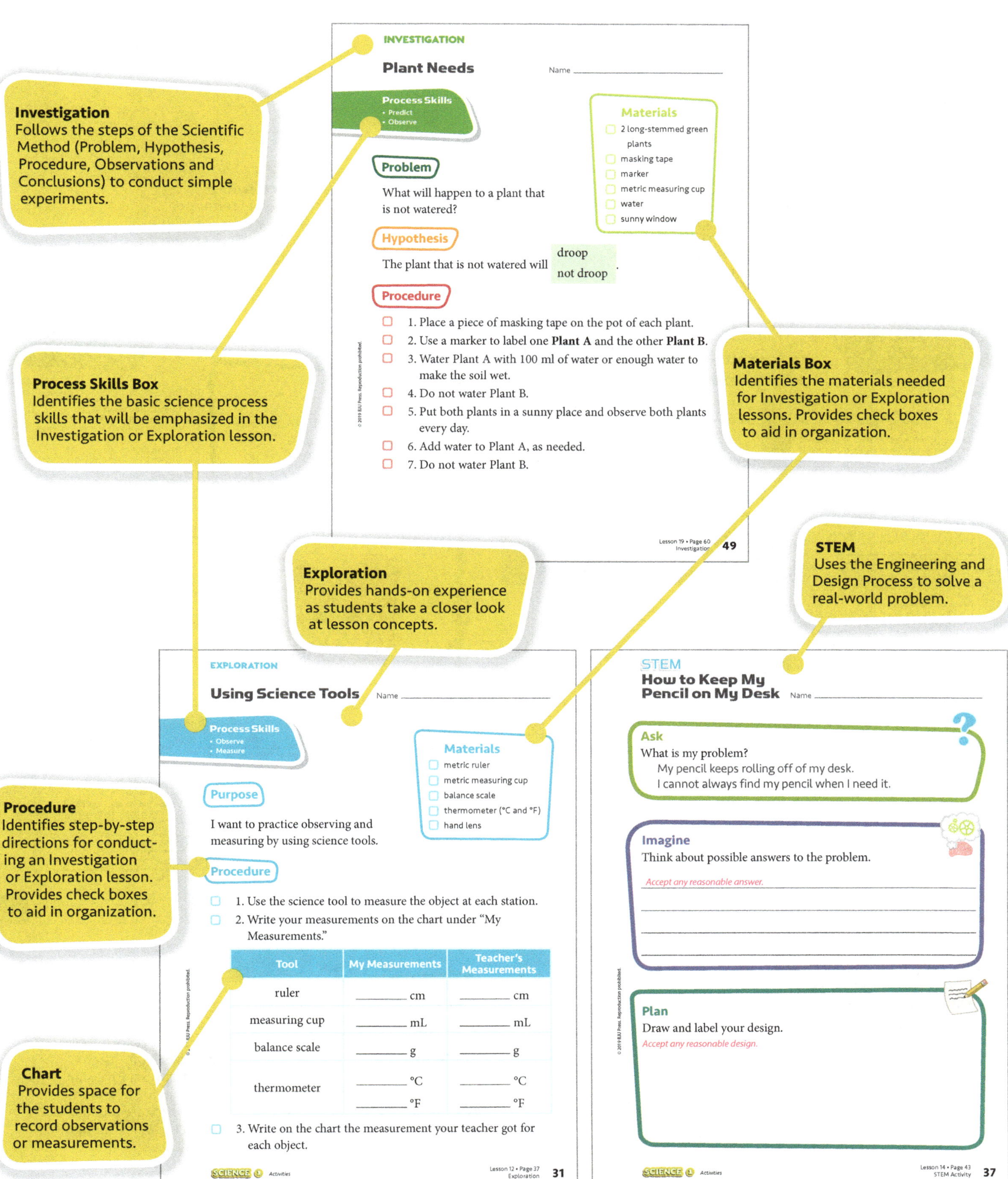

Investigation
Follows the steps of the Scientific Method (Problem, Hypothesis, Procedure, Observations and Conclusions) to conduct simple experiments.

Process Skills Box
Identifies the basic science process skills that will be emphasized in the Investigation or Exploration lesson.

Procedure
Identifies step-by-step directions for conducting an Investigation or Exploration lesson. Provides check boxes to aid in organization.

Chart
Provides space for the students to record observations or measurements.

Exploration
Provides hands-on experience as students take a closer look at lesson concepts.

Materials Box
Identifies the materials needed for Investigation or Exploration lessons. Provides check boxes to aid in organization.

STEM
Uses the Engineering and Design Process to solve a real-world problem.

INVESTIGATION
Plant Needs
Name

Process Skills
• Predict
• Observe

Materials
2 long-stemmed green plants
masking tape
marker
metric measuring cup
water
sunny window

Problem
What will happen to a plant that is not watered?

Hypothesis
The plant that is not watered will droop / not droop.

Procedure
1. Place a piece of masking tape on the pot of each plant.
2. Use a marker to label one Plant A and the other Plant B.
3. Water Plant A with 100 ml of water or enough water to make the soil wet.
4. Do not water Plant B.
5. Put both plants in a sunny place and observe both plants every day.
6. Add water to Plant A, as needed.
7. Do not water Plant B.

Lesson 19 • Page 60
Investigation   49

EXPLORATION
Using Science Tools   Name

Process Skills
• Observe
• Measure

Materials
metric ruler
metric measuring cup
balance scale
thermometer (°C and °F)
hand lens

Purpose
I want to practice observing and measuring by using science tools.

Procedure
1. Use the science tool to measure the object at each station.
2. Write your measurements on the chart under "My Measurements."

Tool | My Measurements | Teacher's Measurements
ruler | cm | cm
measuring cup | mL | mL
balance scale | g | g
thermometer | °C | °C
 | °F | °F

3. Write on the chart the measurement your teacher got for each object.

SCIENCE 1 Activities

Lesson 12 • Page 37
Exploration   31

STEM
How to Keep My Pencil on My Desk   Name

Ask
What is my problem?
My pencil keeps rolling off of my desk.
I cannot always find my pencil when I need it.

Imagine
Think about possible answers to the problem.
Accept any reasonable answer.

Plan
Draw and label your design.
Accept any reasonable design.

SCIENCE 1 Activities

Lesson 14 • Page 43
STEM Activity   37

Teacher Edition Features

Lesson Resource Box
Shows the lesson number and corresponding Student Edition and Activities pages.

BWS Icon
Identifies a Biblical Worldview Shaping objective. The BWS objectives apply biblical worldview truths and principles.

Preparation for Reading
Identifies vocabulary terms and introduces new or more difficult words. Gives the students a purpose for reading.

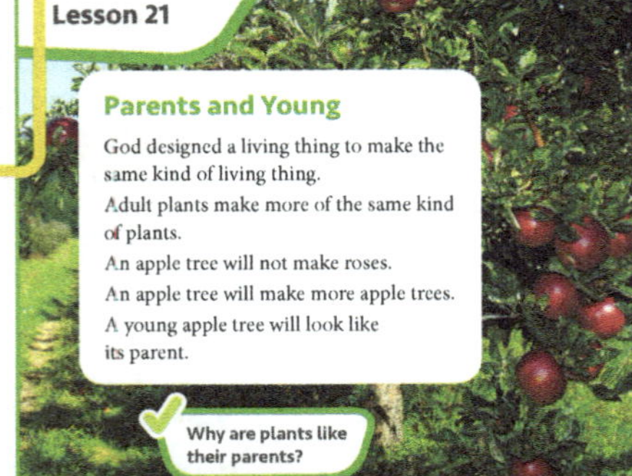

Teach for Understanding
Guides the teacher in modeling and explaining concepts. Engages the students in discussion to evaluate their understanding of science concepts.

Activities
Provides independent practice of concepts learned. Reinforces, enriches, and reviews student learning with pages from the Activities book.

Helps
Provides tips and ideas to help with the lesson.

Background
Clarifies and enhances the lesson with additional information.

Rubric
Provides a tailor-made assessment for each Investigation, Exploration, and STEM lesson. The rubrics are located in *SCIENCE 1* Assessments.

Materials

Identifies materials that need to be gathered to teach the lesson.

Teacher Resources

Lists the Instructional Aids and Visuals needed to teach the lesson. These are located in the Teacher Resources in the back of this book as well as at TeacherToolsOnline.com.

Big Question

Begins each chapter and focuses attention on what the chapter will cover.

Gear Icon

Indicates a higher-order question. Requires some analysis, synthesis, or evaluation of the lesson content. Can be supported with prompts, scaffolding, or background as needed to guide the students to the answer.

Vocabulary

Identifies boldfaced vocabulary terms in the lesson.

Looking Ahead

Alerts the teacher to materials and preparation needed for an upcoming lesson.

Activities

Provides additional hands-on activities and demonstrations to enhance student understanding of concepts taught.

Sample page (left)

LESSON 64 — Student Edition pages 164–71 — Activities pages 153–58

Chapter Objectives
- Identify the components (elements) of weather
- Sequence the water cycle
- Explain the role of a meteorologist in forecasting weather
- Use the science process skills of observe, measure and use numbers, and predict to collect and interpret weather data
- Evaluate a claim about the trustworthiness of science
- Relate a weather prediction to Proverbs 22:3

Lesson Objectives
- Define *weather*
- Recall what temperature is
- Recall the scientific tool that measures temperature
- Define *wind*
- Identify the appearance of a flag when the wind is calm, light, and strong

Materials
- umbrella
- mittens or gloves
- thermometer, outdoor
- pinwheel
- picture of a sailboat

Teacher Resources
- Visual 9.2: *Bible Verses B*

Vocabulary
- weather
- wind

CHAPTER INTRODUCTION
- Direct attention to pages 164–65.
  What is the title of this chapter? "Weather"
- Direct attention to the picture.
  What is the girl holding? an umbrella
- Display an umbrella.
  When do you need an umbrella? when it is raining, snowing, or stormy
- Invite the students to share their experiences in the rain.
  Does it rain every day? no
  Is it dry every day? no
- Point out that weather is always changing.

Big Question
- Direct attention to the Big Question. Ask a volunteer to read the question aloud.
- Explain that the answer to the Big Question will be found as they read Chapter 9.

Chapter 9

Big Question

How does the weather change from day to day?

164

LOOKING AHEAD

Chapter 9 Science Materials
An outdoor thermometer, a pinwheel, and a small flag will be used in several lessons in this chapter to determine air temperature and the type of wind. Displaying a thermometer outside where it is visible from inside will allow the students to easily participate in recording the temperature.

Chapter 10, Lesson 74 Science Materials
Pinhole boxes are required for the Investigation in Lesson 74. You may prepare these ahead of time. See the illustration on Student Edition page 196. Prepare the short end of a shoebox with a pinhole for observation. Along the long side of the shoebox make a hole the same size as the flashlight lens. Make one pinhole box for each science group or each student. Begin gathering enough objects and flashlights for each group or student. Without telling the students, make one object for each group or student be a glow-in-the-dark object, such as a sticker or glow rock.

Visit TeacherToolsOnline.com for resources to enhance the lessons.

184

Science 1

Sample page (right)

Some people have walked on the moon. These people are called astronauts. An **astronaut** is a person who travels into space.

Why did God make the moon?

Science and History
Alan Shepard was the first American astronaut in space. Shepard went 116 miles above the earth in 1961. His trip into space lasted about 15 minutes. He later led a flight to the moon. He was the fifth man to walk on the moon.

Commander Alan Shepard on the moon

141

ACTIVITIES

Reflecting Light
Materials: flashlight, mirror
Turn on the flashlight and shine it at a wall. Dim or darken the room as needed. Put the mirror in the path of light and show how the direction of the light changes.

Size of the Moon
Materials: paper circle from Lesson 48, one ½" mini marshmallow, one ⅛" button (e.g., a man's shirt-collar button)
Show the paper circle and mini marshmallow. Discuss what each represents. Show the ⅛" button. Explain that the button is representing the moon. Point out that the moon is about one-fourth the size of the earth.

BACKGROUND

Men on the Moon
Between July 1969 and December 1972 several American astronauts walked on the moon. Alan Shepard commanded *Apollo 14* and walked on the moon in February 1971.

Chapter 7: The Earth and its Lights

LESSONS 51 / 52

What does the moon move around? the earth
- Use the *Size of the Moon* activity to compare the sizes of the sun, earth, and moon.
- Direct the students to Activities page 119. Complete the graphic organizer about the moon with the students.

Teaching for page 141 begins here.

Have any people walked on the moon? yes; astronauts
What is an astronaut? a person who travels into space
- Direct attention to the *Science and History* box. Read aloud about Alan Shepard as the students follow along.
  Who was the first American astronaut in space? Alan Shepard
  How many miles above the earth did he travel? 116
- Direct attention to the picture. Ask a volunteer to read the caption.
  Where is Alan Shepard? on the moon
- Explain that Alan Shepard walked on the moon about 10 years after the first American flight into space.
- Point out the astronaut's space suit.
  Why do you think Alan Shepard is wearing a space suit? There is no air to breathe on the moon like there is on earth. The space suit gives the astronaut air to breathe. It helps to keep the astronaut safe.
  How do we know that the moon is made of rock? The astronauts brought rocks and soil back from the moon and told people about the moon.
- Point out the American flag. Explain that the American astronauts were representing *all* Americans when they visited the moon. Explain that they left the flag on the moon when they returned to earth. You may point out that there is no wind on the moon and that a pole across the top is used to hold the flag straight out.

answer

to give the earth light at night

155

"

Assessment and Grading

SCIENCE 1 provides a variety of tools that teachers may use for assessment and grading. Frequent assessment enables you to adjust your instructional plan to better meet the needs of each student. The chart identifies suggested percentages to use for calculating each student's total grade.

Tests	Written Work	Investigations, Explorations, & STEM	Participation
50%	25%	25%	
34%	33%	33%	
50%	20%	20%	10%
40%	25%	25%	10%

Assessments

The *SCIENCE 1* Assessments can serve as the objective part of the evaluation of a student's progress. The most effective assessments are an outgrowth of the teaching process. Accordingly, these tests should not replace your individual assessment of a student's understanding and application. The Assessments can be adapted to meet the teaching emphasis and direction as well as the student's maturation level. You may find it necessary to eliminate some items, provide additional test items, or do both. Students will find the mastery-level information for the *SCIENCE 1* Assessments on the Study Guide pages in the Activities book.

Written Work

Written work may include daily assignments, Quick Check answers, Study Guide pages in the Activities book, essay and application questions, and journal entries. Scores may be given based on the completion of some assignments and the accuracy of content on others.

Investigations, Explorations, and STEM Activities

A rubric is a useful tool for assessing work that is not objective, such as the Investigation, Exploration, and STEM activity lessons. Specific rubrics for each Investigation, Exploration, and STEM lesson are located in the Assessments. In a group activity it is often beneficial to give a grade not only to the group as a whole, but to each student, individually, as well. Your scores and comments on a rubric allow the student to see why he or she received a particular grade and what areas need improvement.

Participation

You may choose to evaluate each student's participation during lesson discussions. This subjective assessment should be tracked. You could use checklists, rubrics, point values, or letter grades.

Investigation Rubric: Plant Needs

Name ________________

	Excellent (4)	Very Good (3)	Needs Improvement (2)	Unsatisfactory (1-0)	Points Earned
Use and Understanding of the Scientific Method	Always uses and correctly understands the scientific method to test a hypothesis.	Usually uses and correctly understands the scientific method to test a hypothesis.	Sometimes uses and correctly understands the scientific method to test a hypothesis.	Rarely or never uses and has little or no understanding of the scientific method to test a hypothesis.	
Process Skills: • Predict • Observe	Always uses the science process skills correctly to predict and observe the needs of plants.	Usually uses the science process skills correctly to predict and observe the needs of plants.	Sometimes uses the science process skills correctly to predict and observe the needs of plants.	Rarely or never uses the science process skills correctly to predict and observe the needs of plants.	
Procedure	Always follows the procedure correctly.	Usually follows the procedure correctly.	Sometimes follows the procedure correctly.	Rarely or never follows the procedure correctly.	
Participation	Actively participates during the Investigation.	Usually participates during the Investigation.	Sometimes participates during the Investigation.	Rarely or never participates during the Investigation.	
Attitude	Always displays a positive attitude.	Usually displays a positive attitude.	Sometimes displays a positive attitude.	Rarely or never displays a positive attitude.	
Teamwork	Always listens to and encourages others.	Usually listens to and encourages others.	Sometimes listens to and encourages others.	Rarely or never listens to or encourages others. Usually is a distraction.	
				Total	

Comments

Science 1 Rubrics • For use with Lesson 19

Unit 1: Let's Learn About Science
Chapter 1: Science and Scientists

Lesson	Teacher Edition	Student Edition	Activities	Objectives
1	2–3	1		• Identify and locate the key text features • Infer from key text features the topics of Unit 1
2	4–9	2–7	1–6	**Exploration: Looking at God's World** • Infer from key features the topics for Chapter 1 • Define *science* • Explain from biblical truth why science is important `BWS` • Distinguish science activities from activities that are not science
3	10–14	8–12	1–2, 5–8	• Recall the word *science* • Infer the five senses and the body part used with each sense • Define *senses* • Identify the reason God gave people five senses `BWS`
4	15–18	13–16	1–2, 9–11	• Recall the reason God gave people five senses `BWS` • Describe what scientists do • Explain from the Bible the importance of what scientists do `BWS` • Create a list of ways that students can use science to help others • Classify an engineer as having a STEM career
5–6	19–23	17–21	13–18	• Define *worldview* `BWS` • Identify that every scientist has a worldview `BWS` • Identify that God is the Creator of all things `BWS` • Identify that God designed everything to work together `BWS` • Identify that God made people in His own image to care for the earth `BWS` • Infer that people learn science to take care of the earth and to help others `BWS`
7	24	1–21	1–18	**Review** • Recall terms and concepts from Chapter 1
8	25			**Assessment** • Recall and apply terms and concepts from Chapter 1

Chapter 2: What Scientists Do

Lesson	Teacher Edition	Student Edition	Activities	Objectives
9	26–31	22–27	19–22	• Recall what science is and what scientists do • Define *science process skill* • Observe an object using the five senses • Classify objects based on a chosen criteria • Measure an object using a non-standard unit • Classify science process skills as *observe, classify*, and *measure*
10	32–34	28–30	23–26	• Recall that the science process skills of observing, classifying, and measuring are ways people learn about God's world BWS • Define *inference* as a science process skill • Infer the cause from an effect • Predict the outcome of a certain action • Define what a *scientific prediction* is • Identify *communicate* as a science process skill
11	35–40	31–36	19, 27–28	• Identify science tools and their uses • Measure length using non-standard and standard units • Infer reasons for using standard units of measurement • Explain how people learn about God's world BWS • Explain from Genesis 1:28 why accurate measurement is important BWS
12	41	37	29–32	**Exploration: Using Science Tools** • Measure objects using age-appropriate science tools • Record observations • Compare and contrast observations • Infer steps needed to determine accurate measurements
13	42–46	38–42	33–36	• Identify the purpose for an investigation • Identify the steps of the scientific method • Explain the purpose for the problem and hypothesis in a scientific investigation • Create a hypothesis
14	47	43	37–38	**STEM Activity: How to Keep My Pencil on My Desk** • Recall what an engineer does • Identify the steps of the engineering design process • Apply the engineering design process to solve a real life problem • Relate the work of engineering to the commands of Genesis 1:28 BWS
15	48	22–43	19–38	**Review** • Recall terms and concepts from Chapter 2
16	49			**Assessment** • Recall and apply terms and concepts from Chapter 2

Unit 2: Let's Learn About Living Things
Chapter 3: Plants

Lesson	Teacher Edition	Student Edition	Activities	Objectives
17	50–59	44–53	39–42	• Identify the characteristics of living and nonliving things • Classify items as living or nonliving • Identify the needs of plants • Identify ways people use plants • Explain from Genesis 3:17–18 how the Fall affected plants `BWS`
18	60–65	54–59	43–48	• Identify each part of a plant and its function • Relate plant survival and growth to God's creational design `BWS`
19	66	60	49–50	**Investigation: Plant Needs** • Predict the effects on the growth and survival of a plant when its needs are not met • Observe and describe parts of a plant • Draw a conclusion about plant needs (about the growth and survival of plants) based on observations • Draw a conclusion from the investigation about God's creational design of plants `BWS`
20	67–69	61–63	51	• Define *life cycle* • Identify and describe the stages of the life cycle of a plant • Sequence stages of a plant's life cycle
21	70	64	39, 53–56	• Compare and contrast a seedling with an adult plant • Explain that young plants are like the parent plants because God made plants to reproduce after their kind (Genesis 1:11) `BWS` • Compare and contrast the same kind of plant to show that they are recognized as similar but can also vary
22	71	65	40, 57–58	**STEM Activity: Unwanted Plants** • Design a solution to prevent unwanted plants • Draw and label the design • Explain how the design solves the problem • Relate the growth of weeds and other unwanted plants to Genesis 3:17–18 and how the Fall affected plants `BWS`
23	72	44–65	39–58	**Review** • Recall terms and concepts from Chapter 3
24	73			**Assessment** • Recall and apply terms and concepts from Chapter 3

Chapter 4: Animals

Lesson	Teacher Edition	Student Edition	Activities	Objectives
25	74–79	66–71	59–61	• Infer from key text features the topic for Chapter 4 • Distinguish the identity of living and nonliving things in an environment • Identify the needs of animals • Explain that God designed animals and their environments to work together so they can survive and grow **BWS**
26	80–83	72–75	63–66	• Identify external characteristics of mammals, birds, and fish • Classify animals as mammals, birds, and fish based on similar external characteristics • Classify a zoologist as a scientist
27	84–87	76–79	67–68	• Relate the function of animal body parts to the survival and growth of animals
28	88–93	80–85	69–70	• Identify and sequence the stages of the life cycle of an animal • Name ways that animals care for their offspring • Compare and contrast animals of the same kind • Compare and contrast animals and their offspring • Identify the Bible's explanation for animal death **BWS**
29	94–95	86–87	71–72	**STEM Activity: Copying God's Design** • Identify a real-life human problem • Design a solution to a human problem by using biomimicry • Draw and label the design • Explain how the design solves the problem
30	96	66–87	59–72	**Review** • Recall terms and concepts from Chapter 4
31	97			**Assessment** • Recall and apply terms and concepts from Chapter 4

Unit 3: Let's Learn About Our Bodies
Chapter 5: The Human Body

Lesson	Teacher Edition	Student Edition	Activities	Objectives
32	98–104	88–94	73–75	• Infer the topic of the unit and the chapter based on the pictures and headings • Compare and contrast the needs of animals to the needs of people • Explain how God created the first man and woman **BWS** • Evaluate the statement that people are no different from animals **BWS**
33	105	95	77–78	**Exploration: My Head** • Observe the human head • Identify body parts found on the head • Identify purposes for why God designed the body parts located on the head **BWS** • Associate each of four senses with the correct body part • Apply knowledge of a human body part to give praise to God **BWS**
34	106–10	96–100	74, 79–80	• Recall and describe the body parts of the head • Describe the head, arm, and leg • Label the head, arm, and leg • Explain ways that God's design of the human outside body parts helps people survive and grow (Psalm 139:14) **BWS**
35	111–16	101–6	73–74, 81–82	• Describe the function of the brain, lungs, heart, stomach, bones, and muscles • Label the brain, lungs, heart, stomach, bones, and muscles on a diagram • Explain ways that God's design of the human body parts helps people survive and grow **BWS**
36	117	107	83–89	**Exploration: How My Lungs Work** • Assemble internal body parts to show location • Construct a model that shows how the lungs work • Explain ways that God's design of the lungs helps people survive and grow **BWS**
37	118	88–107	73–89	**Review** • Recall terms and concepts from Chapter 5
38	119			**Assessment** • Recall and apply terms and concepts from Chapter 5

Chapter 6: Care for the Human Body

Lesson	Teacher Edition	Student Edition	Activities	Objectives
39	120–24	108–12	91–94	• Identify kind and respectful behavior • Explain why we should treat other people with kindness and respect `BWS` • Formulate a plan to show how to treat another person with love, care, and respect `BWS` • Identify healthy habits for a strong body
40	125–28	113–16	95–100	• Identify ways to prevent the spread of germs • Identify healthy habits for strong teeth • Explain the importance of developing healthy habits • Practice healthy habits
41	129	117	101–2	**Investigation: Clean Hands** • Formulate a hypothesis to determine the effect that washing hands has on germs • Record observations • Draw conclusions from data collected
42	130–31	118–19	103	• Identify safe habits when at play and in the car • Explain the importance of safe habits
43	132–34	120–22	104–6	• Identify safe habits at home and in the community • Identify fire hazards • Explain the proper response in an emergency • Identify trustworthy adults to go to in a dangerous situation
44	135	123	107–8	**STEM Activity: Safe Shoes** • Propose a possible solution to the real-life problem of slick-soled shoes • Construct a design to solve the problem • Communicate to others how the design solves the problem
45	136	108–23	91–108	**Review** • Recall terms and concepts from Chapter 6
46	137			**Assessment** • Recall and apply terms and concepts from Chapter 6

Unit 4: Let's Learn About Earth and Space
Chapter 7: The Earth and Its Lights

Lesson	Teacher Edition	Student Edition	Activities	Objectives
47	138–44	124–30	109, 111	• Infer topics by previewing the unit and chapter • Explain from Genesis 1 how the earth, sun, moon, and stars were formed **BWS** • Evaluate from the Bible an opposing view of how the earth, sun, moon, and stars formed **BWS**
48	145–49	131–35	113–14	• Describe the earth's daily motion • Identify the sun as a star • Identify the beneficial properties of the sun • Explain from Genesis 1 why God made the sun **BWS** • Describe and predict the sun's pattern across the sky
49	150	136	115–16	**Investigation: Stars in the Day** • Formulate a hypothesis for why it is hard to see stars during the daytime • Observe simulated stars in various lighting • Infer why it is hard to see stars, other than our sun, during the daytime
50	151–53	137–39	117	• Identify the characteristics of stars other than the sun • Identify the telescope as a magnifying tool to observe stars other than the sun • Identify the groups of stars called the Big Dipper and the Little Dipper • Identify the North Star
51–52	154–58	140–44	109, 119–22	• Identify the characteristics of the moon • Identify what an astronaut does • Identify the changes in the shape of the moon over the course of a month • Predict the phases of the moon over the course of a month • Explain from Genesis 1 why God made the moon **BWS** • Explain how the sky changes each day
53	159	145	123–27	**Exploration: Changes in the Sky** • Compare and contrast the nighttime sky with the daytime sky • Predict the moon's phase • Infer the cause for the changes in the sky each day • Apply our knowledge of the earth, sun, moon, and stars to praising God for His greatness and goodness **BWS**
54	160	124–45	109–27	**Review** • Recall terms and concepts from Chapter 7
55	161			**Assessment** • Recall and apply terms and concepts from Chapter 7

Chapter 8: Seasons

Lesson	Teacher Edition	Student Edition	Activities	Objectives
56	162–67	146–51	131–34	• Recall that the earth rotates once each day • Identify that the earth revolves around the sun • Identify that one complete revolution around the sun is equal to one year • Identify the two things that cause the seasons • Sequence the cycle of the seasons
57	168	152	135–36	**Exploration: Using a Thermometer** • Recall two things that cause the seasons • Recall the thermometer as a scientific tool used to measure temperature • Relate the movement of the red line on the thermometer to changes in temperature • Measure temperature to record information • Record temperature using a thermometer
58	169–70		137–40	• Recall the cycle of the seasons by singing a song • Compare and contrast temperature and amount of daylight among the seasons • Infer the temperature and length of daylight hours for each season
59	171–75	153–57	129, 141–42	• Recall the cycle of seasons by singing a song • Explain, using Scripture, that seasonal patterns exist by God's design **BWS** • Identify characteristics of winter and spring
60	176–80	158–62	129, 141–44	• Recall the cycle of seasons by singing a song • Explain what a landscape architect does • Identify characteristics of summer and fall • Defend, using Scripture, that seasonal patterns exist by God's design **BWS**
61	181	163	145–51	**Exploration: Seasons Where I Live** • Compare and contrast the characteristics of seasons with the seasons in your area • Communicate by constructing a booklet that represents the seasons in your area
62	182	146–63	129–51	**Review** • Recall terms and concepts from Chapter 8
63	183			**Assessment** • Recall and apply terms and concepts from Chapter 8

Chapter 9: Weather

Lesson	Teacher Edition	Student Edition	Activities	Objectives
64	184–91	164–71	154–58	• Define *weather* • Recall what temperature is • Recall the scientific tool that measures temperature • Define *wind* • Identify the appearance of a flag when the wind is calm, light, and strong
65	192–95	172–75	153, 159–60	• Define *water cycle* • Sequence the movement of water in the water cycle • Identify the appearance of the sky on clear, partly cloudy, and cloudy days • Identify types of precipitation • Explain how the weather changes from day to day
66	196–97	176–77	154, 161–65	• Define *meteorologist* • Explain what a meteorologist does • Contrast the trustworthiness of Bible promises with the trustworthiness of scientific predictions **BWS** • Evaluate the statement that science gives us the most trustworthy information about our world **BWS** • Practice using tools of a meteorologist
67–68	198–99	178–79	153, 167–73	**Exploration: Weather Watching** • Recall what a weather prediction is • Infer from Proverbs 22:3 that weather predictions help us to prepare for the future **BWS** • Observe, collect, record, and report weather data using tools of a meteorologist • Identify weather patterns in data collected to predict the weather • Compare and contrast weather predictions with actual observations
69	200	164–79	153–73	**Review** • Recall terms and concepts from Chapter 9
70	201			**Assessment** • Recall and apply terms and concepts from Chapter 9

Unit 5: Let's Learn About Energy
Chapter 10: Light

Lesson	Teacher Edition	Student Edition	Activities	Objectives
71	202–10	180–88	175–78	• Identify what energy is • Identify light as energy • Defend, using Scripture, the statement that God created light **BWS** • Describe sources of light as natural or manmade • Identify cause-and-effect energy relationships
72	211	189	179–81	**Investigation: Observing Light** • Predict the amount of light that travels through different objects • Record observations • Graph data from observations • Draw conclusions from the data
73	212–17	190–95	183–85	• Differentiate between objects that are transparent, translucent, and opaque • Recognize that a shadow forms when light is blocked • Explain that a shadow changes when a light source moves
74	218	196	187–89	**Investigation: Illuminate Objects** • Predict whether objects can be seen if light is available to illuminate them or if they give off their own light • Observe objects in a pinhole box • Infer that objects can be seen if light is available to illuminate them or if they give off their own light
75	219–21	197–99	191–92	• Recall that objects can be seen if light is available to illuminate them or if they give off their own light • Identify that light travels in a straight line • Infer that mirrors reflect light
76	222	180–99	175–92	**Review** • Recall terms and concepts from Chapter 10
77	223			**Assessment** • Recall and apply terms and concepts from Chapter 10

Chapter 11: Sound

Lesson	Teacher Edition	Student Edition	Activities	Objectives
78	224–27	200–203	195–96	• Recall hearing as one of the five senses • Identify sound as a form of energy • Identify sound as a vibration that can be heard • Infer different ways sound can be made
79	228–31	204–7	193, 197–198	• Identify that sound travels in waves • Observe that sound travels in all directions • Observe that sound travels through matter • Relate sound and the human ear to God's creational design **BWS** • Relate sound to the vibration of materials
80	232–35	208–11	199–200	• Identify the characteristics of volume • List examples of loud and soft sound • Identify the characteristics of pitch • List examples of sound with high and low pitch • Explain two ways that sound changes
81	236	212	201–3	**Investigation: Hearing Pitch** • Formulate a hypothesis for how the thickness of a rubber band will affect pitch • Measure with numbers the length of a stretched rubber band • Observe that the pitch of a sound is affected by the thickness of a rubber band when the rubber band is plucked • Infer that the thickness of a rubber band influences the pitch of the sound the rubber band produces • Explain how the pitch of a stringed instrument can be changed
82	237	213	205–6	**STEM Activity: Making Music** • Design a musical instrument with four strings of varying pitch • Draw and label the design of the stringed musical instrument • Make a model of the stringed musical instrument • Test and improve the stringed instrument model • Explain how the design of the musical instrument solved the problem of having four strings of varying pitch
83	238	200–213	193–206	**Review** • Recall terms and concepts from Chapter 11
84	239			**Assessment** • Recall and apply terms and concepts from Chapter 11

Chapter 12: Communicating with Light and Sound

Lesson	Teacher Edition	Student Edition	Activities	Objectives
85	240–47	214–21	208–13	• Identify ways light and sound are used to communicate at home and school • Explain how various sources of light and sound communication at home and school can be used to help people **BWS** • Explain how to determine whether light and sound communication is good or bad **BWS** • Evaluate uses of light and sound communication **BWS**
86	248–51	222–25	208–10, 213–16	• Identify ways light and sound are used in the community to communicate • Explain how various sources of light and sound communication in the community can be used to help other people **BWS** • Explain how to determine whether light or sound communication is good or bad **BWS** • Evaluate uses of light and sound communication **BWS**
87	252	226	217–20	**STEM Activity: Helping with Light or Sound** • Propose possible solutions to a real-life problem using light or sound • Draw a design that uses light or sound to solve a real-life problem • Communicate to others how the design solves the problem
88–89	253–60	227–34	13, 207–10, 221–22	• Recall what a worldview is • Summarize from the Bible where the world came from **BWS** • Construct a response explaining why things work the way they do in our world **BWS** • Determine who we are and why we are here **BWS** • Compare and contrast the importance of science with the importance of the Bible **BWS**
90	261	235	208–9, 223–24	**Exploration: A Song of Praise** • Create a song of praise for God's creation **BWS** • Formulate a sentence explaining how the song of praise will be used **BWS** • Explain how to determine whether the words of the song of praise are good or bad **BWS**

Book Objectives

> These are the objectives for first grade science. The first lesson of each chapter includes a list of chapter objectives. Each chapter objective is part of one of the book objectives.

- Defend the truth that God created all things and sustains their existence **BWS**
- Apply the Bible's teachings to science concepts **BWS**
- Apply foundational science process skills to develop orderly approaches to problem solving
- Explain how the design of living things meets their needs for growth and survival **BWS**
- Analyze the patterns of the earth, sun, moon, and stars **BWS**
- Explain the properties of light and sound and communication using light and sound
- Apply science knowledge through active participation and relevant discussions, showing a positive attitude

Lesson Objectives
- Identify and locate the key text features
- Infer from key text features the topics of Unit 1

Materials
- picture of a Malay lacewing butterfly (optional)
- world map or globe

BOOK INTRODUCTION

- Allow time for the students to explore the different parts of the book before looking at it together.
  Turn to the Contents page in the front of your book.
  Can you read the word written in black next to each of the big numbers? Guide the students in reading the word *unit*.
- Explain that there are five units in the book.
  Put your finger on the words *Let's Learn About Science* written in blue next to the blue line.
  This is the title of the first unit. What will the unit be about? learning about science
  Each unit is divided into chapters. Move your finger down to see what the first chapter is about. "Science and Scientists"
  Is there another chapter in Unit 1? yes What is the second chapter about? "What Scientists Do"
  Put your finger on the title of the first chapter, *Science and Scientists*. Now move your finger toward the edge of your book to find the page number where you can find the first chapter.
  What page does the first chapter begin on? 1
  Put your finger on the last two lines of the page.

Science Notebook
Encourage each student to keep a science notebook for science-related pages. A three-ring binder or other easily accessible notebook would work best. The students may keep Activities pages, such as Investigations, Explorations, and reviews, in their notebooks. They may also keep other science projects and handouts.

Keeping a notebook will help the students develop organizational skills. The notebook can also be used in each student's portfolio to demonstrate his progress and ability to think scientifically. Plan to check the content, organization, and neatness of the notebook at least once per chapter.

Unit Photo
The unit photo is a Malay lacewing butterfly on a flower. The caterpillar is not a replica of the caterpillar that turns into the Malay lacewing butterfly. The butterfly lives from Burma to Singapore. It is found mostly in the forests of nature reserves where its preferred plants and flowers live. The borders of its wings are indented with a white, lace-like pattern.

Have you ever used a magnifying glass or a hand lens? What happens when you look at something through a hand lens? The object looks bigger. What is hanging on the handle of the hand lens? a caterpillar

- Explain that a caterpillar turns into the Malay lacewing butterfly.

Let's look at the first chapter. Turn to page 2 in your book. What does it say in big letters in the box? Elicit that it says "Big Question."

> When you are asked to *elicit*, use the Socratic method. It is a pedagogical technique in which you do not give the students the information directly. You ask a sequence of questions that helps the students come to the desired conclusion.

There will be a big question in each chapter. We will find the answer to the question during the chapter.

Turn to page 4. Find the word that is a darker black than the other words. What is the word? science

You will find other darker black or bold words in the chapters. These words are vocabulary words. We will learn what these words mean.

- Direct attention to page 5.

Find the box at the bottom of the page with a check mark. In this box you will find a Quick Check question. These questions will help you know if you understand what you have read.

- Guide the students in finding page 7.

That big word at the top of the page is *Exploration*. An Exploration is a science activity that we will get to do.

- Guide the students in finding page 15.

The big words at the top of the page say "STEM Careers." Do you know what a career is? Elicit from the students that a career is a job.

On these pages you will learn about some of the jobs that scientists have.

> Allowing the students to look at their Activities books during this lesson will minimize distractions during the teaching of Lesson 2.

- Direct attention to the Activities book.

Turn to page 1. Each chapter will start with a Bible Verses page. The verses will be read while we study the chapter.

Turn to page 5. Do you remember what the big word at the top of the page is? Exploration

What is an exploration? a kind of science activity

- Explain that the students may write in their Activities books but they should not write in their science books. They may tear out pages in the Activities books but not in their science books.

What are the last two sections in the book? Guide the students in reading the words *Glossary* and *Index*.

- Explain that a glossary is like a dictionary. You can find each science vocabulary word and what it means in the glossary.

Put your finger on the word *Glossary*. Move your finger toward the edge of your book to find the page number where you can find the glossary. What page is the glossary found on? 237

- Explain that an index tells you on what page in your book you can find information about different words, subjects, or people.

What page can you find the index on? 253

- Guide the students in finding the glossary and index. *Note*: Time will be allotted in a later lesson to help the students use the glossary.

UNIT INTRODUCTION

Turn to page 1 in your book. What do you see? Possible answers: butterfly, flower, caterpillar, magnifying glass

- Explain that the butterfly is a Malay lacewing. It lives in Asia. Show the students where the butterfly lives on a map. Show the students a picture of the Malay lacewing (optional).

Chapter 1: Science and Scientists

Chapter Objectives

In the first lesson of each chapter, there is a list of objectives for the chapter. Each of the lesson objectives is part of one of the chapter objectives.

- Associate key text features with information found in each feature
- Identify what science is and what scientists do
- Associate the five senses with the body part that God gave to observe and learn about His world **BWS**
- Explain that all people have a worldview **BWS**
- Infer that people learn science to take care of the earth and to help others **BWS**
- Explain why science is important from a biblical worldview **BWS**
- Identify the four characteristics of a biblical world-view **BWS**

Lesson Objectives

- Infer from key features the topics for Chapter 1
- Define *science*
- Explain from biblical truth why science is impor-tant **BWS**
- Distinguish science activities from activities that are not science

Materials

- lab coat
- hand lens

Teacher Resources

For your convenience, the Teacher Resources are available in the back of this Teacher Edition and on TeacherToolsOnline.com.

- Instructional Aid 1.1: *Bookmarks*, copy on cardstock
- Visuals 1.1–1.4: *Bible Verses A*; *Bible Verses B*; *Is It Science? A*; *Is It Science? B*

If you do not have access to a type of multimedia presentation, you may want to make each student or group of students a copy of the visuals that they do not have access to in their books.

Vocabulary

- science

CHAPTER INTRODUCTION

Chapters 1 and 2 contain information founda-tional to *Science 1*.

Chapter Photo

The photo on page 2 is a boy looking for an ocean creature to catch in his net.

Introduction to Lessons

Because first graders are easily distracted, you may choose to complete your introduction before having the students take out their books.

Silent Reading

You may choose to read the text to the students. The goal is for the students to read independently. Even small sections of silent, independent reading will help the students recognize that they can gain information on their own.

Visit TeacherToolsOnline.com for resources to enhance the lessons.

Science and Scientists

God created our world.

He created all that is in it.

Last of all, God created people.

God gave people minds.

He gave us minds to learn about His world.

God created people to be curious.

God created us to ask questions.

What questions do you ask about God's world?

3

- Put on the lab coat and use the hand lens to look around the room. Ask the students if they think you look like a scientist.
- Discuss that some scientists wear lab coats, but many do not.

What do you think makes a person a scientist? Answers will vary.

What is the title of the first chapter? "Science and Scientists"

Where do you think the boy is? at the shore or beach

The gear icon ( ) indicates a higher-order question. These questions are based on information gathered from the text, but they require some analysis, synthesis, or evaluation of the text. Supply prompts or background as needed to guide the students to the answer.

What is he doing? Possible answers: searching for something to catch in his net, looking at something in his net

What things do you think he might catch in his net? Possible answers: fish, crabs, oysters, clams, starfish

What could the boy learn about the things he catches in his net? Possible answers: where they live, what they look like, how they move

Big Question

- Guide the students in reading the question.

Why do you think the question is here? Guide the students to see that while they read through the chapter, they will learn the answer to the question.

- Ask the students to look at the chapter to find out what it is about.

What do you think this chapter is about? As you give your answer, tell what made you think that. Possible answers: science, scientists, frogs, smelling flowers, tasting different foods

PREPARATION FOR READING

Reading for information is an important skill for the students to learn. The Preparation for Reading "find out" or "find where" helps motivate reading and focuses the students' attention. Although it is more useful for silent, independent reading, you may choose to use the "find out" even if you read the text aloud. The "find out" will be answered as the information is discussed.

- Direct the students to read page 3 silently to find out why God gave us minds.

TEACH FOR UNDERSTANDING

What did God create? Possible answers: our world, all that is in the world, people, minds

- Display *Bible Verses A*. Direct attention to Genesis 1:1 on Activities page 1.

At the beginning of each chapter in the Activities, there will be a page with most of the Bible verses used in that chapter. Using this page will allow the students to read the verses with you without having to spend extra time looking up the verses.

- Read Genesis 1:1 as the students follow along.

What does this verse tell us that God created? the heavens and the earth

Why did God give us minds? to learn about His world, to be curious, to ask questions

What does it mean to be curious? to ask a lot of questions, show interest in

What do you think the boy with the net is curious about? what is in his net, what is in the water, what is on the beach

What are you curious about in God's world?

PREPARATION FOR READING

A vocabulary term is a bold word for which the students learn the definition. Write the terms for display and pronounce them with the students. Some lessons include words that are not vocabulary terms but may be difficult for the students to decode and pronounce. Assist the students with any words that may be difficult.

- Preview and pronounce the vocabulary term *science*. Preview and pronounce the word *amazing*.
- Explain that headings are the blue words or bigger words at the top of a section in the book. Direct the students to the heading.
  What is the heading? "What Is Science?"
  A heading will tell you what information you will be reading about.
- Direct the students to read pages 4–5 silently to find out if science can answer all your questions.

TEACH FOR UNDERSTANDING

What is science? the study of God's world

What does it mean to study? learn about, look carefully at

What is God's world? the earth and everything on it

Note: God's world includes more than the earth and everything on it. At the first-grade level *God's world* will be used as a synonym for the earth.

Look at the words with the picture.

- Explain that words, sentences, or questions that are near a picture are called *captions*.
  What is the caption for this picture? What questions do you have about God's world?
- Allow several volunteers to share some of the questions they have about God's world or questions about the picture.
  Can science answer all your questions? no

Name some questions that science can answer. Possible answers: What do you add to the dirt to grow plants better? What happens to ice in the sun? Why is the moon not always a circle? Why do some lizards change color?

Name some questions that science cannot answer. Possible answers: Where did the world come from? Why are you here? Who am I? Am I going to heaven? How does Jesus talk to me?

⚠ HELPS

Open-Ended Questions
To help a first grader relate what he is learning to his own experiences, open-ended questions are included. Answers should not be limited to those given in the Teacher Edition. If an answer seems unusual, question the student as to why he gave that particular answer. This will help distinguish divergent thinking from silliness. A student should never be made to think his answer is silly when he intended to be serious.

Reading Assignments
The reading assignments for each lesson are divided into one- to three-page segments. You may choose to adjust the length of assignments based on the reading ability of the students.

Science can answer some questions.

What do you add to the dirt to grow plants?

What happens to ice in the sun?

Science can answer these questions.

Science cannot answer some questions.

Where did the world come from?

Why are you here?

Science cannot answer these questions.

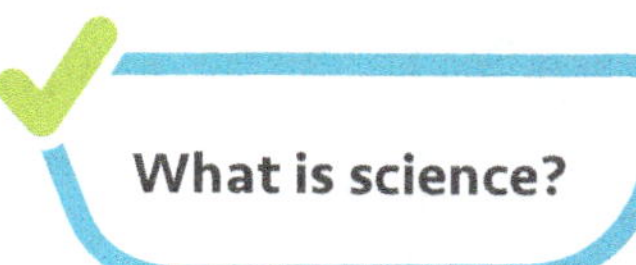

What is science?

5

In later lessons students will learn that some questions the Bible answers and some science answers. Also they will learn that when science disagrees with the Bible, the Bible is always right.

- Direct attention to the picture.
- What is the picture of? a frog sitting on a lily pad What is the caption? Can this frog swim? Can science answer your question?
- Can the frog swim? Accept any reasonable answer.
- Can science answer your question? yes
- Direct attention to the Quick Check question. Remind the students that these questions will help them know if they understand what they have read.

answer

the study of God's world

PREPARATION FOR READING

- Preview and pronounce the word *neighbor*.
- Direct the students to read page 6 silently to find out why you should study science.

TEACH FOR UNDERSTANDING

- Display *Bible Verses B*. Direct attention to Mark 12:30–31 on Activities page 2. Read Mark 12:30–31 aloud as the students follow along.
  In the first line of the verse, Jesus tells us what the first commandment is. What is it? to love God
  How are we to love God? with all our heart, and with all our soul, and with all our mind, and with all our strength
  What is the second commandment? to love your neighbor like yourself
- Ask a volunteer to read the second line on the text page.
  Why do we study science? to show love
  Which commandments tell us to show love? the first and second commandments
- Ask a volunteer to read the third line.
  What is another reason why we study science? to help others
  What example of learning science to help others does the book give us? learning about a plants' needs to help grow a flower to give to a sick neighbor
  Which commandment tells us to show love to our neighbor? the second commandment
  How does giving your neighbor a flower show love? Possible answers: showing you care, showing you are thinking about him
- Display *Bible Verses A* and direct attention to Activities page 1.
- Read Genesis 1:26 aloud as the students follow along.
- Explain that the word *dominion* means to care for, or to rule over. God tells us to care for the earth.
- This verse and the word *image* will be explained more in Lessons 5–6.
- Ask a volunteer to read the second sentence in the second section on the page.
  What is another reason why we study science? to care for the earth
  What are some examples of using science to care for the earth? Possible answers: learning about an animal's needs to help take care of dogs at a shelter; learning about what a plant needs to help grow vegetables or flowers

God tells us to love our neighbors.

People study science to show love.

People study science to help others.

You can learn what a plant needs.

This helps you grow a flower.

You can give it to a neighbor who is sick.

You are using science to help others.

God tells us to care for the earth.

People study science to care for the earth.

You can learn what an animal needs.

This helps you care for animals.

You can care for dogs at a shelter.

You are using science to care for the earth.

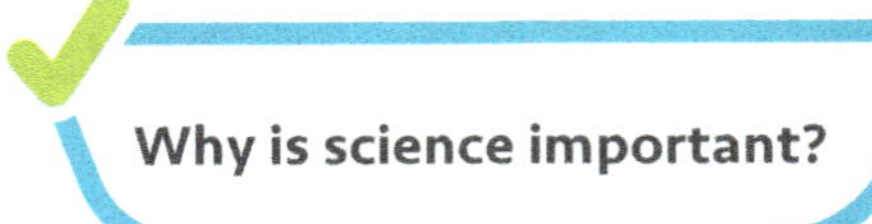

6

- Direct attention to the picture.
  How is the girl using science to care for the earth? She is taking what she learned about animals' needs and using the information to care for a dog at a shelter.
- Display *Is It Science? A*. Direct attention to the first picture.
  What is happening in this picture? A boy is looking at the stars.
  Do you know what he is looking through? a telescope
  Does the picture show science being done? yes
- Direct attention to the second picture.
  What is happening in this picture? A boy is having a birthday party.
  Does the picture show science being done? no
- Follow the same procedure with the pictures on *Is It Science? B*. picture 3: the weather being reported, science being done; picture 4: outside temperature and amount of wind being measured, science being done

EXPLORATION

Looking at God's World

In this Exploration you will look at God's world.

Your home is part of God's world.

Everything around you is part of God's world.

Let's take a closer look at your home.

7

PREPARATION FOR READING

- Direct the students to read page 7 and Activities pages 5–6 silently to find out what they will be doing.
  What will you be doing in the Exploration? taking a closer look at my home
- Explain that it is helpful to look over all the pages before starting an Exploration. It will help the students to see what they are going to do and when they will need to write or record information.

TEACH FOR UNDERSTANDING

- The **Process Skills** will be taught in Chapter 2. In this lesson the students will learn what the process skill *observe* is generally and learn its definition in Lesson 3. This Exploration will be done at home. From the students' answers they will infer the senses in Lesson 3.
  What science process skill will you be using in this Exploration? Observe
  What does *observe* mean? Elicit that it means to look closely at something.
- Direct attention to the **Purpose** on Activities page 5.

What is the purpose of this activity? to look at, or observe, my home as a part of God's world

- Direct attention to the **Materials** box on Activities page 5.
- Explain that the items they need for doing the Exploration are listed here.
- Read the list with the students.
- Direct attention to the **Procedure**. Explain that a procedure is a list of steps that tells them how to do the activity. Each step is to be followed. Encourage the students to check off each step as it is completed.
- Explain that the students will take this page home to complete. They may ask someone to help them read the steps. Instruct them to return the page tomorrow. It will be used in the next science lesson.
- In this Exploration the **Conclusions** section will be discussed in the next lesson.

ACTIVITIES

Bible Verses, pages 1–2
These Bible verses will be used in the teaching of this chapter.

What Is Science?, page 3
This page reinforces the concepts taught in this lesson.

- Direct attention to the footer on page 3. Point out the Student Edition page numbers and explain that they are the pages used to complete the exercise.
- To help students, you may choose to distribute the prepared bookmarks (Instructional Aid 1.1) to each student.
- Instruct the students to mark the beginning of a lesson with the green bookmark and the end with the red bookmark. This will help the students keep track of the page ranges as they complete their Activities assignments.

You may choose to read the directions and questions on the Activities pages to your students, pausing for them to answer each question. Or you may assist the students as they complete the pages independently.

Exploration: Looking at God's World, pages 5–6
This activity prepares the students for studying the senses in the next lesson. It is worked at home.

Objectives
- Recall the word *science*
- Infer the five senses and the body part used with each sense
- Define *senses*
- Identify the reason God gave people five senses **BWS**

Materials
- several sound clips, such as a bell ringing, dog barking, whistle blowing, cow mooing, or train whistle

Teacher Resources
- Visuals 1.1–1.2: *Bible Verses A*; *Bible Verses B*

Vocabulary
- senses
- observe

INTRODUCTION

What is the study of God's world called? science
Why is it important to study science? to show love, to help others, to care for the earth

- Direct attention to Activities pages 5–6 (pages completed at home).
  Look at question 1. What color is your front door? Allow several students to share their colors.
- What body part did you use to tell what color your front door is? eyes
- What do we call it when you use your eyes to look at something? seeing or sight
  Look at question 2. What does your living room floor feel like? Allow several volunteers to tell about the feel of their living room floor.
- What body parts did you use to feel the floor? hands and fingers
- What are you doing when you feel something? Elicit that you are touching it.
- Follow the same procedure with questions 3–5. Elicit the body part used in each and the word used to name what is being done. 3: nose, smell; 4: mouth or tongue, taste; 5: ear, hear
- Direct attention to **Conclusions** question 1.
  What body parts did you use to answer the questions? Allow volunteers to share what body parts they used.
  Today we will learn about using these body parts to find out about God's world.

PREPARATION FOR READING

- Preview and pronounce the vocabulary term *senses*.
- Direct attention to the blue heading on Student Edition page 8.

Lesson 3

What Are the Five Senses?

God gave you ways to find out about His world.

Senses are the ways God gave you to find out about the world.

Your five senses are sight, touch, smell, taste, and hearing.

Sight

You see with your eyes.

You can see shapes or sizes of things.

You see many different things.

Your sense of sight helps you find out about God's world.

8

What is the heading? "What Are the Five Senses?"
- Preview and pronounce the word *discover*.
- Direct the students to read pages 8–9 silently to find out what the word *senses* means.

TEACH FOR UNDERSTANDING

What do we call the ways God gave you to find out about the world? senses
What are the five senses? sight, touch, smell, taste, and hearing
What body part do you use with the sense of sight? eyes
 What are some of the things you see with your sense of sight? Possible answers: shapes, sizes, colors, flowers, desks, tables, people
Look at the picture at the top of the page. What is it? jelly beans
What does your sense of sight tell you about the jelly beans? Possible answers: what they are, the colors, the shapes
Look at the picture at the bottom of the page. What does your sense of sight tell you about the picture? Possible answers: The picture is in the shape of a heart. The candy is heart shaped. The candy is different colors.

Touch

You use your hands and fingers to touch.

You touch bumpy or smooth things.

You touch many different things.

Your sense of touch helps you find out about God's world.

Smell

You use your nose to smell.

You can discover good smells or stinky smells.

You smell many different things.

Your sense of smell helps you find out about God's world.

What are the five senses?

9

What body parts do you use with your sense of touch? hands and fingers

What are some of the things you can feel with your sense of touch? Encourage the students to answer with words that describe touch, such as "soft." Possible answers: bumpy things, smooth things, rough things, wet things

Can you feel the clothes you are wearing? yes

- Explain that every part of our bodies can touch things, but usually our hands and fingers are what we use to touch things around us.

What body part do you use with your sense of smell? nose

What are some of the things you can smell with your sense of smell? Encourage the students to answer with words that describe smell, such as "sweet." Possible answers: stinky smells, flowery smells, fruity smells

What are some things that smell good? Possible answers: candles, food, bath soap, perfume, flowers

What are some things that smell bad? Possible answers: a skunk, dirty socks, tennis shoes, garbage

- Explain that sometimes your sense of smell helps protect you. Food that is no longer good to eat may smell bad, or you may smell smoke, which could mean a fire is nearby and something is burning.

Look at the picture. Which of the senses is the girl using? smell and touch

How is she using the sense of smell? smelling the flower

How is she using the sense of touch? feeling the flower

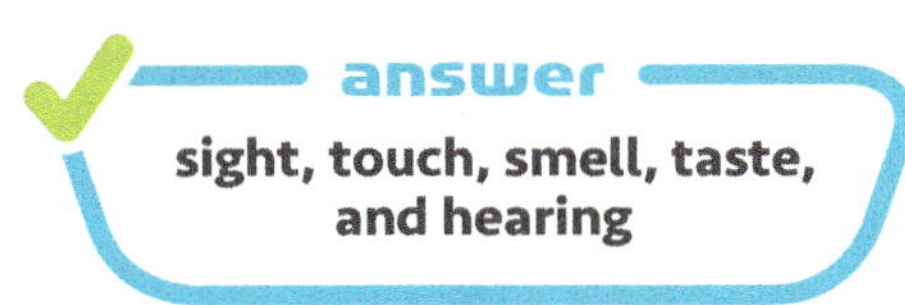

answer

sight, touch, smell, taste, and hearing

ACTIVITIES

Play "I Spy"

Choose an object in the room and say "I spy a(n) _____ object." Describe the object using its size, shape, or color. Encourage the students to name things in the room that fit your description. Then allow a student to choose an object and describe it.

Touch-Related Words

Materials: golf ball, table tennis ball, smooth rubber ball, tennis ball, softball

Hand one of the balls to a student. Encourage him to describe the ball using touch-related words. Repeat for each of the other balls.

Distinguish Smells

Materials for each station: 3 paper cups; 3 scented items such as lemon slice, peppermint candy, orange slice; 3 rubber bands; aluminum foil; pencil; paper

Prepare three cups, each with a different scented item. Use a rubber band and aluminum foil to cover each cup. Punch a few small holes in the foil. Number the cups 1–3.

Direct the students to write the numbers 1–3 on their papers. Tell them to sniff each cup and write down what they think is in each cup next to the number.

PREPARATION FOR READING

- Preview and pronounce the word *tongue*.
- Direct the students to read pages 10–11 silently to find out what kinds of things you can hear.

TEACH FOR UNDERSTANDING

Which of your senses do you use your tongue for? sense of taste

What is the picture at the top of the page? lemons

How does a lemon taste? sour

What are some things you like to eat that are sour? Possible answers: pickles, limes, sour cream

What is in the red cup? hot chocolate with marshmallows

How does hot chocolate with marshmallows taste? Possible answers: hot, sweet, soft

What is the boy eating? ice cream

How does ice cream taste? Possible answers: sweet, crunchy, soft

- Explain that in Psalm 119:103 David says that God's words are sweeter than honey. In Bible times people used honey as we use sugar today. David loved God's words more than he loved honey.

Taste

You use your tongue to taste.

You taste things that are sweet.

You taste things that are sour.

You taste many different things.

Your sense of taste helps you find out about God's world.

10

ACTIVITIES

Identify Tastes

Materials for each student: napkin, jelly bean, unsweetened chocolate, pretzel, sour candy

Place each of the different foods on the napkin for the students. Instruct them to taste the jelly bean. Ask them whether the candy is sweet, sour, salty, or bitter. Repeat with the other food items.

Make Faces

Tell the students to think of what kind of face they might make as they taste something sweet, sour, salty, or bitter. Allow them to demonstrate each face.

Variation: Allow a volunteer to make a face for sweet, sour, salty, or bitter. Ask the other students to guess which taste the student is showing.

HELPS

You may choose to play recordings of sounds. Visit TeacherToolsOnline.com for resources to enhance the lessons.

Hearing

You use your ears to hear.

You hear loud things and soft things.

You hear bangs and whistles.

You hear many different things.

Your sense of hearing helps you find out about God's world.

What do you use to taste?

11

- Play several sound clips, asking the students to identify each sound.

Which one of the senses did you use to tell what the sounds were? sense of hearing

What body part do you use with the sense of hearing? ears

Look at the picture of the children. How is the girl in the pink shirt making sounds? by blowing into the instrument

Do you think the sound that she is making is loud or soft? Accept either answer.

What are some loud sounds? Possible answers: whistle blowing, band playing, dog barking, ambulance siren

What are some soft sounds? Possible answers: a whisper, kitten crying, dripping water, ticking clock

How is the boy who is seated making sounds? hitting the triangle with a stick

Do you think the sound would be loud or soft? Accept either answer.

- Direct attention to Proverbs 20:12 on *Bible Verses B*. Read the verse aloud.

What senses does the verse talk about? sense of sight and sense of hearing

Who does the verse say made both of these senses? the Lord

Senses Song
Sing to the tune of "The Farmer in the Dell."

I use my eyes to see.
I use my eyes to see.
I find out more about God's world
When I use my eyes to see.

I use my hands to touch.
I use my hands to touch.
I find out more about God's world
When I use my hands to touch.

I use my nose to smell.
I use my nose to smell.
I find out more about God's world
When I use my nose to smell.

I use my tongue to taste.
I use my tongue to taste.
I find out more about God's world
When I use my tongue to taste.

I use my ears to hear.
I use my ears to hear.
I find out more about God's world
When I use my ears to hear.

PREPARATION FOR READING

- Preview and pronounce the vocabulary term *observe*.
- Direct the students to read page 12 silently to find out how you can find out about God's world.

TEACH FOR UNDERSTANDING

How can you find out about God's world? use your senses, observe

What does *observe* mean? to use your senses to find out something

- Sing together "O Be Careful Little Eyes" or the *Senses Song* on Teacher Edition page 13.

What are our five senses? sight, touch, smell, taste, and hearing

Where did our senses come from? Let's read a Bible verse to find out.

- Direct attention to Psalm 139:14 on *Bible Verses B*. Read the verse aloud as the students follow along on Activities page 2.

Who made you? God

The verse says "marvellous are thy works." Marvellous means amazing, or fabulous. He made us amazing.

Are your senses part of your body? yes

Did He also make your senses? yes

This verse says we should praise God for making us wonderfully. To *praise* God means to thank God for who He is and what He does.

- Direct attention to the Big Question on page 2. Ask a volunteer to read the question.

Why did God give people five senses? to find out about God's world

- Direct attention to the picture on page 12.

How are the senses being used? Possible answers: tasting ice cream, smelling the hamburger, feeling the hamburger, hearing the bird singing, hearing the speakers playing music, seeing the menu, feeling the dog

answer

to use your senses to find out something

ACTIVITIES

Exploration: Looking at God's World, pages 5–6

These pages were used during the lesson to infer the five senses.

Study Guide, pages 7–8

These pages review the concepts in Lessons 2–3. The students will need to include the corrected pages in their science notebooks.

The Study Guide pages review the information in smaller segments. The material for each chapter test comes from the Study Guides. A student who understands the material (and knows more than just the answers) covered on the chapter Study Guides will be adequately prepared to take the test.

ASSESSMENT

Rubric

A rubric is a useful tool for assessing work that is not objective, such as the Exploration, Investigation, and STEM activity lessons. Specific rubrics for each Exploration, Investigation, and STEM activity are located in the Assessment Packet. However, because this Exploration was worked at home, it does not need a rubric.

Lesson 4

What Is a Scientist?

A **scientist** is a person who studies God's world.

Scientists use their senses to observe the world.

Scientists are curious.

They ask questions.

They try to find answers to help others.

13

LESSON
4

Objectives
- Recall the reason God gave people five senses **BWS**
- Describe what scientists do
- Explain from the Bible the importance of what scientists do **BWS**
- Create a list of ways that students can use science to help others
- Classify an engineer as having a STEM career

Materials
- object placed in a small, wrapped gift box

Teacher Resources
- Instructional Aid 1.2: *STEM The Engineering Design Process*, one per student
- Visuals 1.1–1.2: *Bible Verses A*; *Bible Verses B*

Vocabulary
- scientist
- engineer
- design

INTRODUCTION

- Direct students to the Big Question on page 2.
  Why did God give people five senses? to observe and learn about His world
  What are your five senses? sight, touch, taste, smell, and hearing
  Today we will learn about people who use their senses to study God's world.

PREPARATION FOR READING

- Preview and pronounce the vocabulary term *scientist*.
- Preview and pronounce the word *curious*.
  What is the heading? "What Is a Scientist?"
- Allow time for several volunteers to tell what they think a scientist is.
- Direct the students to read page 13 silently to find out what a scientist is.

TEACH FOR UNDERSTANDING

What is a scientist? a person who studies God's world

Why do scientists try to solve problems? to help others

Why do scientists ask questions? They are curious; they are trying to find answers to problems.

What does it mean to be curious? Elicit that it means to be eager to learn or to have questions.

- Show the students the wrapped gift box.
  What does this box look like? a present
  Are you curious to know what is inside the box?
  What do you do if you are curious? Elicit that the students can ask questions.
- Allow time for the students to ask questions before opening the gift.
  Look at the top picture. What do you see? scientists looking through microscopes
- Explain that microscopes are science tools that make things look bigger.
  Look at the middle picture. What do you think the scientist is doing? Possible answer: collecting a sample of water
  What do you think he will do with the water? Possible answers: study it, test it, compare it to the water in another stream
  Look at the bottom picture. What do you think the scientist is doing? Possible answers: observing the plants, recording what he observes
  What senses are the scientists in these pictures using? Possible answers: sight and touch

PREPARATION FOR READING

- Direct the students to read page 14 silently to find out why a scientist's job is important.

TEACH FOR UNDERSTANDING

- Read together Genesis 1:1 from *Bible Verses A*.
  Who does the world belong to? God
  Why does the world belong to God? because He made it
- Direct attention to Genesis 1:28 from *Bible Verses A*. Ask the students to follow along on Activities page 1 as you read the verse aloud.
  What job did God give to people? to have dominion over the earth

> Some Bible translations use the word *dominion* in Genesis 1:28. If the translation your students are using does, be sure to explain that the word *dominion* means to rule over, to care for, and to manage.

- Point out that God has given people the job of ruling over, or caring for, the earth and all the plants and animals that He has made.
  When you care for the earth, who are you obeying? God
- Direct attention to Mark 12:30–31 on *Bible Verses B*. Ask the students to follow along on Activities page 11 as you read the verses aloud.

> The students will take *Bible Verses*, Activities page 11, home tonight. See the explanation at the end of the lesson.

What does the Bible say is the first commandment? to love God
What does the Bible say is the second commandment? to love our neighbor just like we love ourselves; to love other people
How does a scientist obey God's commandment to love other people? by helping them
How can a scientist help other people? Possible answers: keep water clean; finds ways to grow more food by making plants stronger; finds ways to make sick people well
Why is a scientist's job very important? Elicit that a scientist is following God's commandments to love others by helping them and that he is having dominion over, or caring for, the earth.

God made the world.
The world is God's world.
He tells people to care for the earth.
He tells us to love our neighbors.
A scientist shows love when he helps other people.

What does a scientist do?

14

Can you be a scientist this year? yes
Name ways you can use science to help other people today or when you get older. Possible answers: learn about what makes people sick; learn about animals to raise animals for food or to take care of animals; learn about plants to grow a garden or farm
- As the students name ways, write a list for display.
- Direct the students to Activities page 9. Read the scenarios aloud.
- Allow the students to choose a story and discuss it with a science partner. Encourage them to think like a scientist.
- Allow the teams to share their ideas orally with the class.

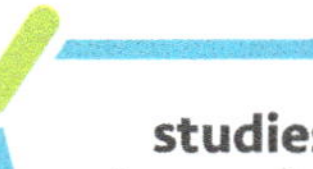

answer
studies God's world; uses his senses to learn about God's world; asks questions; tries to find answers; helps other people

STEMCareers

Engineer

An **engineer** uses science and math to find answers to problems.

Engineers **design**, or plan, something.

They use a design process.

The design process has steps.

There are different types of engineers.

Some engineers design cars.

Others design buildings.

Some engineers design robots.

Do you like to find answers to problems?

Maybe you can be an engineer.

15

Science Groups

Science groups are a good way to encourage cooperative learning. Using groups also helps limit needed supplies while giving each student an opportunity to participate. Groups of two or three students are best for most activities. Assign tasks to the group members, such as getting the supplies, measuring, and recording information.

STEM Career pages will appear throughout the Student Edition. These pages are designed to explain various STEM careers and develop student interest in careers that require STEM (science, technology, engineering, and mathematics) skills.

PREPARATION FOR READING

- Preview and pronounce the vocabulary terms *engineer* and *design*.
- Direct the students to read page 15 silently to find out what an engineer uses to find the answer to a problem.

TEACH FOR UNDERSTANDING

- Display the acronym STEM.
  Science
  Technology
  Engineering
  Math
- Explain that each letter stands for a different word. Read aloud each word and pronounce it together.
  What does the *S* in STEM stand for? science
  What is science? the study of God's world
  How do we study God's world? by using the five senses
  What does the *E* stand for? engineering
- Underline *engineer* in Engineering. Point out that the word *engineer* is part of the word *engineering*.
  What do you think the word *career* means? Elicit that it is a person's job, or work.
  What does an engineer do? uses science and math to solve or find the answer to problems
- Explain that STEM skills are used by people every day to solve or find the answers to problems. Some careers, such as engineering, use more than one STEM skill.
  What does the word *design* mean? plan
  What do we call the plan that is used by an engineer to finds the answers to a problem? design process
  Have you ever made your own paper airplane or planned a way to make a fort out of blankets? Ask for a show of hands.
- Point out that when the students designed a paper airplane or made a blanket fort, they did the work of an engineer.
  What kinds of things do engineers design? Possible answers: cars, buildings, robots, bridges

- Direct attention to The Engineering Design Process. Explain that these are the basic steps an engineer uses to find answers to everyday problems.

While the engineering design process that is used in *Science 1* has six steps, there are variations that include fewer or more than six steps. Often the steps are illustrated cyclically to indicate that the process or parts of the process may need to be repeated many times to reach a workable solution.

Count the steps. How many steps does The Engineering Design Process have? six

What is the first step in the design process? Ask

- Explain that the first thing an engineer asks is "What is the problem?" Point out that engineers try to find answers to everyday problems.

Once an engineer has identified the problem, what is his next step? Imagine

What do you think the word *imagine* means? Elicit that it means to think, or to form a picture in your mind.

- Explain that when an engineer imagines, he thinks about a possible answer to his problem.

What is the third step in the design process? Plan

- Point out that when an engineer plans, he writes down what he is going to do to find the answers to the problem. He will draw a design or picture and label the parts of his design.

Look at the top picture on page 15. What do you think the engineer is doing? Elicit that he is drawing plans of the car he is going to make.

What is the next step in the design process? Make

- Explain that the engineer will make a model of his design or picture. A model is an example of the answer the engineer has imagined.

Why do you think the engineer makes a model? Possible answers: to see if it will work; to make sure it will answer the problem

Look at the bottom picture on page 15. What step has the engineer finished? Make

In step five, what do you think the engineer will make better? Possible answers: any problems found in the model; anything that did not work right

Why do you think the engineer tests his model? If the model does not find the answer to the problem, the engineer will make it better until the problem is solved.

What is the last thing an engineer does? Share

- Explain that an engineer tells others how his design answered the problem.
- Point out that in *Science 1* this year the students will have the opportunity to use the design process and become engineers.

- Give the students a copy of *STEM The Engineering Design Process* to keep in their science notebooks.

In Lesson 16, the students will use the The Engineering Design Process to conduct their first STEM activity.

What Is a Scientist?, page 10
This page reinforces the concepts taught in this lesson.

Bible Verses, page 11
These Bible verses will be used throughout *Science 1*. Send the pages with the verses home to be read each night with an adult or other family member. By reading these verses for several nights, the students will become more familiar with them. Ask the students to fill in the blank in the directions with the number of nights that you would like them to read the verses.

Lessons 5–6

What Is a Scientist's Worldview?

You study science to learn about God's world.

Scientists study God's world too.

The way a person thinks about the world is called a **worldview**.

All scientists have a worldview.

A scientist's worldview helps him.

It helps him explain what he finds.

Not all scientists believe what the Bible teaches.

Let's take a look at a biblical worldview of science.

17

Objectives

- Define *worldview* BWS
- Identify that every scientist has a worldview BWS
- Identify that God is the Creator of all things BWS
- Identify that God designed everything to work together BWS
- Identify that God made people in His own image to care for the earth BWS
- Infer that people learn science to take care of the earth and to help others BWS

Materials

- container
- video of gears working together (optional)
- picture of people digging a ditch
- picture of a field being plowed

Teacher Resources

- Visuals 1.1–1.2; 1.5–1.7: *Bible Verses A*; *Bible Verses B*; *Worldview*; *Worldview K-L Chart*; *A Biblical Worldview*

Cut apart the strips on the *Worldview* page. Fold each and place in a container. The strips will be used to illustrate a worldview.

Vocabulary
- worldview
- image

Two days have been alloted for this lesson. You may divide the lesson where it works best for your instruction time.

INTRODUCTION

- Generate interest and help the students recognize that they already have a point of view, or a worldview.
- Direct a volunteer to draw a strip from the container and read it aloud. Assist as needed.
- Ask the students to vote on whether they agree or disagree with the statement that has been read. After they have voted, say something similar to this statement.

  You have voted on how you think about the world. Fourteen of you think that God created the world in six days.

- Repeat with the remaining strips.

  Each of these statements tells how you think about the world. When you agreed or disagreed with a statement, you were telling everyone what you think about the world.

  Today you will learn about how a person thinks about the world.

PREPARATION FOR READING

- Preview and pronounce the vocabulary term *worldview*.
- Direct attention to the blue heading. Guide the students in reading it aloud together.

  What is a scientist? a person who studies God's world

- Preview and pronounce the word *biblical*. Write the word *Bible* beside *biblical*. Point out that *Bible* is part of the word *biblical*.

  When you see the word *biblical*, it is talking about something found in the Bible.

- Direct the students to read page 17 silently to find out if all scientists have the same worldview.

Teaching for Student Edition page 17 continues on the next page.

TEACH FOR UNDERSTANDING

What is a worldview? the way a person thinks about the world

- Relate the word *worldview* to the Introduction activity.

  Remember the activity that we did at the beginning of the lesson. Do you have a certain way that you think about the world? yes

- Allow several students to tell how they think about the world. After each one, tell him that the way he thinks about the world is his worldview.

  The way you think about the world is your worldview. Does every person have a worldview? yes

  Do all scientists have a worldview? yes

  Do all scientists believe what the Bible says? no

- What do you think that a biblical worldview is? Elicit that a biblical worldview is looking at the world through what the Bible says.

- Look at the picture at the bottom of page 17. What do you see? Elicit that a scientist is looking at a dinosaur bone.

- How does a worldview help this scientist? It helps him explain what he finds.

  If a scientist believes that the dinosaur and man were created on the same day by God, the scientist will explain that the dinosaur and man lived at the same time.

  If a scientist does not believe that the dinosaur and man were created at the same time, then he will explain that the dinosaur lived millions of years before man.

- Display *Worldview K-L Chart*. Write these words for display: *animals, care, create, creation, design, earth, glory, image, learn, others,* and *world.* Explain to the students that they may find these words helpful as they answer the four questions on Activities page 13.

- Read the four questions aloud. Direct the students to write what they know about each question under the word *Know.* Explain that if they do not have an answer, it is acceptable to leave it blank.

- Invite students to share some of their answers.

> Collect the Activities pages and save them. In Lesson 88 the students will answer the questions again, showing what they have learned since the beginning of the school year.

PREPARATION FOR READING

- Direct attention to the headings on pages 18–19. Guide the students in reading them aloud together.

- Direct the students to read pages 18–19 silently to find out how God is an Engineer.

Where did our world come from?

God created the earth.

God created the land and plants.

God created the sun, moon, and stars.

God created the animals.

He created man and woman.

God created all things in six days.

God created a perfect world.

18

TEACH FOR UNDERSTANDING

- A graphic organizer helps students make pictures that show relationships and connections between facts, information, and terms. In this web the center oval shows the topic—worldview. The next set of colored shapes lists some fact or trait about the topic. In this web the next set of shapes lists the four biblical worldview questions. The last set of shapes lists the answers to those four questions.

- Display *A Biblical Worldview* web. Guide the students as they complete the definition for *worldview* in the center of Activities page 15.
  Note: Activities page 15 is intended to be completed during the discussion. Students may complete the web organizer as a reinforcement exercise.

- Genesis 1:1 says, "In the beginning God created the heaven and the earth." What does the word *create* mean in this verse? Elicit that it means that God spoke and made all things.

- What did God make the heavens and earth from? from nothing

- Name some things God created. Possible answers: earth, light, land, plants, sun, moon, stars, water, animals, man, and woman

- Read Genesis 1:31 aloud.

  How long did it take God to create everything? six days

Why do things in our world work together the way they do?

God is the Engineer of the earth.

He designed the earth.

God's design is beautiful.

God is the Artist of the earth.

He designed the parts of the earth to work together.

Each part works with the other parts.

They work together like gears in a machine.

How is God an engineer?

19

Worldview Terms

The following vocabulary may be helpful when discussing aspects of a biblical worldview.

Creation is the six-day event described in Genesis 1 when God created all things very good.

The **Curse** is the punishment described in Genesis 3 that God placed on all of His creation because of Adam's sin.

Dominion is the job God gave to people to manage the world for the benefit of others and the glory of God. Terms used in the Student Edition in conjunction with this concept include *care for*, *manage*, and *rule over*.

The **Fall** is the event described in Genesis 3 when the first man sinned and brought death into the world.

The **Flood** is the event described in Genesis 6–9 when God caused the entire earth to be covered by water.

Redemption is Christ's act of rescuing and freeing from sin.

What is the answer to the first biblical worldview question, "Where did our world come from?" God created a perfect world.

- Direct attention to the pictures that begin on this page and flow onto the next page. Discuss what is shown on the six days of Creation.
- Guide the students as they complete the answer to the first biblical worldview question on Activities page 15.
- Ask a volunteer to read the second biblical worldview question at the top of page 19.

What does an engineer do? Possible answers: uses science and math to answer problems; uses the design plan to answer problems

What does the word *designed* mean? planned

Why is God an Engineer? because He designed the earth

- Direct attention to the picture at the top of the page.

What is the picture showing us? the gears of a machine

Does each gear of the machine work by itself? no

How do the gears work? Possible answers: They work together. They help each other turn together.

- Explain that each cog, or tooth, of the gear fits in between a cog from another gear. The gears work together to turn and help the machine work. Show a video of gears working together.

How did God design the parts of the earth? for each part to work together

- Point out that God is the marvelous Engineer of the earth. He designed a beautiful world where all the parts work together.
- Mention that an example of the parts of the world working together can be found in Genesis 1:29–30. In these verses God says that He gave all the green plants and the fruit of the trees to be food for the birds and animals.

How would you answer the question "Why do things work together the way they do?" God designed everything to work together.

- Guide the students as they complete the answer to the second biblical worldview question on Activities page 15.

answer

An engineer designs to answer problems, and God has designed the parts of the earth to work together.

PREPARATION FOR READING

- Direct attention to the headings on pages 20–21. Guide the students in reading them together.
- Preview and pronounce the vocabulary term *image*. Direct the students to locate the bold word and then to look for its meaning as they read.
- Direct the students to read pages 20–21 silently to find out what job God gave people.
  Note: The students do not need to read *Meet the Scientist* on page 21.

TEACH FOR UNDERSTANDING

Who is the most important part of creation? people

- Ask a volunteer to read Genesis 1:26 from the page.
  How did God make man? in His image
  What does the word *image* mean? likeness
  God made man in His image. The word *man* means people.
  God made all of the people in His image. You do not look like God. But you are made like God. God thinks, loves, and talks. Like God you can think, love, and talk.
- Point out that animals are not made in God's image. Only people are. That is why animals cannot do all the things people can do.
  What is another way we are like God? by our work
- Display *Bible Verses A* and read Genesis 1:28.
  God told people their job was to rule over the earth. What do you think it means to rule over the earth? Accept any reasonable answer.
- Explain that this does not mean telling other people what to do, like a queen, or a king, or a president does.
  God wanted people to rule over His world by caring for the earth and the animals. He wanted Adam and Eve to be in charge of all the things He had made. He wanted them to manage the things He had made. It was their job.
- Direct attention to the picture.
  What is happening in the picture? The boy is helping his father fix a bike.
  How does this picture tell you why you are here? Possible answers: God told people to care for the earth. God told people to help others. God told people to manage the earth.
  What is the answer to the third biblical worldview question, "Who are we? Why are we here?" We are made in God's image to care for, or rule over, the earth.
- Guide the students as they complete the answer to the third biblical worldview question on *Activities* page 15.

Who are we? Why are we here?

People are the most important part of God's creation.

In Genesis 1:26 God said, "Let us make man in our image, after our likeness."

God created you in His **image**, or likeness.

God created all people in His image.

You are like God in many ways.

You are like God through your work.

God tells people to rule over the earth.

God tells you to care for the earth.

God tells you to help others.

20

ACTIVITY

See Your Image

Pass around a hand mirror. Invite each student to look at himself in the mirror. Point out that when he looks in the mirror he is looking at his image. Remind each student that God made him in God's image, or likeness. Each student can think, love, and talk, just as God can.

HELPS

Inference

To answer the fourth biblical worldview question, the students should use inference. An inference is an idea or conclusion that's drawn from evidence and reasoning. It is an educated guess, or the process of inferring things from what is already known. The questions for the fourth question are an example of using inference. You may need to ask some different questions, depending on what answer the student gives. Your questions should lead the student to the conclusion. *Note:* The student does not need to know the terms *inference* or *infer* at this time.

Why is science important?

Science is helpful in your life.

You use science to see God's glory.

The world shows you that God is great.

Science helps you learn how nature works.

Then you can use nature to help people.

You use science to care for the earth.

✓ **How can you use nature?**

Meet the Scientist

Sir Francis Bacon was a scientist from long ago. He believed in God and the Bible. He found a way to study science. He started with facts. He used tests to answer problems. His plan is known as the scientific method.

21

Scientist: Francis Bacon

Sir Francis Bacon (1561–1626) was born to a well-established couple in London, England. His parents were Sir Nicholas Bacon and Lady Anne Cooke. He became Sir Francis Bacon after being knighted by King James I in 1603. He developed the inductive version of the scientific method. We know that Bacon believed in the Bible and God but cannot be certain he accepted Christ as his savior.

See Helps on page 22 before beginning the discussion of page 21.

* What is science? the study of God's world
  How can science be helpful? to see God's glory, to see how great God is, to learn how nature works, to make changes
* Direct attention to Genesis 1:28 on Activities page 1 or *Bible Verses A*.
  What does Genesis 1:28 teach us about our job? Possible answers: People are to rule over, care for, or manage the earth.
  Who has given us the job of taking care of the earth? God

---

* What are some ways we can care for the earth? Possible answers: take care of the plants; take care of people; take care of animals
* How can we use nature to help others? We can use nature to make changes and then use these changes to help others.
* Display the pictures of digging a ditch and plowing a field. Elicit that nature is being changed in each picture and ask how the change is being used to help others. by providing pipes for water or gas; providing food for people or cotton for clothes
* Direct attention to Mark 12:30–31 on Activities page 2 or on *Bible Verses B*.
  What does Mark teach us is the first commandment? to love God
  What is the second commandment? to love our neighbor like we love ourselves; to love other people
* How can you show love to other people? by helping them
* Direct the students to think about what two things the verses in Genesis and Mark teach us. Then direct them to use these two things to answer this question:
* Why is science important? We learn science to take care of the earth and to help others.
* Guide the students as they complete the answer to the fourth biblical worldview question on Activities page 15.
* Direct attention to *Meet the Scientist*. Read aloud about Francis Bacon as the students follow along.
* Why do you think Francis Bacon has the "Sir" in front of his name? He was knighted by the king.
* Why is Bacon's plan called the *scientific method*? It starts with facts and uses tests to solve problems.
  Note: The students will learn about the scientific method in the next chapter.

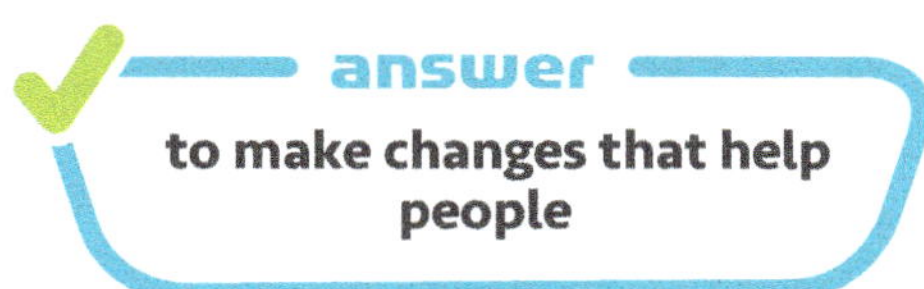

✓ **answer**
to make changes that help people

Worldview K-L Chart, page 13
The students will complete the Know column for each question. Save this page for Lessons 88–89.

A Biblical Worldview, page 15
Complete the web as part of the lesson, or instruct the students to complete the page as reinforcement.

Study Guide, pages 17–18
These pages review Lessons 4–6. Direct the students to keep the finished pages for review for Test 1.

Objective

- Recall terms and concepts from Chapter 1

REVIEW

- Material for Test 1 will come from the Study Guides on Activities pages 7–8 and 17–18. The same Write About It from the Study Guide will appear on each chapter test. You may review any or all of the material during the lesson.
- You may choose to review Chapter 1 by playing the *Dress the Scientist* review game or a game from the Game Bank. For your convenience, the Game Bank is available in the back of this Teacher Edition and on TeacherToolsOnline.com.

♛ REVIEW GAME

Dress the Scientist

Materials: *Dress the Scientist* (Instructional Aid 1.3), one copy per team; magnetic board; magnets

Cut apart the scientists and the pieces of science equipment. Display the scientists on a magnetic board. Divide the students into teams. Alternate asking each team review questions, using the questions from the Activities Study Guides. If a student's answer is correct, he attaches a piece of science equipment with a magnet to his team's scientist. If the answer is incorrect, the question is asked to the opposing team. The team that dresses their scientist first wins.

NOTES

Objective
- Recall and apply terms and concepts from Chapter 1

ASSESSMENT

- Administer Test 1.

Tests can serve as the objective part of the evaluation of a student's progress. The most effective tests are an outgrowth of the teaching process. Accordingly, these tests should not replace your individual assessment of a student's understanding and application. The tests can be adapted to meet the teaching emphasis and direction as well as the student's maturation level. You may find it necessary to eliminate some items, provide additional test items, or do both.

NOTES

Chapter Objectives

- Use science process skills and age-appropriate tools to gather information and reach conclusions
- Explain how people use their senses and other tools to learn about God's world **BWS**
- Explain how Genesis 1:28 states one of the main purposes for what scientists do **BWS**
- Apply the basic steps of the scientific method
- Apply the engineering design process to solve a real life problem

Lesson Objectives

- Recall what science is and what scientists do
- Define *science process skill*
- Observe an object using the five senses
- Classify objects based on a chosen criteria
- Measure an object using a non-standard unit
- Classify science process skills as *observe, classify*, and *measure*

Materials

- pretzel, one per student *Note:* A cracker or similar crisp item may be substituted for the pretzel.
- several pretzels, in a variety of shapes and sizes, in a small self-closing plastic bag, one bag per student

Teacher Resources

- Instructional Aid 2.1: *Science Process Skills*, for display and one per student
- Visual 2.1: *Using My Senses to Observe*

> For your convenience, the Teacher Resources are available in the back of this Teacher Edition and on TeacherToolsOnline.com.

Vocabulary

- science process skill
- classify
- measure

CHAPTER INTRODUCTION

> Chapters 1 and 2 contain information that is foundational to *SCIENCE 1*. Chapters 1 and 2 should each be taught in their entirety.

- Direct attention to the chapter title on page 23.
  What is the title of this chapter? "What Scientists Do"
  Look at the picture on page 22.

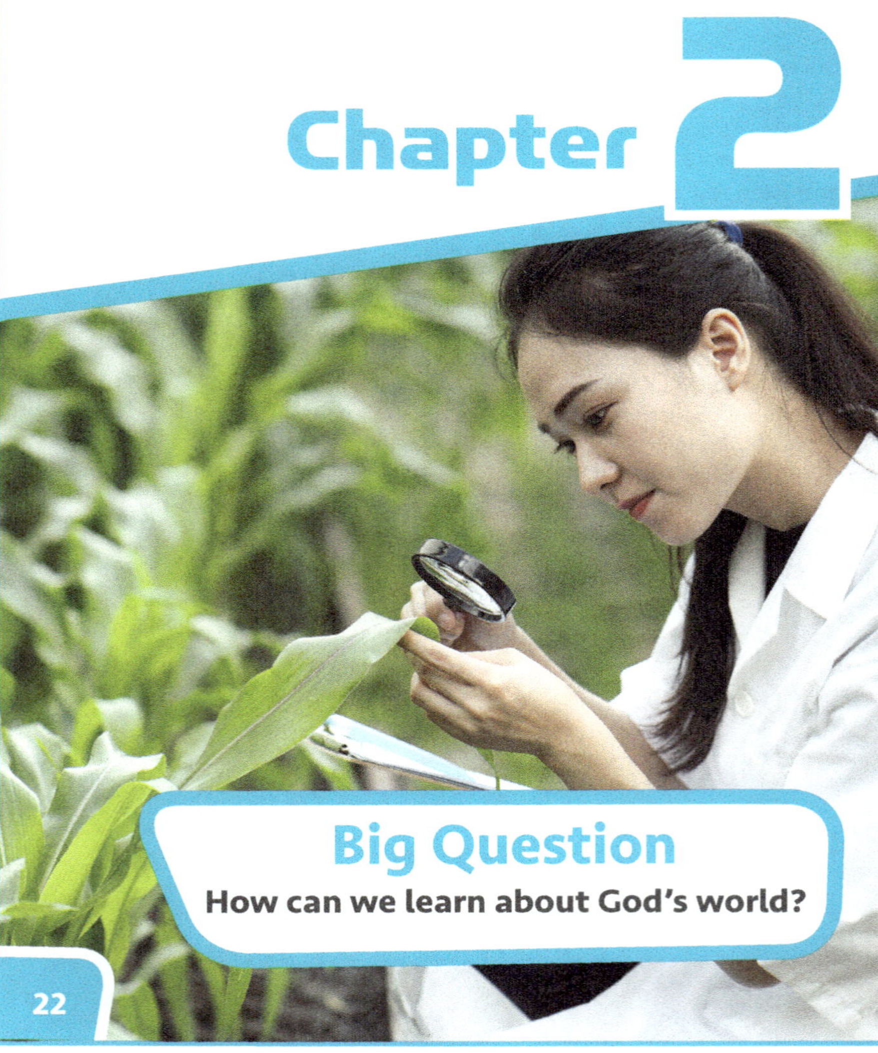

- What is this a picture of? a woman looking at a plant leaf; a scientist studying a leaf
- Why do you think this woman is a scientist? Possible answers: she is wearing a lab coat; she is using a hand lens, a science tool, to observe a leaf
- Do all scientists wear a lab coat? no
- Discuss that some scientists wear a lab coat, but many do not. Explain that the students will learn about scientists who wear lab coats and about some who do not.

Big Question

- Direct attention to the Big Question on page 22. Read the question aloud.
- How do you think we can learn about God's world? Answers will vary.
- Explain that the answer to the Big Question will be found as the students read Chapter 2.

What Scientists Do

Scientists study God's world.

Scientists use skills to help them.

They also use tools to help them.

They do tests to find answers to problems.

You can be like a scientist.

You can learn about God's world.

23

- Direct the students to read page 23 silently to find out what scientists use to help them study God's world.

- Remind the students that in Chapter 1 they learned what science is and what scientists do.

  What is science? the study of God's world

  What do scientists do? Answers should include that scientists study God's world; they use their senses to observe the world.

  What do scientists use to help them study or observe God's world? They use skills and tools.

  What do scientists do to find answers to problems? They do tests.

- How can you be like a scientist? Possible Answers: I can use my senses to learn about God's world; I can learn about God's world by using skills and tools and by doing tests.

Chapter Photo
The photo on page 22 is of a scientist using a hand lens to study the leaf of a corn plant.

Visit TeacherToolsOnline.com for resources to enhance the lessons.

- Preview and pronounce the vocabulary term *science process skill*.
- Direct the students to read pages 24–25 silently to find out why they observe.

- Explain that a skill is the ability to do something.
- Discuss the kinds of skills the students may use, such as reading, writing, hitting a ball, riding a bike, drawing a picture, or playing an instrument.

What do we call the kinds of skills that scientists use? science process skills

What are science process skills? skills that help a scientist discover and use facts

What can science process skills help you to do? learn about God's world

Lesson 9

Science Process Skills

Scientists use different skills to study the world.

These are called science process skills.

A **science process skill** helps a scientist discover and use facts.

You can use science process skills to learn about God's world.

Observe

You observe to learn about God's world.

You see, hear, taste, smell, and touch.

You use your senses to observe things.

✓ **Why do we use science process skills?**

25

⚠ **HELPS**

Science Notebook

You may choose for each student to keep a science notebook to provide a place to save science-related pages throughout the course. In addition to the Activities pages such as Bible Verses, Investigations, Explorations, and Study Guides, they may also place Instructional Aids in their science notebooks for reference. A three-ring binder works well for a science notebook.

ACTIVITY

Spy with My Little Eye

Choose an object in the classroom and describe how it looks, feels, sounds, or smells. When a child correctly guesses the item that has been described, invite the student to share which sense(s) he used to identify the object. Repeat the game two or three more times.

What does *observe* mean? To observe means to use your senses to find out something.

Why do you observe? to learn about God's world

What do you use to observe? your senses

What are your five senses? sight, touch, smell, taste, and hearing

- Display the *Science Process Skills* chart and distribute one copy to each student. Direct attention to the word *observe* and its definition. Ask a volunteer to read the word and definition aloud.
- Direct attention to the pictures on pages 24–25.

Which picture shows someone using the sense of sight? the girl looking at the leaf

Which picture shows someone using the sense of touch? the girl hugging the kitten

Which picture shows someone using the sense of smell? the girl smelling the flower

Which picture shows someone using the sense of taste? the boy eating the watermelon

Which picture shows someone using the sense of hearing? the boy with the shell up to his ear

- Direct attention to the Big Question on page 22.

How can we learn about God's world? We can use our senses to see, touch, smell, taste, and hear things.

- Display *Using My Senses to Observe*. Point out that the object being observed is a pretzel. Distribute one pretzel to each student.
- Direct the students to remove Activities pages 21–22 from their books.
- Guide the students as they use their senses to observe the pretzel and complete the chart. Be sure that taste is the last sense used for observation.
- You may choose to reinforce the science process skill of *observe* with the *I Spy with My Little Eye* activity.

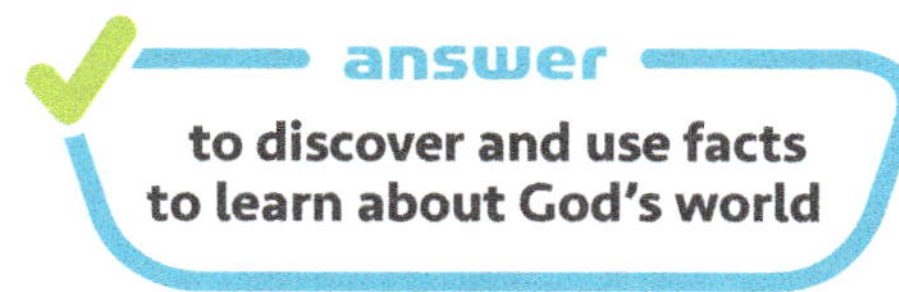

✓ **answer**
to discover and use facts to learn about God's world

- Preview and pronounce the vocabulary terms *classify* and *measure*.
- Direct the students to read pages 26–27 silently to find out what they do when they classify.

What do you do when you classify? group things that are alike

- Direct the students to form groups according to their hair color.

What are some other ways we could group everyone? Possible answers: by eye color, by boys or girls, by height, by shoe style, by clothing color

When you were grouped by hair color, what science process skill did we use? classify

- Point out that scientists separate things into groups to help scientists understand things better.
- Direct attention to the picture.

How are the things in the picture classified? by shoes and blocks

What are some ways the shoes and blocks can be classified? shoes: by kind (girls, boys, tie, do not tie) or size (large or small); blocks: by color or shape

- Direct attention to *classify* and its definition on the *Science Process Skills* chart. Ask a volunteer to read the word and definition aloud.

What are some other ways the shoes could be classified? Possible answers: by dress or play shoes; by color

- To practice the science process skill of classify, use the bag of pretzels from the materials list and the *Classifying Pretzels* activity.

Classifying Pretzels
Using the small bag of pretzels, instruct the students to classify the pretzels using any criteria. Invite the students to share with a partner how they classified their pretzels. Criteria may include shape, size, salted, or unsalted.

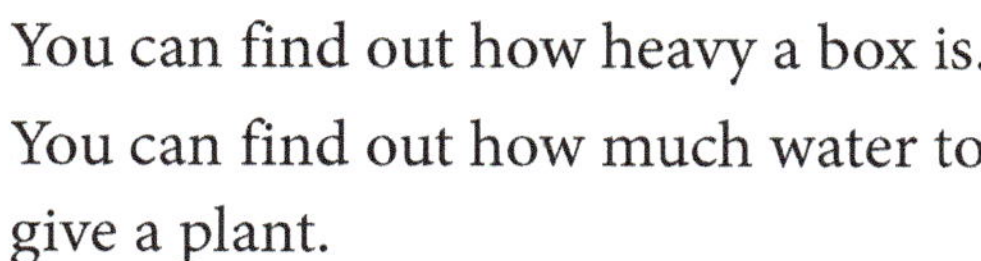

Measure

To **measure** means to find the size, number, or amount of something.

You can measure how long a pencil is.

You can find out the number of crayons in a box.

You can find out how heavy a box is.

You can find out how much water to give a plant.

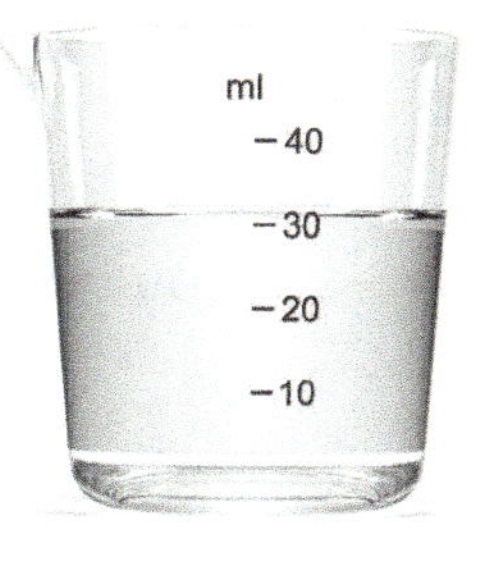

27

Why do we measure things? to find the size, number, or amount of something

- Direct attention to the word *Measure* and its definition on the *Science Process Skills* chart. Ask a volunteer to read the word and definition aloud.
- Direct attention to the pictures.

What is being measured in the top picture? how long a pencil is

Can you use your science book to measure how wide your desk is? Answers will vary.

- Direct the students to use the *Science 1* Student Edition to measure the width of their desks. Demonstrate this skill using the book and your desk.

How wide is your desk in science books? Since all students will have the same size science book, and if the students have the same desk style, answers should be the same.

What is being measured in the middle picture? how much water to give a plant

What is being measured in the bottom picture? how heavy a box is

- You may use the *What Can Be Measured?* activity to practice identifying the ways things can be measured.

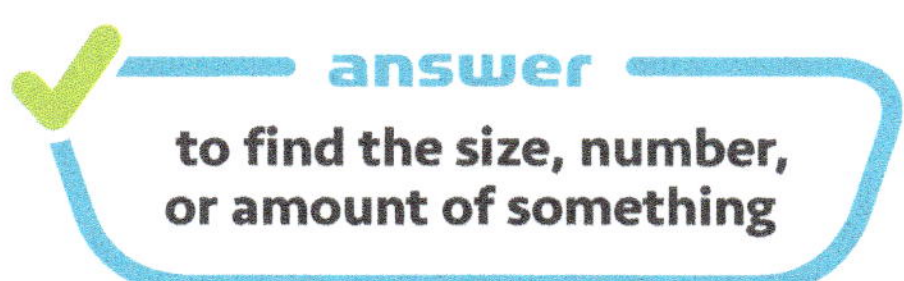

Direct the students to place the *Science Process Skills* chart in their science notebooks. Explain that they may refer to this chart as they need to throughout the course.

⚠ HELPS

Using Measuring Tools

This discussion introduces the science process skill of measuring rather than the use of measuring tools. Specific science tools are discussed in Lesson 11. Students will use the science tools in Lesson 12.

ACTIVITY

What Can Be Measured?

Materials: pictures of things that can be measured

Display pictures of common items and have the students think of different measurements that could be taken of that item. For example, a handful of jelly beans could be counted (how many) and weighed (how much). A potted plant could be weighed (how heavy) and its height measured (how long).

ACTIVITIES

Using My Senses to Observe, page 21

This page was completed during the lesson.

Science Process Skills, page 22

This page reinforces the concepts in the lesson.

Objectives
- Recall that the science process skills of observing, classifying, and measuring are ways people learn about God's world **BWS**
- Define *inference* as a science process skill
- Infer the cause from an effect
- Predict the outcome of a certain action
- Define what a *scientific prediction* is
- Identify *communicate* as a science process skill

Materials
- video of a weather prediction from a local newscast

Teacher Resources
- Instructional Aid 2.1: *Science Process Skills*
- Visual 2.2: *Cause and Effect*

INTRODUCTION

What are three of the science process skills people use to learn about God's world? observe, classify, measure

What do we use to observe God's world? our five senses

What do we do when we classify? group things that are alike

What is one way we could classify the students in this classroom? Possible answers: by hair color, height, eye color, boys or girls

Why do we measure? to find the size, number, or amount of something

In this lesson you will read about three more science process skills.

PREPARATION FOR READING

- Introduce and pronounce the words *infer*, *predict*, *guess*, and *communicate*.

> A vocabulary term is a bold word for which the students learn the definition. Some lessons include words that are not vocabulary terms but that may be difficult for the students to decode and pronounce. Assist the students with any words that may be difficult.

- Direct the students to read pages 28–30 silently to find out what you do when you communicate.

TEACH FOR UNDERSTANDING

- Direct attention to the picture.

What is this a picture of? a glass of water and ice

- Explain that the glass, ice, and water are things we know from looking at the picture.

Lesson 10

Infer

To infer means to use what you know to tell why things happen.

You infer after you observe.

You know that ice is cold.

You infer what ice will do to water.

28

- Read the caption aloud and ask a volunteer to answer the question. Elicit that the water is cold.

Why do you think the water is cold? There is ice in the glass and ice will make the water cold.

- Explain that recognizing that the glass of water is cold because there is ice in the water is called *inferring*.

What does it mean to *infer*? to use what you know to tell why things happen

When do you infer? after you observe

Suppose you look outside and see people wearing coats, hats, and gloves. What can you infer about the weather? It is cold outside.

- Display the *Science Process Skills* chart and read the definition for *infer* aloud.

- Explain that although the students may not be able to feel the weather, they can infer that it is cold outside by observing. They know that people wear clothing such as coats, hats, and gloves when it is cold outside.

- Direct attention to page 23 in the Activities book. Guide the students as they infer what is happening in each picture.

- Display the *Cause and Effect* visual and discuss the chart.

What is an effect? something that happens

What question is asked to find an effect? "What happened?"

Predict

To predict means to make a careful guess at what may happen.

You predict after you observe.

You observe that the sun's light is warm.

You can predict what ice will do in the sun's light.

What does it mean to predict?

29

Weather Forecasting
While weather forecasting introduces the science process skill of predicting, the specific job skills of a meteorologist will be discussed in Lesson 66.

What is a cause? why something happens
What question is asked to find a cause? "Why did it happen?"

- Explain that we can infer the *cause* from an *effect*.
- Explain that the *effect* in the first question on page 23 is that the boy is hot and sweaty. Point out that the *cause* is that he has been playing with his friends.

What is happening in picture number two? The girl is tasting a lemon.

Why does her face look that way? She is tasting something sour.

What is the *effect* in picture number two? The girl's face is scrunched up.

What is the *cause* that you can infer from the look on the girl's face? She is tasting something sour.

How do you know the lemon is sour and not sweet? by the look on the girl's face

- Discuss other examples of *infer* as time allows.

Teaching for Student Edition page 29 begins here.

- Invite students to tell about watching or hearing a weather report.
- Play a short video of a weather prediction from a local newscast.
- Explain that people who give weather reports use the science process skill called predicting. Scientists gather information about the weather and then use that information to predict what weather may happen in the future.

What does it mean to *predict*? to make a careful guess at what may happen

What do you do before you predict? observe

- Read the caption aloud and ask a volunteer to answer the question. Elicit that the ice will melt.

Why will ice melt when it is in the sun? because the sun's light is warm

- Display the *Science Process Skills* chart or direct the students to the copy in their science notebooks. Read the definition for *predict* aloud.
- Direct attention to page 24 in the Activities book. Guide the students as they predict what will happen next in each picture.
- Discuss other examples of *predict* as time allows.

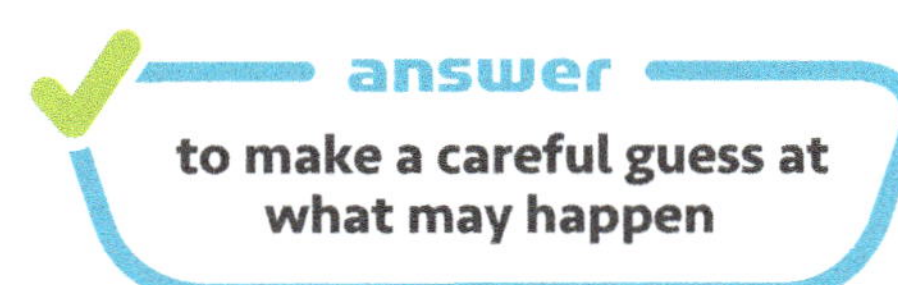

Communicate is another science process skill. What do you do when you communicate? **share with others what you know**

Suppose you saw a colorful butterfly in your garden. You could not catch the butterfly, but you wanted others to know what you observed. How could you communicate about what you saw? **tell about it, write about it, take a photo, draw a picture**

- Explain that scientists communicate, or share, what they have learned. Scientists will often write papers and scientific journal articles to communicate the results of their research. You may choose to display a scientific journal and allow the students to scan it.
- Display the *Science Process Skills* chart or direct the students to the copy in their science notebooks. Read the definition for *communicate* aloud.
- Discuss other examples of *communicate* as time allows.
- Direct attention to the Big Question on page 22.

How can we learn about God's world? **We can use the science process skills to observe, classify, measure, infer, predict, and communicate.**

ACTIVITIES

Science Process Skills, pages 23–24
These pages were completed during the lesson.

Study Guide, pages 25–26
These pages review Lessons 9–10. After completion, direct the students to keep them for review for Test 2.

Communicate

You communicate when you share what you know.

You may tell others what you know.

You may show others what you know.

30

☐ ACTIVITY

Practice Communicating

Materials: several small, interesting objects; shoebox with a lid

Place an object in the shoebox. Choose a student to look inside the box without letting anyone else see inside. Direct him to communicate what he saw without telling the name of the object. He can describe it, draw it, or use any other communication method. Allow the other students to guess what the object is. Repeat with other objects as time allows.

Lesson 11

Science Tools

A tool makes work easier.

A **science tool** is an object that helps scientists work.

They use tools to find out about God's world.

Tools help them observe and measure.

You can use science tools to observe and measure.

Why do we use science tools?

31

Thermometer
When a thermometer is referenced, it will be an alcohol or bulb thermometer, displaying degrees Celsius and Fahrenheit, as opposed to a digital thermometer, unless otherwise stated.

Objectives
- Identify science tools and their uses
- Measure length using non-standard and standard units
- Infer reasons for using standard units of measurement
- Explain how people learn about God's world **BWS**
- Explain from Genesis 1:28 why accurate measurement is important **BWS**

Materials
- hand lens
- previously used crayon, for each student
- ruler with metric and standard marks, for each student and teacher
- clear glass with water
- beakers or measuring cups with metric marks
- balance scale and weights
- thermometer with °C and °F markings

Teacher Resources
- Visual 2.3: *Bible Verse*

Vocabulary
- science tool
- temperature

INTRODUCTION

Scientists use tools. In this lesson you will learn about the tools that a scientist uses.

PREPARATION FOR READING

- Preview the vocabulary term *science tool* and pronounce it together.
- Direct the students to read page 31 silently to find out how science tools can be helpful.

TEACH FOR UNDERSTANDING

- Direct attention to the picture.
- What is the scientist doing? looking at red water; using a science tool to measure red water

  What is a science tool? an object that helps scientists work

  Why do scientists use tools? to find out about God's world; to observe and measure
- How can you be like a scientist? I can find out about God's world by using science tools to observe and measure.

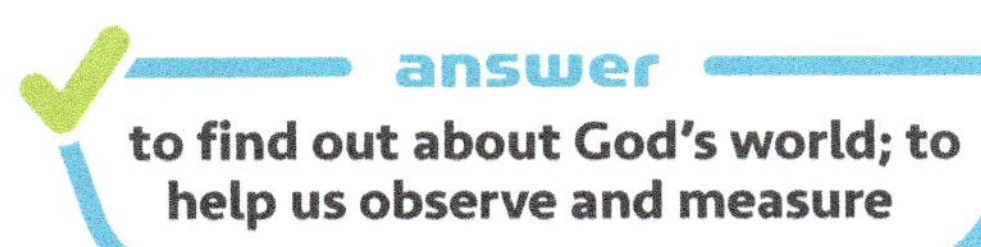

answer
to find out about God's world; to help us observe and measure

- Preview and pronounce the words *hand lens*, *ruler*, and *centimeters*.
- Direct the students to read pages 32–33 silently to find out what a ruler is used for.

- Show a hand lens.
  What do you use a hand lens for? to observe small things; to help make things look bigger
- When using a hand lens, how can you make things look larger and clearer? by moving the hand lens
- Demonstrate the proper use of a hand lens. Explain that you make the object look clearer by moving the lens closer to or farther away from the object you are looking at. Point out that when the lens is farther away from the object, the object looks bigger.

Hand Lens

You use a hand lens to observe small things.

A hand lens makes things look bigger.

Moving the lens makes things look larger.

Moving the lens makes things look clearer.

32

⚠ HELPS

Science Tools

This discussion introduces the science tools. Students will practice using each tool in Lesson 12. As each tool is introduced, emphasize the proper way to handle, care for, and store it. In this way the tool will be ready and available for the next time it is needed. Explain that caring for our tools is part of how we care for and manage the earth.

- Display a ruler with standard and metric marks. Make sure each student has a ruler with standard and metric marks for this discussion.
  What do you use a ruler to do? to measure how long something is
  What do rulers measure in? some measure in inches, some measure in centimeters, and some measure in both inches and centimeters
- Explain that inches are standard units of measurement and centimeters are metric units of measurement.
  What do scientists measure in? centimeters
  Look at your ruler. Look at the long lines.
  What numbers do you see marked at each of the long lines on your ruler? 0 through 12 (inches) and 0 through 30 (centimeters)
- Explain that the lines are like the ends of the crayon: the space from one long line to the next is one inch or one centimeter.
  Do you think an inch is the same on each ruler? yes
  Do you think a centimeter is the same on each ruler? yes
  Do you think an inch (or a centimeter) is the same length in Alaska as it is in New York? yes
- Point out that the lines for measuring on every ruler are the same distance apart. Explain that each ruler will give the same measurement no matter who uses it or where it is used.
- Explain that in science class the students will measure in centimeters.
- Using a metric ruler, demonstrate how to measure the length of a book.
- Direct the students to use their metric rulers to measure the width of their science books in centimeters. Instruct the students to write down the number.
- Allow several students to share their measurements in centimeters. Display the measurements.
  Were your measurements mostly the same or mostly different? the same
  Is it better to measure with a crayon or a ruler? ruler Why? The measurement will be the same even if someone different measures it.
- Explain that measuring with crayons can tell how long something is, but using crayons as a measurement unit can be confusing because not all crayons are exactly the same length.
- Point out that exact measurements for how long something is are important in science.

- Explain that the students are to measure the width of their *Science 1* Student Edition using a crayon. Students should not use a new crayon for this activity.
- Using a crayon, demonstrate how to measure the width of the science book.
- Instruct the students to write down the number of crayon lengths for how wide their science books are.
- Invite several students to share their crayon measurements. Display the measurements.
  Are your measurements mostly the same or mostly different? different
  Are your science books the same size or different sizes? same size
  What do you think might be a problem with using a crayon as a measuring tool? The crayons are not all the same size.
- Show the students several different sizes of crayons.
  What would happen if I measured my desk using different crayons? The measurements would be different each time.

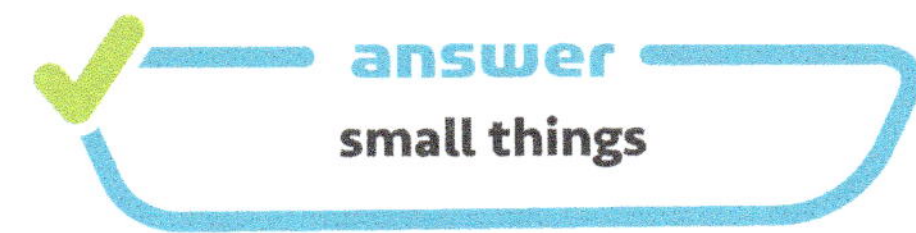

- Preview and pronounce the words *measuring cup*, *milliliters*, and *balance scale*.
- Direct the students to read pages 34–35 silently to find out what a beaker is.

What do you use a measuring cup to measure? how much space something takes up

What is a beaker? a measuring cup that scientists use

What do beakers measure in? milliliters

- Explain that everything takes up space. Point to several objects in the classroom and explain that each object takes up space.

 Do you take up space? yes

 Does water take up space? yes

- Direct attention to the beakers of green water on page 34. Point out the milliliter units. Explain that beakers measure the amount of space water takes up using milliliters.
- Display a clear glass with some water in it. Explain that we can use a measuring cup or beaker to find out how much space the water takes up.
- Direct attention to the beaker.
- Pour the water from the glass into the beaker. Be sure the amount of water is less than the size of your beaker.

To measure the amount of space the water takes up, you must look at the milliliter lines on the beaker at eye level.

- Demonstrate reading the measurement at eye level.
- Ask a volunteer to read how much space the water takes up. Encourage the student to express the answer in milliliters.

Measuring Cup

You use a measuring cup to measure how much space something takes up.

Beakers are measuring cups that scientists use.

Beakers measure in milliliters.

Volume
The definition for *volume,* not the term, will be used in this chapter.

Balance Scale

You use a balance scale to measure the weight of something.

A balance scale weighs objects to find out how heavy they are.

A balance scale weighs objects in grams or pounds.

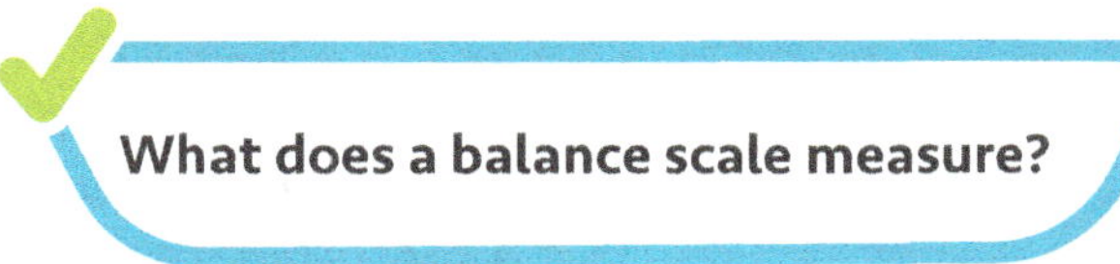

35

What does a balance scale measure? the weight of something

Why do we use a balance scale to weigh objects? to find out how heavy an object is

What does a balance scale weigh objects in? grams or pounds

- Direct attention to the picture.

What is the balance scale weighing? an apple

- Point out that the blue weights on the right arm of the scale measure weight in grams. Explain that the scale is balanced when the beam is level, or not tipped up or down.

How do you know the stack of weights weighs the same as the apple? because the balance is level

- Display your balance scale and demonstrate how to use it.

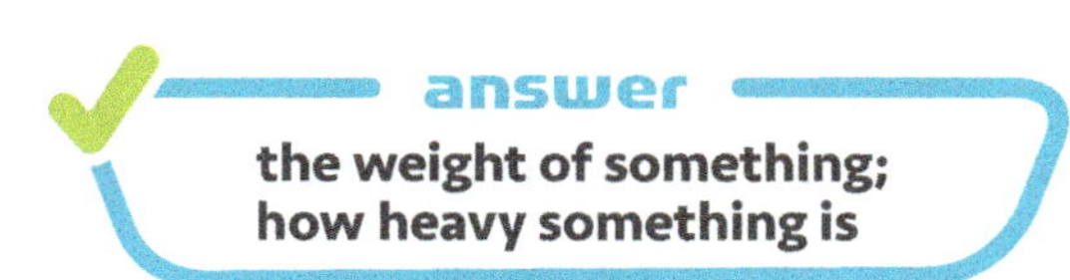

BACKGROUND

Balance Scale

A balance scale weighs an object against a standard weight. It has a beam that balances on a fulcrum. An object is placed on one end of the beam. The beam is balanced by placing standard units of measurement, in either pounds or grams, on the other end of the beam. When the beam is level, the weight of the object can be determined.

Grams or Pounds

A gram is a commonly used metric unit of mass measurement. A pound is the unit of mass in the English Engineering System.

⚠ HELPS

The word *matter* will be introduced in Chapter 11 and *mass* will be introduced in *SCIENCE 2*.

- Preview the vocabulary term *temperature*. Preview and pronounce the words *thermometer* and *Celsius*.
- Direct the students to read page 36 silently to find out what temperature is.

TEACH FOR UNDERSTANDING

What tool do we use to measure temperature? a thermometer

What is temperature? how hot or cold something is

- Display a thermometer that shows the temperature in °C and °F. Point out the Celsius and Fahrenheit markings and explain how to read a thermometer.

What do most scientists measure temperature in? degrees Celsius

- Explain that, while most other countries measure temperature in Celsius, the United States uses Fahrenheit.
- Ask volunteers to read a thermometer sitting at room temperature and then after sitting in the beaker of water for about a minute. Discuss how the temperature changed.

How do science tools help us to learn about God's world? We can use science tools to observe and measure things in God's world.

- Display *Bible Verse*. Direct attention to Activities page 19. Read Genesis 1:28 aloud.

What does it mean to "have dominion over"? to manage, care for, rule over

- Explain that science tools help people manage, care for, or rule over the earth. Point out that making careful measurements helps other people.

Explain from Genesis 1:28 why it is important to learn to measure correctly using science tools.

- Point out that God made people to rule over the earth. Science tools help people care for, or manage, the earth.

ACTIVITIES

Bible Verses, page 19
This Bible verse will be used in the teaching of this chapter.

Science Tools, pages 27–28
These pages reinforce the concepts taught in this lesson.

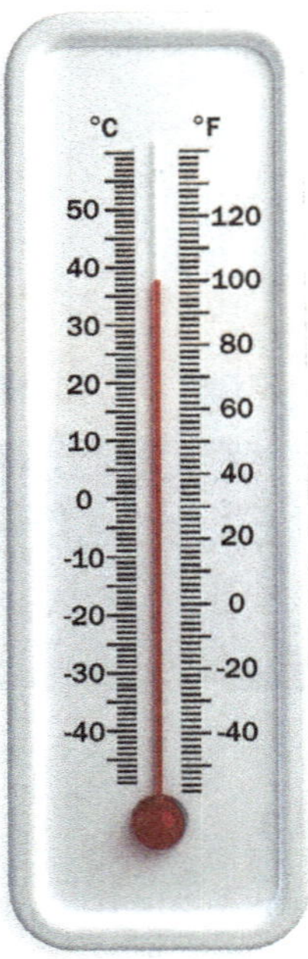

Thermometer

You use a thermometer to measure temperature.

Temperature tells how hot or cold something is.

Most scientists measure in degrees Celsius.

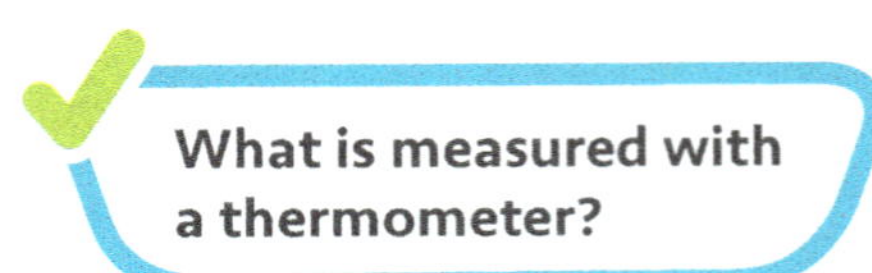

36

Measuring Stations
To limit needed supplies, you may choose to set up one measuring station for each tool used in Lesson 12.

Science Groups
Science groups are a good way to encourage cooperative learning. Using groups also helps limit needed supplies while giving each student an opportunity to participate. Groups of two or three students are best for most activities. You may need to assign tasks to the group members, such as getting the supplies, measuring, and recording information. A scoring rubric has been provided with the Assessments for each Exploration, Investigation, and STEM activity.

Units of Measurement
Most measurements will be made using metric units. Temperature may be recorded in Celsius or Fahrenheit.

Teacher's Measurements and Drawing
For Lesson 12, the students will compare their measurements and drawings to the teacher's. You may choose to do the Exploration ahead of time and have your measurements and drawing ready for display.

EXPLORATION

Process Skills
Observe
Measure

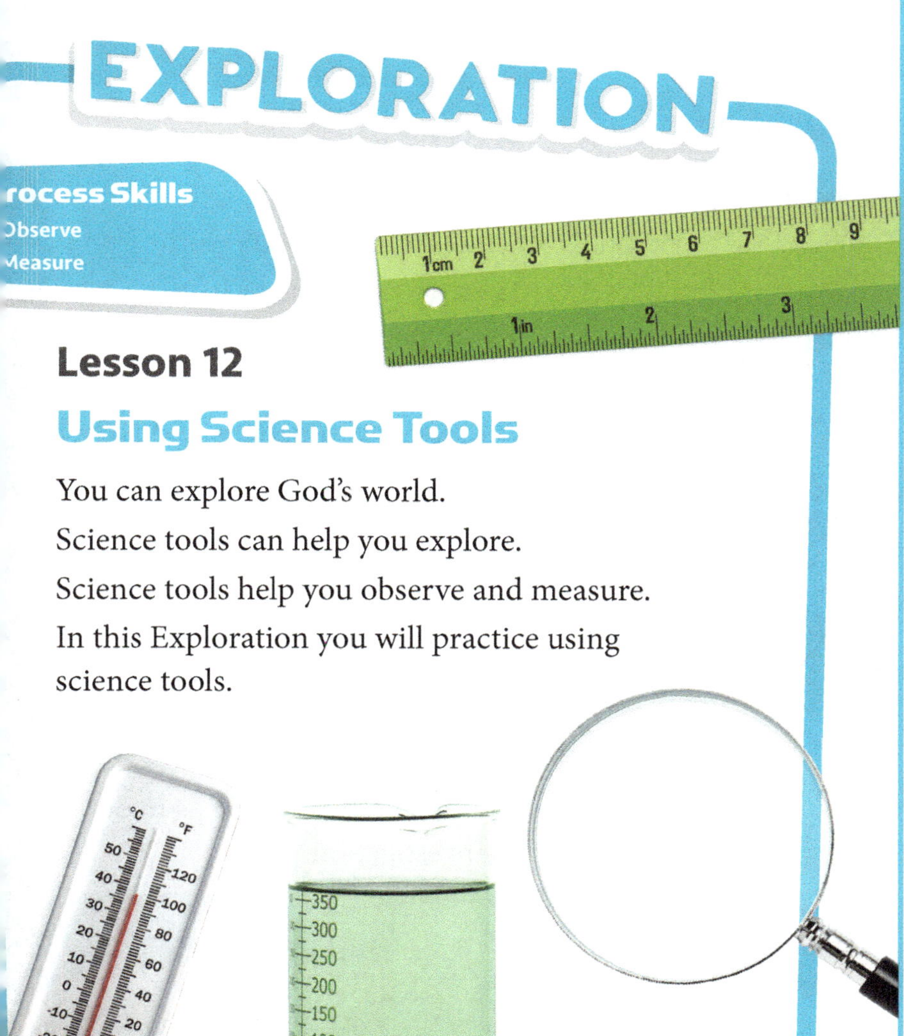

Lesson 12

Using Science Tools

You can explore God's world.

Science tools can help you explore.

Science tools help you observe and measure.

In this Exploration you will practice using science tools.

**LESSON
12**

Objectives
- Measure objects using age-appropriate science tools
- Record observations
- Compare and contrast observations
- Infer steps needed to determine accurate measurements

Materials
- See Activities page 31

Teacher Resources
- Visuals 2.4–2.5: *Science Safety Tips*; *Using Science Tools Chart*

INTRODUCTION

- Explain that we have different tools that we can use to help us observe and study God's world. Remind the students that learning how to use these tools will help us to take better care of the world.

- Direct attention to Activities page 29. Display and discuss the *Science Safety Tips*. Demonstrate proper use of each tool. Explain the purpose of the safety goggles and demonstrate their proper use. Safety goggles may not be necessary for each Exploration or Investigation.

- Set up measuring stations for Step 2. For example, students may measure the length of a piece of string, how much water is in a beaker, the weight of a small object, and the temperature of the air. Provide an object to be observed for Step 6.

PREPARATION FOR READING

- Direct the students to remove pages 31–32 from their Activities books.
- Direct the students to read page 37 and Activities pages 31–32 silently before beginning.
- Explain that it is helpful to look over all the pages before starting an Exploration. It will help the students to see what they are going to do and when they will need to write and record information.

TEACH FOR UNDERSTANDING

Process Skills
What science process skills will you be using in this Exploration? Observe and Measure
- Review the process skills *observe* and *measure*.

Purpose
- Direct attention to the Purpose.
What is the purpose of this Exploration? to practice observing and measuring by using science tools

Procedure
- Direct attention to the list of procedures. Explain that each step is to be followed. Encourage the students to check off each step as it is completed.
- For student comparisons, display *Using Science Tools Chart* with your measurements. Display your drawing.

Conclusions
- Discuss reasons for and ways to correct differences in the measurements or drawing.
- Why is it important to measure carefully? People can help care for the earth by measuring carefully.

ACTIVITIES

Science Safety Tips, page 29
Exploration: Using Science Tools, pages 31–32
The students will explore observing and measuring by using science tools.

ASSESSMENT

Rubric
Select the prepared rubric or design a rubric to include your chosen criteria.

Objectives
- Identify the purpose for an investigation
- Identify the steps of the scientific method
- Explain the purpose for the problem and hypothesis in a scientific investigation
- Create a hypothesis

Materials
- See Activities page 33

Teacher Resources
- Instructional Aid 2.2: *Steps of the Scientific Method*, for display and one per student

Vocabulary
- investigation
- hypothesis

INTRODUCTION

- Display the insulated lunch bag. Invite students to tell about events they have used a bag like this for.
- Do you think food will stay cooler inside or outside an insulated lunch bag? Accept any answer.
- How could we find out if the bag actually keeps food cooler? Possible answers: put something in the bag and use a thermometer to see if it gets warm; see whether something in the bag melts more slowly
- Explain that today you are going to be scientists and find out something by doing an investigation.

PREPARATION FOR READING

- Display the vocabulary terms *investigation* and *hypothesis*. Pronounce the words together.
- Direct the students to read pages 38–39 silently to find out what a hypothesis is.

TEACH FOR UNDERSTANDING

- What do scientists do? Answer should include that they study God's world; try to answer questions; try to solve problems.
- What do they use to find answers to questions? investigations
- What is an investigation? a test to find an answer to a question
- Explain that a test is the process or steps the scientist follows.
- The steps that an investigation follows are part of what? the scientific method
- Display *Steps of the Scientific Method* as this lesson is taught. Distribute one copy to each student.
- Explain that these are the steps they will follow this year as they do an investigation.

Problem

An investigation begins with a problem.

First decide what you want to know.

Then write a question about it.

You ask a question if you want an answer to a problem.

What question do you want to answer?

Hypothesis

Make a hypothesis.

A **hypothesis** is a possible answer to the question.

39

⚠ **HELPS**

Scientific Method Visual

You may choose to display a copy of *Steps of the Scientific Method* as a poster.

Investigation or Experiment

SCIENCE 1 uses the term *investigation* rather than *experiment* due to the introductory level of each investigation.

Set Up a Second Test

The investigation used in this lesson requires time for ice to melt to a noticeable amount. You may set up a second test of materials at least an hour before the lesson so that ice can begin melting. This will allow the students to make observations and draw conclusions during the lesson.

- Direct the students to remove pages 33–34 from their Activities books.
- Guide the students as they complete the pages during the lesson.

Problem

- Direct attention to Step 1 of *Steps of the Scientific Method*.

  What does an investigation begin with? a problem

  What is the problem? a question that needs to be answered

  ❋ At the beginning of the lesson, what question did we want to know the answer to? Will something stay cooler in an insulated lunch bag or outside the lunch bag?

  This question is the problem of the investigation you will conduct today.

- Direct attention to the Problem on Activities page 33. Read the problem aloud. Point out that the problem is a question.

- Explain that for this investigation we will use two cups of ice. One cup will be inside the insulated lunch bag and one will be outside the bag. Explain that the problem is to see which cup of ice will melt slower.

Hypothesis

- Direct attention to Step 2 of *Steps of the Scientific Method*.

  What is a hypothesis? a possible answer to the problem, or question

  What does an investigation test? It tests the hypothesis to find out whether it is true.

- Explain that when we make or form a hypothesis, we predict a possible answer to the problem, or question, by using the scientific method.

- Allow the students to share what they already know about ice, such as the fact that ice stays frozen in a freezer but melts in warm places.

- Direct the students to complete the hypothesis on Activities page 33 by circling either *faster* or *slower*. Point out that the hypothesis is written as a statement and there are no wrong answers for the hypothesis.

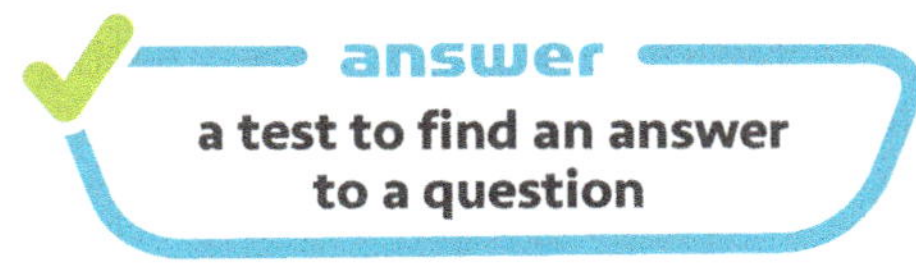

PREPARATION FOR READING

- Preview and pronounce the words *materials*, *procedure*, *observations*, and *conclusions*.
- Direct the students to read pages 40–42 silently to find out what the procedure is.

TEACH FOR UNDERSTANDING

Materials

- Direct attention to Step 3 of *Steps of the Scientific Method*.

  What are the materials? the supplies needed to do the investigation

- Direct attention to the list of materials on Activities page 33. Choose a student to read aloud the materials list.

  Look at the picture on page 40.

  Can you find all the materials listed in the picture? yes

Procedure

- Direct attention to Step 4 of *Steps of the Scientific Method*.

  What is the procedure? the steps of the investigation

- Point out that we will test the hypothesis by following the procedure.
- Direct attention to the procedure on Activities pages 33–34. Explain that each step is numbered and each step needs to be followed carefully as we test the hypothesis. Point out the check boxes and encourage the students to check off each step of the procedure as it is completed.
- Follow the procedure on Activities pages 33–34. Read Steps 1–3 aloud and ask for volunteers to help with each step.

  Why do you think it is important to fill each cup with the same number of ice cubes? to make sure the test is fair; if the number of ice cubes is not the same, the test will not be fair

  Suppose I put one cup of ice on the windowsill in the sun and the insulated bag with the other cup of ice somewhere else in the room. Would that be a fair test? No; the cup inside the insulated bag and the cup outside the bag need to be in the same place for the test to be fair.

Guide the students in supporting "yes" or "no" responses.

Materials

Gather the materials.

The materials are the supplies needed to do the investigation.

Procedure

The procedure is the steps of the investigation.

Follow the steps carefully to test your hypothesis.

40

- Ask a volunteer to read aloud Step 4 of the procedure.

You may finish this lesson after one hour, or you may choose to use the second test to make the observations immediately. If you use the second test, explain that you prepared a second test one hour earlier to save time during the lesson.

Observations

Observe what happens.

Use your five senses to observe.

Write down what you observe.

Draw what you observe.

41

Observations

- Direct attention to Step 5 of the procedure on Activities page 34. After the set amount of time, allow the students to observe what happened to the ice.
- Direct attention to Step 5 of *Steps of the Scientific Method*.

  What do you use to observe? five senses

  Scientists observe what happens when they do an investigation.

- Discuss the importance of observing carefully when testing the hypothesis.
- Explain that scientists need to write down or draw what they observe to help them remember what has happened during the test.

  Which cup has more melted ice? the cup that was outside the bag

  If the ice melted more in the cup outside the bag, does it mean the ice in the cup inside the bag melted faster or slower? slower

- Direct the students to circle the answer choice that matches their observation.

Conclusions

- Direct attention to Step 6 of *Steps of the Scientific Method*.

  What are the conclusions in a science investigation? the answer to the test
- Explain that to "draw conclusions" means to "make conclusions."

  How do you draw or make conclusions? you read your hypothesis and study your observations
- Allow time for the students to write the answer to the three questions under *Conclusions*.
- Poll the students, by a raise of hands, to determine how many students predicted in the hypothesis that the cup of ice inside the insulated lunch bag would melt slower than the cup outside the bag.
- Point out that the conclusions tell a scientist whether the hypothesis was right or wrong. The conclusions tell what was learned from the investigation.
- Go over the answers to the conclusions on Activities page 34.

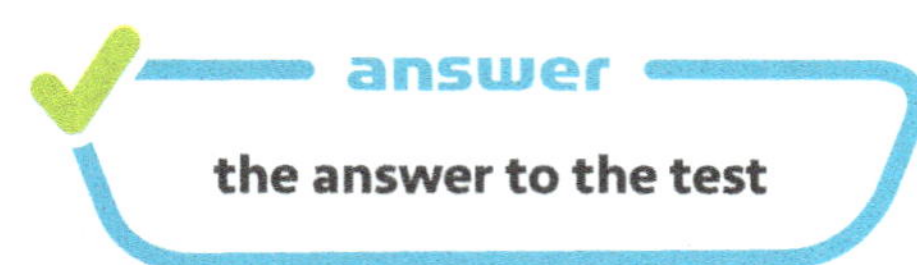

ACTIVITIES

An Investigation, pages 33–34
This first Investigation will be completed with the students during the lesson.

Study Guide, pages 35–36
These pages review Lessons 11–13. Direct the students to keep them for review for Test 2.

Conclusions

Read your hypothesis.

Study your observations.

Do your observations match your hypothesis?

Draw conclusions to tell what you learned.

Conclusions are the answer to your test.

42

STEM

Lesson 14
How to Keep My Pencil on My Desk

An engineer finds answers to everyday problems.

When he answers problems he is obeying God.

In Genesis 1:28 God tells people to rule over the earth.

An engineer uses math and science to rule over the earth.

An engineer follows a plan with steps.

The engineering design process is the plan an engineer follows.

In this STEM activity you will find an answer to an everyday problem.

You will use the engineering design process.

43

LESSON 14

Objectives
- Recall what an engineer does
- Identify the steps of the engineering design process
- Apply the engineering design process to solve a real life problem
- Relate the work of engineering to the commands of Genesis 1:28 **BWS**

Materials
- construction paper, one sheet per student
- pipe cleaner, one per student
- clear adhesive tape

Teacher Resources
- Visuals 2.3, 2.6: *Bible Verse*; *STEM The Engineering Design Process*

INTRODUCTION

Invite students to tell about a problem they needed to find the answer to.

Today you will find an answer to an everyday problem.

PREPARATION FOR READING

- Preview and pronounce the word *imagine*.
- Direct the students to remove pages 37–38 from their Activities books. Direct them to read page 43 and Activities pages 37–38 silently before beginning.
- Explain to the students that it is helpful to look over all the pages before starting a STEM Activity. It will help them to see what they are going to do and when they will need to write information.

TEACH FOR UNDERSTANDING

What does an engineer do? An engineer finds answers to everyday problems; he uses science and math to rule over the earth; he follows the engineering design process to find answers to problems.

- Point out that when engineers find answers to everyday problems they are obeying God.
- Display *Bible Verse* or direct attention to Activities page 19. Read aloud Genesis 1:28.

  In Genesis 1:28, what did God tell people to do? to rule, or have dominion, over the earth

  How does an engineer rule over the earth? Elicit that an engineer uses the engineering design process to find answers to everyday problems.

  What are the six steps in the engineering design process? Ask, Imagine, Plan, Make, Test and Make Better, and Share

- Display *STEM The Engineering Design Process*. Explain that to *imagine* means to think of a possible answer to a problem. An engineer uses his imagination to think of an answer to the problem.
- Distribute the materials and explain that they will use *The Engineering Design Process* and the materials provided to find an answer to the problem.
- This activity is intended for each student to design his own solution to the problem. However, you may want to group students to collaborate and share materials.
- Invite the students to share their designs with their science group or a partner.

ACTIVITIES

STEM Activity, pages 37–38
The students will design a way to keep their pencils from rolling off their desks in this STEM activity.

ASSESSMENT

Rubric
Use the prepared rubric or design a rubric to include your chosen criteria.

Objective
- Recall terms and concepts from Chapter 2

REVIEW

- Material for Test 2 will be taken from Activities Study Guide pages 25–26 and 35–36. You may review any or all of the material during the lesson.
- You may choose to review Chapter 2 by playing a review game from the Game Bank. For your convenience, the Game Bank is available in the back of this Teacher Edition and on TeacherToolsOnline.com.

NOTES

Objective
- Recall and apply terms and concepts from Chapter 2

ASSESSMENT
- Administer Test 2.

NOTES

Materials
- Audio of Boreal Owl calls

UNIT INTRODUCTION

- Direct attention to pages 44 and 45.

  What is the title of Unit 2? "Let's Learn About Living Things"

  What do you see in the picture? tree, owl, mouse

  What kind of animal is an owl? a bird

  Where do you think owls make their homes? in trees, barns, bushes, caves

- Invite students to tell about having seen or heard an owl when they were outside.

- Explain that the owl in the picture is a Boreal Owl. You may share the Background information about the owl.

- Play *Boreal Owl Calls* (TeacherToolsOnline.com). Encourage the students to mimic the call of the Boreal Owl.

  Let's look at Unit 2.

- Direct the students to look through Chapters 3 and 4, pages 46–87. Guide the students through several pages, stopping to look at headings and pictures.

  What do you think we are going to be talking about in Unit 2? living things, plants, animals

44

LOOKING AHEAD

In Lesson 29, the following materials will be needed for each student: one empty paper towel tube; one large brown paper grocery bag.

Visit TeacherToolsOnline.com for resources to enhance the lessons.

NOTES

BACKGROUND

Unit Photo

The owl pictured on page 45 is an adult Boreal Owl. In the Northern Hemisphere, the Boreal Owl can be found year round in the boreal forest or taiga of Alaska, of Canada, and the western mountains of the lower forty-eight states of the United States. While this owl was photographed in a deciduous tree, most Boreal Owls make their homes in coniferous forests. It is a small owl (approximately 8–10 inches in length; wingspan of 20–24 inches). Like most owls, it hunts mainly at night and feeds mostly on insects and small mammals, including mice.

Chapter Objectives
- Classify living and nonliving things
- Identify plant needs, parts, and plant part functions
- Explain from Genesis 3:17–18 how the Fall affected plants **BWS**
- Relate plant survival and growth to God's creational design **BWS**
- Apply science process skills of observing, predicting, and drawing conclusions
- Sequence and describe stages of the plant life cycle
- Compare and contrast like kinds of plants with each other and parent plants with their young
- Explain that young plants are like the parent plants because God made plants to reproduce after their kind (Genesis 1:11) **BWS**
- Apply steps of the engineering design process to solve a real life problem

Lesson Objectives
- Identify the characteristics of living and nonliving things
- Classify items as living or nonliving
- Identify the needs of plants
- Identify ways people use plants
- Explain from Genesis 3:17–18 how the Fall affected plants **BWS**

Materials
- houseplant
- cut flower
- vegetable
- rock

Teacher Resources
- Visuals 3.1–3.2: *Bible Verses A; Bible Verses B*

Vocabulary
- living
- nonliving
- need
- energy

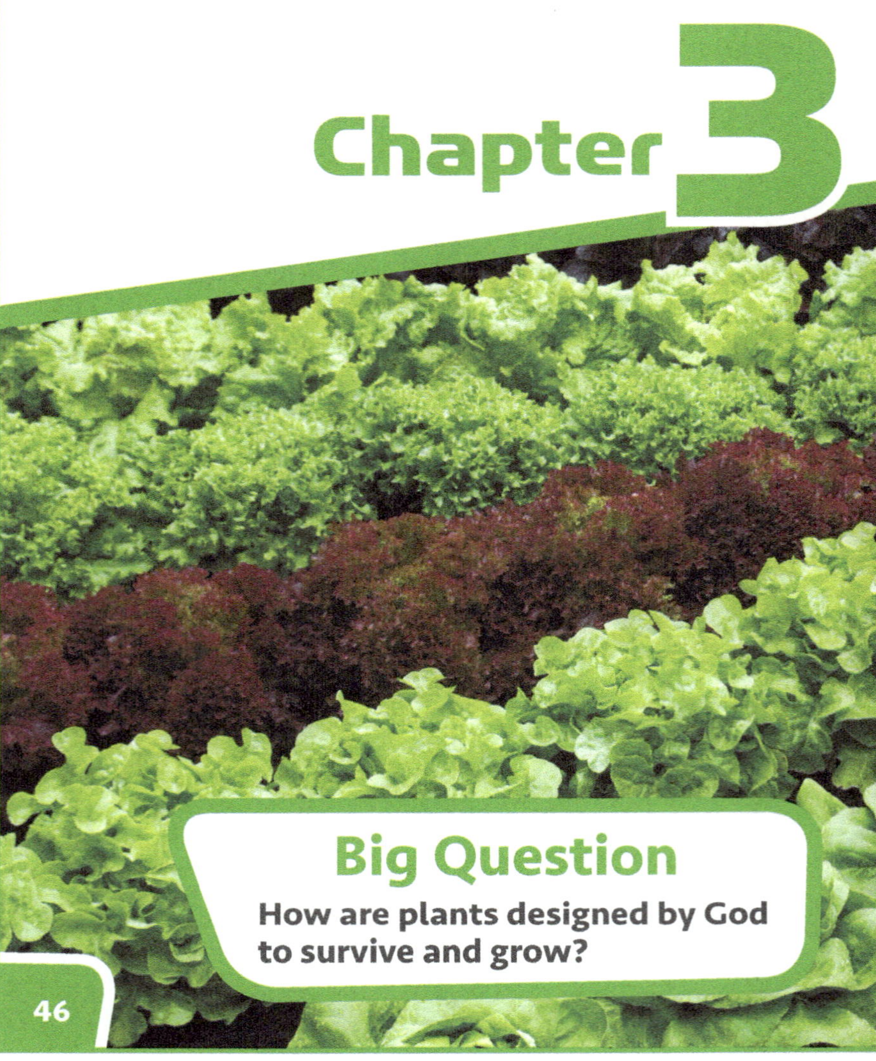

Chapter 3

Big Question
How are plants designed by God to survive and grow?

46

CHAPTER INTRODUCTION

- Direct attention to pages 46–47.
  What is the title of Chapter 3? "Plants"
- Direct attention to the picture.
- What is this a picture of? a vegetable garden
- Are all the plants in this garden the same? no
- How do you know? They look different; they are different colors; the leaves have different shapes.

  Guide the students in supporting their "yes" or "no" responses.

- Poll the students to find out how many have grown vegetables or flowers in a garden.
  Today you will begin learning about plants and what they need as they grow and change.

Big Question
- Direct attention to the Big Question. Read the question aloud.
- Explain that the answer to the Big Question will be found as the students read Chapter 3.

Plants

God made our world.

God made plants on the third day of Creation.

God made plants for people to use.

God designed plants to grow.

God designed plants to make the same kind of plants.

47

- Direct the students to read page 47 silently to find out on which day of Creation God made plants.

Who made plants? God

On which day of Creation did God make plants? the third day

- Display *Bible Verses A*. Direct attention to Activities page 39 and read Genesis 1:11–13 aloud.

What plants are mentioned in these verses? grass, herbs, fruit trees

What is one reason God made plants? Plants were created for people to use.

- Display the houseplant, cut flower, and vegetable.

How are these all alike? They are all plants or parts of plants.

What might you use the cut flower for? Possible answers: to cheer up someone; to show someone love

What could you use the vegetable for? eating

According to Genesis 1:11–13, what did God design plants to do? to grow; to make the same kind of plants

Which of these do you think can still grow? the houseplant

- Point out that God made many different kinds of plants for us to use. Explain that in this chapter the students will learn about the parts of plants and some things made from plants.

Chapter Photo

The photo on pages 46–47 is of a vegetable garden.

Classroom Garden

Plant a small garden or terrarium. Allow the students to help care for the plants. Tending a garden teaches responsibility and has many worldview applications.

PREPARATION FOR READING

- Preview and pronounce the vocabulary terms *living* and *nonliving*.
- Direct the students to read pages 48–49 silently to find out what living things need.

TEACH FOR UNDERSTANDING

Who made living things? God
What do living things need? food, water, air, and space

- Explain that *space* means "room."
What do living things do? They grow and change.
Who made you? God
Are you living? yes
How do you know you are living? I need food, water, air, and space, and I am growing and changing.
- Direct attention to the picture on pages 48–49. Ask volunteers to identify some of the living things in the picture. boy, man, lady, dog, squirrel, birds, trees, bushes, grass
- Invite the students to name some other living things that are not pictured on pages 48–49.

Did God make only living things? no

What do you call things that do not have to have food, water, and air? nonliving

- You may choose to explain that *non-* means "not" and that *nonliving* means "not living."

Do nonliving things grow and change? no

- Display the rock.

Did God make rocks? yes

What are some other nonliving things that God made? light, dirt, water, air

- Explain that God has not made *all* nonliving things, but He *has* made everything people need to make other nonliving things.

What are some nonliving things in the picture on pages 48 and 49 that God did not make? buildings, an airplane, kite, string, a bicycle, park benches, lamp posts, a bridge, a leash, clothing, and a stroller.

- Point out that the bridge looks to be made of rock, and that people make bridges like this out of the rock that God made.

answer

Living things must have food, water, air, and space. Living things grow and change. Nonliving things do not.

Nonliving Things May Move
The students may consider movement to be an indication of whether or not something is alive. Explain that though nonliving things may move, they cannot move on their own. Something else must make them move.

PREPARATION FOR READING

- Preview and pronounce the vocabulary terms *need* and *energy*.
- Direct the students to read pages 50–51 silently to find what a plant needs to make food.

TEACH FOR UNDERSTANDING

Who made plants? God

How many living things have needs? all

What is a need? what a living thing must have to stay alive

What does a plant need to make food? light and air

What part of the plant uses light and air to make food? the leaves

What does food give to a plant? energy

What is energy? what is needed to cause change

What does energy from the sun's light help a plant do? grow and change

- Explain that different plants need different amounts of sunlight to grow and change. Some grow best where there is a lot of sunlight. Some grow best in shade, or where there is little sunlight.
- Direct attention to the picture.

Can you see light in the picture? yes

Can you see air in the picture? no

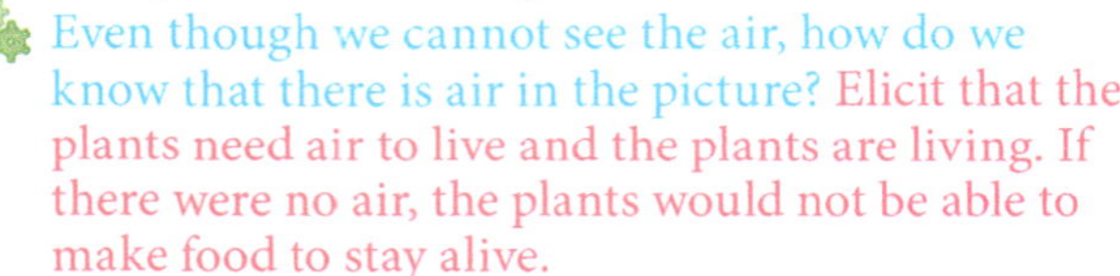 Even though we cannot see the air, how do we know that there is air in the picture? Elicit that the plants need air to live and the plants are living. If there were no air, the plants would not be able to make food to stay alive.

Needs of Plants

Plants are living things made by God.

All living things have needs.

A **need** is what a living thing must have to stay alive.

A plant has five main needs to live and grow.

Light and Air

Two of a plant's needs are *light* and *air*.

The plant's leaves use light and air to make food.

Food gives energy.

Energy is what is needed to cause change.

Energy from the sun's light helps a plant grow and change.

50

Water, Dirt, and Space

Water and *dirt* help a plant live and grow.

The plant gets water that it needs from the dirt.

Some plants need lots of water.

Some plants need only a little water to live and grow.

The last need of a plant is *space*.

A plant needs space to grow larger.

Some plants need lots of space to live and grow.

Some plants need only a little space.

What are the five needs of plants?

51

Besides light and air, what other needs do plants have? water, dirt, and space

Where does a plant get the water it needs? from the dirt

Where is dirt found? on the ground

• Mention that another word for *dirt* is "soil." Point out that dirt is what plants grow in. Explain that dirt holds the water that plants need to live and grow.

Do all plants need the same amount of water to live and grow? No; some plants need lots of water, and some plants need only a little water to live and grow.

• Direct attention to the picture.

Which plants in the picture do you think need the most water? the trees

Why do trees need the most water? Because the trees are larger than the other plants in the picture, they need more water to live and grow.

Why does a plant need space, or room? so it can grow larger

Would a small plant need more space or less space to live and grow than a large plant? less

• Direct attention to the picture on page 50. Point out the small yellow flowers. Discuss that one of the yellow flowers would need less space than one of the big trees or bushes.

What do you think would happen if a plant did not have light, air, water, dirt, and space? It would not live; it would not grow and change.

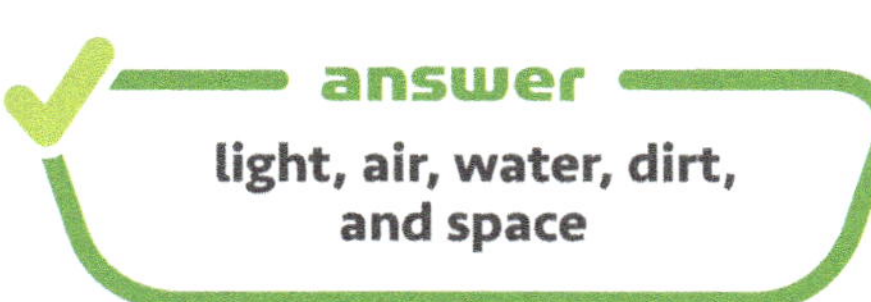

answer

light, air, water, dirt, and space

Plants and Soil

Some plants, such as some orchids (a type of epiphyte), grow on other plants and do not get nutrients directly from the soil. At this grade level, however, it is better for students to generalize that plants need soil.

PREPARATION FOR READING

- Preview the words *thorns* and *thistles.* The students may need help pronouncing these additional words.
- Direct the students to read pages 52–53 silently to find some ways that people use plants.

TEACH FOR UNDERSTANDING

Why did God create plants? for people to use
Did God make all plants the same? no
What are some ways people use plants? for food, clothing, to build houses

✦ What foods are made from plants? Possible answers: peanut butter, jam, bread, juice, cereal, salad, pasta

✦ What plants were used to make the food pictured? Possible answers: peanuts (peanut butter), strawberries (jam), wheat (bread)

✦ What plant do you think the girl's dress is made from? cotton

- Direct attention to the picture of the cotton boll. Explain that the soft cotton fibers grow in a boll, or a protective covering. Point out that many of our clothes are made from cotton.

✦ What plants are used to build houses? trees

- Explain that lumber is used to build some houses and that lumber comes from trees. Explain that trees are also used to make paper and furniture. Point out that the paper in their books is made from trees, which are plants.

Ways People Use Plants

God created plants for people to use.
God made different kinds of plants.
Some plants are used for food.
Other plants are used for clothing.
Some plants are used to build houses.

52

Make a Paper Fan
Materials: half sheet of paper per student

Give each student a half sheet of paper and demonstrate folding it to make a fan. Allow the students time to enjoy waving their fans. Discuss that paper is made from plants and that they have made something useful out of a plant.

Result of Sin

God created plants.

They were perfect.

The Bible says in Genesis 3:17–18 that plants changed.

Plants changed when Adam sinned.

Plants are no longer perfect.

There are now weeds, thorns, and thistles.

Weeds take up space and make it hard for other plants to grow.

Thorns and thistles make some plants hard for people to use.

What happened to plants after Adam sinned?

53

When God created plants, what were they like? They were perfect.

- Explain that *perfect* means the plants were the best; there was nothing wrong with them.

What does the Bible say happened to plants when Adam sinned? They changed; they are no longer perfect.

- Display *Bible Verses B*. Direct attention to Activities page 40 and read Genesis 3:17–18 aloud.

What are some ways plants changed when Adam sinned? There are now weeds, thorns, and thistles.

How do weeds make it hard for plants to grow? They take up space, or room, that plants need.

- Direct attention to the pictures. Explain that the top right picture is showing thistles and the bottom picture is showing plants with thorns. Point out the insert showing the closeup of thorns.

What do thorns and thistles have in common? They are sharp; they have points that can prick or stick people or animals.

How do thorns and thistles make it hard to use some plants? The sharp points can hurt people who try to use the plants.

- Point out that thorns and thistles show that plants changed because of sin and they have made the job to care for the earth much harder.

answer
They are no longer perfect; there are now weeds, thorns, and thistles.

ACTIVITIES

Bible Verses, pages 39–40
These Bible verses will be used in the teaching of this chapter.

Living and Nonliving Things, pages 41–42
These pages reinforce the concepts taught in Lesson 17.

Objectives
- Identify each part of a plant and its function
- Relate plant survival and growth to God's creational design **BWS**

Materials
- toy car
- time-lapse video of a growing plant
- celery stalk
- sunflower seeds
- leaf, one per student
- hand lens, one per student

Teacher Resources
- Visuals 3.1; 3.3: *Bible Verses A*; *Plant Parts Song*

Vocabulary
- survive
- roots
- stem
- leaves
- flower

INTRODUCTION

- Display the toy car and point out its parts, such as the wheels, roof, hood, trunk, and doors.
- Explain that the toy car has many parts that work together to make a whole car.
-  If we took the wheels off, would the car work properly? no
- Point out that all the parts are needed for the car to look and work the way it should.
- Explain that plants have different parts, too. Point out that God designed the parts of a plant to work together so the plant can live and grow.

PREPARATION FOR READING

- Preview and pronounce the vocabulary term *survive*.
- Direct the students to read pages 54–55 silently to find out what God designed plants to do.

TEACH FOR UNDERSTANDING

Do plants all look the same? no
In what ways can plants be different? They can be different sizes, shapes, and colors.
Name one big plant. Possible answers: a full-grown tree, a full-grown cactus
Name one small plant. Possible answers: grass, a dandelion, a buttercup
Look at the picture. Do all plants in the picture have the same shape? no

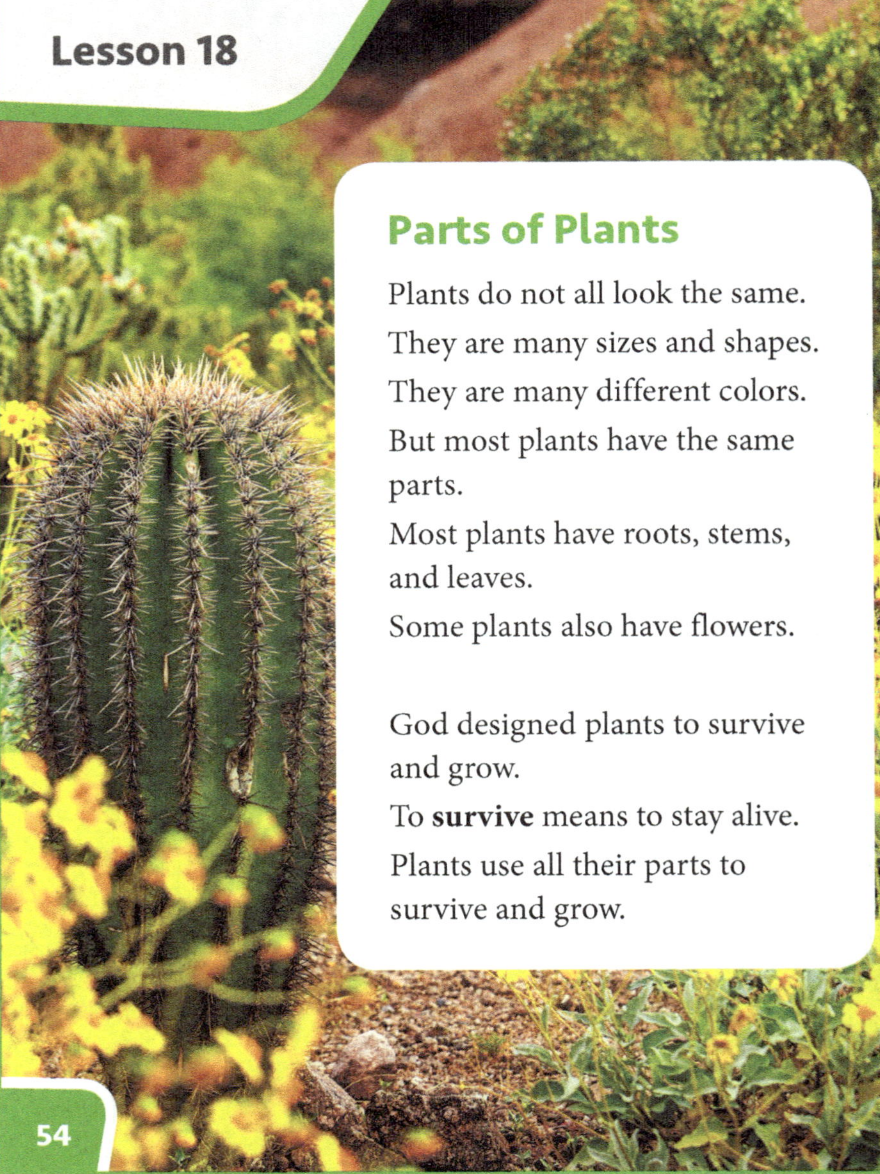

Parts of Plants

Plants do not all look the same.

They are many sizes and shapes.

They are many different colors.

But most plants have the same parts.

Most plants have roots, stems, and leaves.

Some plants also have flowers.

God designed plants to survive and grow.

To **survive** means to stay alive.

Plants use all their parts to survive and grow.

Are the plants the same color? no
What sentence tells what most plants have that is the same? "But most plants have the same parts."
What parts do *most* plants have? roots, stems, and leaves
What plant part do *some* plants have? flowers
What did God design plants to do? to survive and grow
What does *survive* mean? to stay alive
What does a plant do when it grows? It gets bigger.
- Show a time-lapse video of a plant growing. Discuss that time-lapse video takes what would normally be a slow process and speeds it up. Point out the changes in the plant as it grows, especially the change in size.
What do plants use to survive, or stay alive, and grow? all their parts
- Point out that if one part of the plant is missing, the plant cannot survive and grow. God designed all parts of the plant to work together.

What are the four parts of a plant?

55

- Introduce the words *roots*, *stems*, *leaves*, and *flowers*.
  Look at the picture of the plant. What parts of the plant are shown here? flowers, leaves, stems, and roots
- Which part of the plant is in the dirt? the roots
- Direct the students to put their finger on the roots.
- Which part of the plant connects the roots to the leaves and flowers? the stem
- Direct the students to put their finger on a stem and follow the stem from the roots to a flower.
- What part of the plant makes food? the leaves
- Direct the students to put their finger on a leaf.
- Do most plants have roots, stems, and leaves? yes
  Put your finger on a flower of this plant.
- Do all plants have flowers? no

answer
flowers, leaves, stems,
and roots

BACKGROUND

Barrel Cactus
In the foreground of the picture on page 54 is a barrel cactus. The barrel cactus comes in different shapes and sizes. Some can reach heights up to ten feet.

Zinnia
The flowering plant on page 55 is a zinnia.

ACTIVITY

Identify Parts of a Plant
Materials: small flowering plant
You may show a small flowering plant and allow the students to touch and iden-ify the leaves, stems, roots, and flowers.

Note: You could use a cutting that has been rooted in water, a young bean plant, or any flowering houseplant. If you do not want to have the plant out of the soil, you may use a plant whose seed has been planted against the side of a clear container. The students can see the roots without removing the plant from the soil.

PREPARATION FOR READING

- Preview and pronounce the vocabulary terms *roots* and *stem*.
- Direct the students to read pages 56–57 silently to find out what moves up the stem from the roots.

TEACH FOR UNDERSTANDING

What part of the plant is under the ground? roots
What two things do roots do? hold the plant in the ground and take in water
Where do roots get water from? the ground
How do you think water gets in the ground? Water falls from the sky (rain, sleet, hail, snow) and soaks into the ground.

- Remind the students that God designed plants to need space, or room, to survive and grow. Point out that the plant roots need space to take in the water that the plant needs to survive and grow.
- Direct attention to the picture and caption.

What is the boy holding? carrots
Which part of the carrot is the root? the orange part
What roots do people eat? carrots, radishes, turnips, beets

- Use Visual 3.3 to introduce the *Plant Parts Song* activity.

Roots

Roots hold the plant in the ground and take in water.

Roots are under the ground.

The roots take in water from the ground.

Roots help the plant survive and grow.

56

ACTIVITY

Plant Parts Song

Invite the students to sing the *Plant Parts Song* (Visual 3.3) to the tune of "The Muffin Man." Display and sing one verse of the song at a time as each part of the plant is taught. After all parts of the plant have been taught, sing all the verses together.

What stems do people eat?

Stems

The **stem** holds up the plant.

Water from the roots moves up the stem.

A tree has a large stem.

A dandelion has a small stem.

The stem helps the plant survive and grow.

What is the job of the roots?

57

What holds up a plant? the stem

What moves up the stem from the roots? water

What plant is described as having a large stem? a tree

What do you call the part of a tree that is the stem? the trunk

- Explain that the trunk is the main stem of a tree. The branches of a tree are also part of the stem. They are a branching part of the stem that supports the twigs and leaves.

What is a plant with a small stem? a dandelion

- Direct attention to the large background picture.

Put your finger on the tree stem.

Put your finger on a dandelion stem.

How does the stem help a plant survive and grow? It moves water from the roots to all parts of the plant.

- Direct attention to the picture insert and caption.

What stems do people eat? celery, asparagus, rhubarb

What stem is the girl in the picture eating? celery

- Display a celery stalk. Pull it slightly apart at the end to show the "strings." Explain that the strings are actually little tubes and that the water for the plant moves from the roots up through these little tubes in the stem. Explain that inside trees there are larger tubes that carry water.

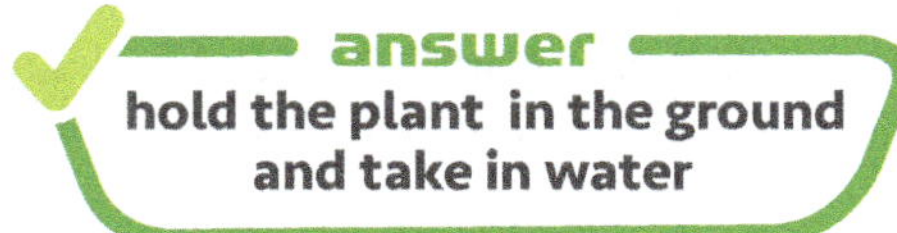

hold the plant in the ground and take in water

BACKGROUND

Edible Parts

Plants make food for their own energy needs and store the extra food. People and animals eat the parts of some plants to get the stored energy. Although plants store food in many places, not all stored food is good for humans to eat. For example, while rhubarb stems are edible, rhubarb leaves are poisonous. Remind the students to never eat any part of a plant without first checking with an adult.

ACTIVITY

Grow Roots from a Stem

Materials: 4"–6" cutting from a coleus or climbing philodendron plant, transparent glass jar, water

Use scissors to cut a 4"–6" section of a stem, just above a node. Remove the leaves from the bottom two inches of the stem and place the stem in a glass jar or vase filled with water. A transparent container allows the students to observe the roots as they grow. Place the jar or vase in a sunny location. Keep the water level filled and check daily for root formation. Once the roots are established, you may plant the cutting in soil.

PREPARATION FOR READING

- Preview and pronounce the vocabulary terms *leaves* and *flower*.
- Direct the students to read pages 58–59 silently to find out what leaves use to make food for the plant.

TEACH FOR UNDERSTANDING

What did God design leaves to do for a plant? to make food

What do leaves use to make food for the plant? light, air, and water

How do plant leaves get the water they need? Elicit that the roots bring water from the ground, up the stem, and to the leaves.

- Display *Bible Verses A* and read Genesis 1:11–13 aloud.

Who made the plants? God

How long do you think it took God to do this? God made plants in one day, on the third day of Creation.

How long would it take you to design a plant and its parts? Answers will vary, but elicit that only God could design a plant and its parts and that God created plants in just one day.

- Remind the students that God is the Engineer who designed all parts of the plant to work together. God made each part, and He designed all parts to work together so that the plant can survive and grow.
- Direct attention to the caption.

Do all leaves look the same? no

Look at the pictures of the leaves. Have you seen any of these leaves before? The students may or may not recognize any leaves. Moving clockwise from the left, the leaves are cedar, oak, palm, and caladium.

Look at the picture of the girl. What is she holding? lettuce

What part of the lettuce plant do we eat? leaves

What are some other leaves people eat? cabbage, kale, spinach, collards

Leaves

Plants make their own food.

Leaves make food for the plant.

Leaves use light, air, and water to make food.

Leaves help the plant survive and grow.

⚗ ACTIVITY

Grow Leaves

Materials: seed trays or large containers, potting soil, lettuce seeds, water, spray bottle, paper (optional)

Fill the trays or containers with soil. Leave some space between the surface of the soil and the tops of the containers. Sprinkle a few lettuce seeds on top of the soil. Cover with a very thin layer of soil. Place the containers in a sunny place and use a spray bottle to keep the soil lightly watered. If using seed trays, cover the top with paper until the seeds sprout and reach up to the paper. Thin the seedlings and transplant as needed.

Note: If using seed trays or shallow containers, you will need to transplant the seedlings to a garden later for full growth. Large, deeper containers can be used to grow the lettuce to maturity.

Flowers

Many plants have flowers.

The **flower** makes seeds.

A new plant will grow from a seed.

Flowers also add color and beauty to plants.

God designed all parts of the plant.

God designed the plant parts to work together.

Plants survive and grow because the roots, stem, leaves, and flowers work together.

What flowers do people eat?

✓ **What is the flower's job?**

59

What do many plants have? flowers
What do flowers make for plants? seeds

- Direct attention to the picture of the sunflower. Explain that the seeds will grow in the dark center of the sunflower. Show the students sunflower seeds.
  What will grow from a seed? a new plant
- Explain that, when planted, most seeds will grow into new plants.
  What else do flowers do for plants? They add color and beauty.
- Invite students to share the color of their favorite flower.
- Ask volunteers to identify a flower they think is beautiful.
- Direct attention to the picture of the girl.
  What is the girl holding? a cauliflower
  What flowers do people eat? cauliflower, broccoli, artichokes, squash blossoms
  Who designed all parts of the plant? God
  What did God design the plant parts to do? to work together
- Direct attention to the Big Question on page 46.
  How are plants designed by God to survive and grow? Plants need food and water to survive and grow. Elicit that God designed all the parts of the plant to work together. The roots take in water from the ground, and God designed the stem to move the water up from the roots to the other parts of the plant. God designed the leaves to make food for the plant, and He designed the flower to make seeds so there will be more plants.
- Display the *Plant Parts Song* and review the parts of a plant by singing all the verses.

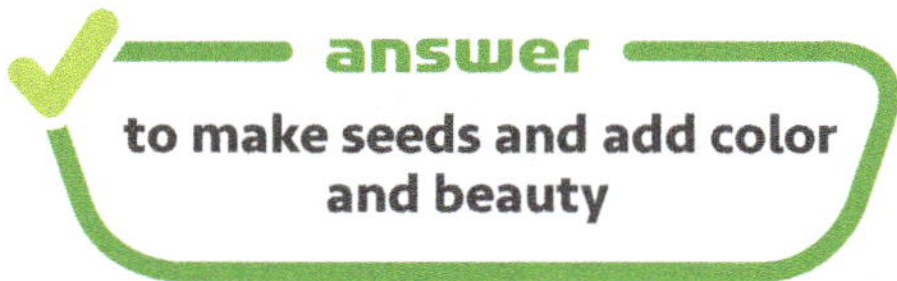

✓ **answer**
to make seeds and add color and beauty

ONLINE ACTIVITY

Hunt for Flowers

Materials: an electronic device for each pair of students

Pair the students. Instruct them to share the devices as needed for research. Instruct the students to conduct a keyword search of images using "flowers." Direct the students to locate flowers with color and beauty. Encourage each student to print a color image of his favorite flower. Write the student's name on the page and make a class scrapbook. Display the completed book.

ACTIVITIES

Study Guide, pages 43–44
These pages review the concepts taught in Lessons 17–18. After completion, direct the students to keep them for review for Test 3.

Hand Lens Observation, pages 45–46
This Enrichment activity may be used to reinforce the concepts taught in Lesson 18.

Eating Plants, pages 47–48
This Enrichment activity may be used to reinforce the concepts taught in Lesson 18.

Objectives

- Predict the effects on the growth and survival of a plant when its needs are not met
- Observe and describe parts of a plant
- Draw a conclusion about plant needs (about the growth and survival of plants) based on observations
- Draw a conclusion from the investigation about God's creational design of plants **BWS**

Materials

- See Activities page 49. *Note:* To limit plant materials, this Investigation is designed to be completed as a class. For Plant B to show visible effects of no water after seven days, the plants should be thin, long-stemmed plants, such as the basil plants shown in the picture.

INTRODUCTION

What are the needs of plants? light, air, water, dirt, and space

Have you ever wondered what would happen to a plant if just one of its needs was not met?

Today you are going to investigate what happens to a plant that does not get water.

PREPARATION FOR READING

- Direct the students to remove pages 49–50 from their Activities books. Direct them to read page 60 and Activities pages 49–50 silently before beginning.

TEACH FOR UNDERSTANDING

- Direct attention to and discuss the **Process Skills**, using information from Chapter 2.
- Choose a student to read the **Problem** aloud. *Note:* This Investigation may be changed or expanded to include needs other than water. The experimental variable could be the lack of air (place one plant in a sealed clear plastic bag) or the lack of light (place one plant in a closed cupboard or cardboard box).
- Direct the students to complete the **Hypothesis**.
- Instruct the students to follow along with you as you complete Steps 1–7 of the **Procedure**. Point out that "as needed" in Step 6 means that Plant A will receive enough water to keep the soil moist.
- Direct the students to place Activities pages 49–50 in their science notebooks, or you may choose to collect the pages and return them after seven days.
- Make **Observations** of both plants daily. Water Plant A as needed. Discuss any noticeable changes in either plant.

INVESTIGATION

Lesson 19

Plant Needs

All living things have needs.

What do plants need to survive and grow?

In this Investigation you will predict what will happen to a plant that does not get water.

You will make a hypothesis.

You will observe to check your hypothesis.

Let's investigate!

60

- After seven days, instruct the students to complete the **Observations** and **Conclusions**. Remind them to use their observations to draw their conclusions.
- Discuss the answers to Conclusions 1 and 2. Remind the students that there are no right or wrong hypotheses, but that the purpose of the Investigation is to test the hypothesis to see if the answer to the problem is correct.
- Discuss the answers to Conclusions 3 and 4.

ACTIVITIES

Investigation: Plant Needs, pages 49–50
Students will predict the effects on the growth and survival of a plant when its needs are not met.

ASSESSMENT

Rubric
Use the prepared rubric or design a rubric to include your chosen criteria.

Lesson 20

A Plant's Life Cycle

Living things grow and change.

All the steps of the life of a living thing are a **life cycle**.

A plant's life cycle starts with a *seed*.

Most new plants grow from seeds.

Plant a seed and a seedling will grow.

A *seedling* is a young plant.

The seedling has roots.

It also has a thin stem and small leaves.

61

LESSON 20

Objectives
- Define *life cycle*
- Identify and describe the stages of the life cycle of a plant
- Sequence stages of a plant's life cycle

Materials
- packet of seeds

Teacher Resources
- Visual 3.4: *Life Cycle of a Pumpkin Plant*

Vocabulary
- life cycle

INTRODUCTION

- Review the parts of plants.
  Display your packet of seeds.
  What part of the plant makes seeds? the flower
  Today you will find out more about how plants grow from seeds.

PREPARATION FOR READING

- Preview and pronounce the vocabulary term *life cycle* together. Introduce and pronounce the words *seedling* and *pumpkin*.
- Direct attention to the diagram on page 63 and read the labels together. Instruct the students to look for these words and refer to the diagram as they read.
- Direct the students to read page 61 silently to find out what will grow when a seed is planted.

TEACH FOR UNDERSTANDING

When living things grow, they also do what? change

What are some living things in this room? the people and any plants or animals.

What do we call all the steps of the life of a living thing? life cycle

What does a plant's life cycle start with? a seed

Italicized words are not vocabulary terms, but they are important words. Their definitions have been included in the glossary.

- Direct attention to the picture of the seeds. Explain that they are pumpkin seeds.
- Direct attention to the diagram on page 63 and display *Life Cycle of a Pumpkin Plant*. Point out the seed and explain that it has started to grow underground, in dirt. Explain that the roots are the first part of a plant to grow.

Why do you think the roots are the first part of a plant to grow? The roots take in water from the dirt for the plant to survive and grow.

- Point out the arrows on the diagram and explain that the word *cycle* means "circle." Explain that in a cycle one step follows the next step in a circle.
- Instruct the students to put a finger on the seed and follow the arrows until they get back to the seed.

What will grow when a seed is planted? a seedling
What is a seedling? a young plant

- Direct attention to the picture of the seedling on page 61.

What parts of a plant does a seedling have? roots, thin stem, small leaves

- Instruct the students to put a finger on the thin stem and small leaves of the pumpkin seedling. Direct attention to the diagram on page 63 and point out the roots.

How have the roots changed? There are more roots and they are longer.

PREPARATION FOR READING

- Display the word *fruit.* The students may need help pronouncing this additional word.
- Direct attention to the diagram on page 63 and instruct the students to refer to the diagram as they read.
- Direct the students to read pages 62–63 silently to find out what a seedling grows into.

TEACH FOR UNDERSTANDING

What does a seedling grow into? an adult plant
- Explain that an adult plant is a full-grown plant that can make more plants.

What does an adult plant grow? flowers
- Direct attention to the pictures. Instruct the students to put a finger on the flowers of the pumpkin plant.

What do the flowers grow? fruit
- Instruct the students to put a finger on the fruit.

What is inside the fruit? seeds
- Point out the seeds inside the open pumpkins.

What will the seeds make? new plants

Growing a Vegetable Garden
You may read aloud *The Gardener's Gold Ring* by Nancy Bopp. This book is available from JourneyForth Books, a division of BJU Press, at journeyforth .com.

Life Cycle of a Pumpkin Plant

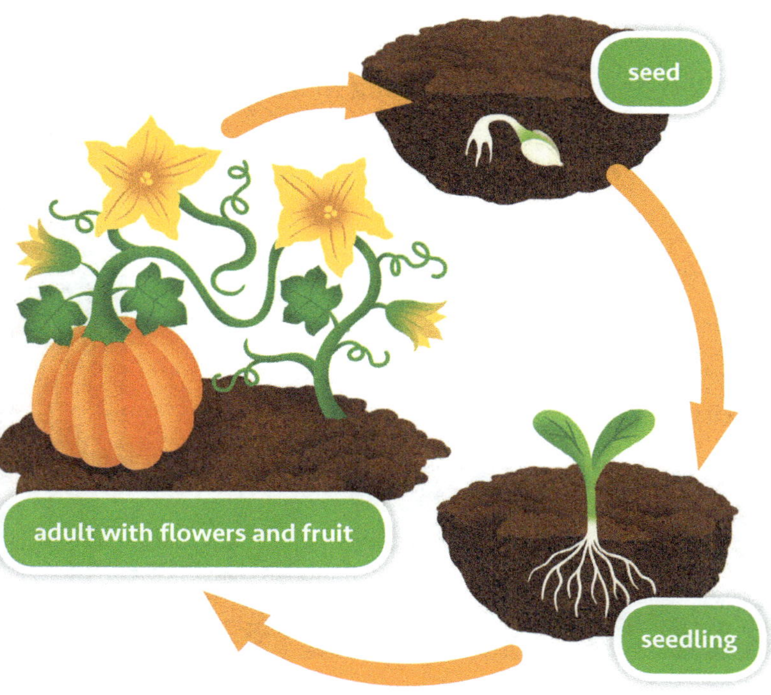

63

- Display the *Life Cycle of a Pumpkin Plant* and direct attention to the diagram. Review the life cycle of a plant with the students.

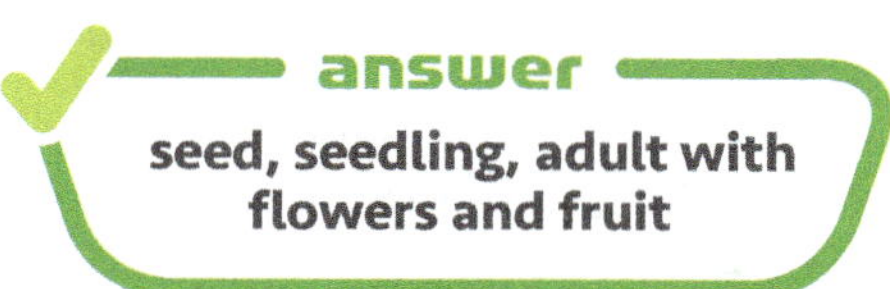

ACTIVITIES

Plant Life Cycle, page 51
This page reinforces the concepts taught in Lesson 20.

Objectives
- Compare and contrast a seedling with an adult plant
- Explain that young plants are like the parent plants because God made plants to reproduce after their kind (Genesis 1:11) **BWS**
- Compare and contrast the same kind of plant to show that they are recognized as similar but can also vary

Teacher Resources
- Visuals 3.1; 3.5–3.6: *Bible Verses*; *Adult Tomato Plant and Seedling Venn Diagram*; *Adult Roses T-Chart*

INTRODUCTION

Today you will observe how adult plants and seedlings are alike and how they are different. You will also observe how plants that are the same kind are alike but can also be different.

PREPARATION FOR READING

- Direct attention to the *Meet the Scientist* feature box. Explain that feature boxes give extra information related to the text. Preview and pronounce the name *George Washington Carver*.
- Direct the students to read page 64 silently to find out what kind of trees an apple tree will make.

TEACH FOR UNDERSTANDING

What did God design a living thing to do? make the same kind of living thing

What do adult plants make more of? the same kind of plants

Will an apple tree make roses? no

What kind of trees will an apple tree make? apple

What will a young apple tree look like? its parent; the adult apple tree

- Display the Venn diagram. Guide the students as they complete the Venn diagram on Activities page 53.
- Discuss ways an adult plant and a seedling are alike and different.
- Display *Bible Verses A* and read Genesis 1:11 aloud. Discuss the truth that God created adult plants to make more of the same kind of plant. Guide the students as they answer the question at the top of Activities page 54.
- Pair the students. Instruct them to observe the pictures of the adult roses on Activities page 54 and to complete the T-chart with their partner.
- Display the *Adult Roses T-Chart* and encourage the students to list ways the roses are alike and different. Point out that even though they have some differences, both plants are still alike and are both roses.

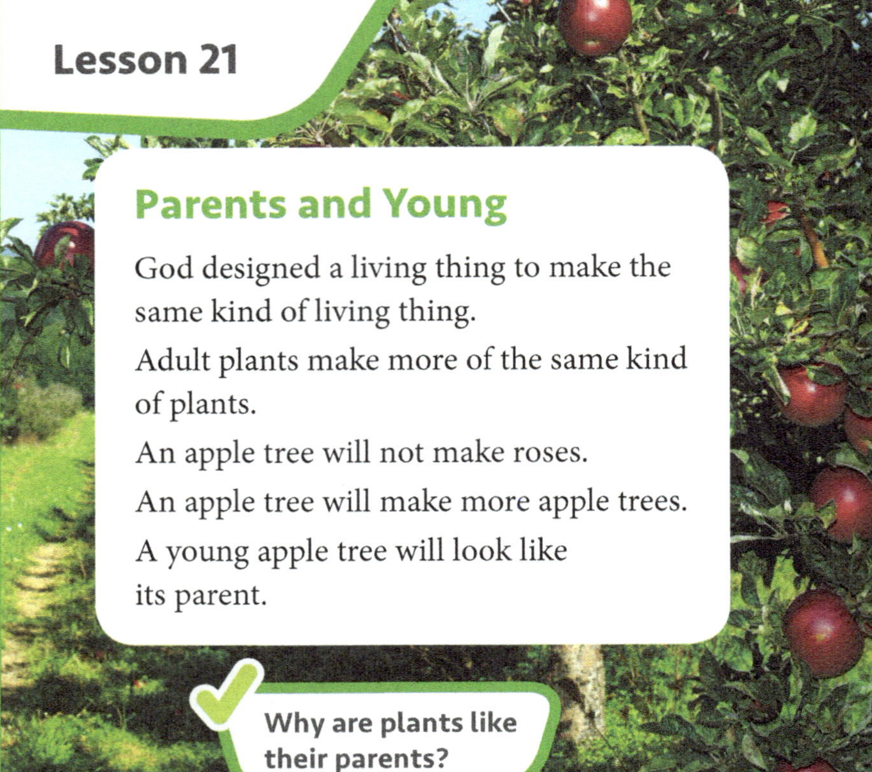

Parents and Young

God designed a living thing to make the same kind of living thing.

Adult plants make more of the same kind of plants.

An apple tree will not make roses.

An apple tree will make more apple trees.

A young apple tree will look like its parent.

Why are plants like their parents?

Meet the Scientist

As a child George Washington Carver was called the plant doctor. He learned to care for the family garden. George was curious. He enjoyed studying plants. He gave advice about plants to his friends. He helped many sick plants survive and grow.

64

- Direct attention to the *Meet the Scientist* box.

What was George Washington Carver called as a child? "the plant doctor"

How did George become interested in plants? He learned to care for the family garden; he was curious.

- Remind the students that to be curious means to ask lots of questions.

Why was George called the plant doctor? He studied plants and gave advice about plants to his friends; he helped sick plants survive and grow.

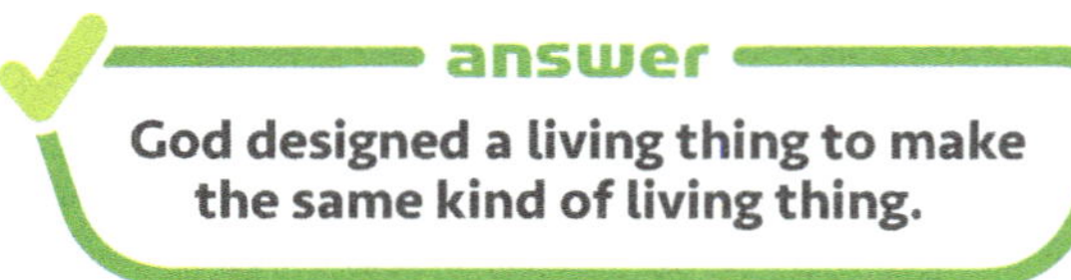

=== answer ===
God designed a living thing to make the same kind of living thing.

ACTIVITIES

Adult Plants and Seedlings, pages 53–54
Study Guide, pages 55–56
These pages review the concepts taught in Lessons 19–21.

STEM

Lesson 22

Unwanted Plants

There are weeds growing in your garden.

The weeds are taking up space that your other plants need to grow.

You do not want the weeds to grow there.

You want to remove the weeds.

You do not want to hurt the other plants.

How can you get rid of the weeds that are growing in your garden?

65

⚠ HELPS

Venn Diagram

A Venn diagram helps students visualize how two or more objects are alike and how they are different. Demonstrate how to complete a Venn diagram. Display two overlapping circles. Label each outside section "Different" and label the overlapping section "Alike." Choose two similar objects (football and baseball or a large block and a small block). Ask the students to identify any characteristics that are alike. List these in the overlapping section. Ask the students to identify any characteristics that are different and list these in the sections that do not overlap. Discuss the ways the objects are alike and the ways they are different using the lists made by the students.

BACKGROUND

Scientist: George Washington Carver

George Washington Carver was born a slave in Missouri. As an adult, he became an important scientist, inventor, and educator. He made many contributions to science that improved farming in the South. He is probably best known for his work with peanuts and is credited with discovering more than three hundred uses for the peanut plant. Carver gave God the credit for what he was able to learn and share with others.

Objectives

- Design a solution to prevent unwanted plants
- Draw and label the design
- Explain how the design solves the problem
- Relate the growth of weeds and other unwanted plants to Genesis 3:17–18 and how the Fall affected plants **BWS**

Teacher Resources

- Visuals 2.6; 3.2: *STEM The Engineering Design Process*; *Bible Verses B*

INTRODUCTION

Today you will think of ways to get rid of weeds growing in your garden.

PREPARATION FOR READING

- Instruct the students to remove pages 57–58 from their Activities books.
- Direct the students to read silently page 65 and Activities pages 57–58 before beginning the STEM activity.

TEACH FOR UNDERSTANDING

Why are weeds "Unwanted Plants" in your garden? They take up space that other plants need to grow.

- Display *Bible Verses B*; read Genesis 3:17–18 aloud. What are weeds the result of? Adam's sin

  When you remove weeds from your garden, what do you *not* want to do? hurt the other plants

- Display *STEM The Engineering Design Process*. Explain that for this STEM activity they will *Imagine*, *Plan*, and *Share* a design to find an answer to the problem. Point out that they will not *Make* or *Make Better* their design.

- For this activity, each student may design his own solution to the problem, or students may work in science groups to collaborate as they design a solution.

- Invite the students to share their designs with a partner or another science group.

ACTIVITIES

Unwanted Plants, pages 57–58

In this STEM activity, the students will design one way to get rid of weeds growing in a garden.

ASSESSMENT

Rubric

Use the prepared rubric or design a rubric to include your chosen criteria.

Objective

- Recall terms and concepts from Chapter 3

REVIEW

- Material for Test 3 will come from the Study Guides on Activities pages 43–44 and 55–56. You may review any or all of the material during the lesson.
- You may choose to review Chapter 3 by playing *Grow a Flower* review game or another game from the Game Bank. For your convenience, the Game Bank is available in the back of this Teacher Edition and on TeacherToolsOnline.com.

🏆 REVIEW GAME

Grow a Flower

Display a horizontal line to represent the soil and a vertical line to divide the area in half.

Divide the students into two teams. Alternate asking review questions to both teams. After each correct answer, allow the team member to draw a part of a plant, starting with the roots, then the stem, several leaves, and finally a flower. You could direct the students to add more than one leaf or a flower with multiple petals.

You may choose to draw a simple plant for the students to follow as a guide. You could also draw the parts ahead of time on separate pieces of paper and let each team assemble its plant like a puzzle.

The first team to have a completed plant or the bigger "garden" wins.

NOTES

Objective

- Recall and apply terms and concepts from Chapter 3

ASSESSMENT

- Administer Test 3.

NOTES

Chapter Objectives
- Explain that God designed animals and their environments to work together to meet their needs to survive and grow **BWS**
- Explain from the Bible why animals die **BWS**
- Classify animals based on similar external characteristics
- Compare and contrast animals of the same kind and animals and their young
- Sequence and describe the stages of the life cycle of an animal
- Apply steps of the engineering design process to solve a real-life problem

Lesson Objectives
- Infer from key text features the topic for Chapter 4
- Distinguish the identity of living and nonliving things in an environment
- Identify the needs of animals
- Explain that God designed animals and their environments to work together so they can survive and grow **BWS**

Materials
- audio of bottlenose dolphin
- video of bottlenose dolphin

Teacher Resources
- Visuals 4.1–4.2: *Bible Verses; Elephant*

Vocabulary
- environment

CHAPTER INTRODUCTION

- Direct attention to the Contents page.
  We are ready to begin Chapter 4. Find what page Chapter 4 begins on. page 66 Turn to page 66.
- Allow time for the students to look through the chapter to decide what the chapter is about.
  What do you think the chapter is about? Possible answers: animals, birds, fish, babies, zoo animals
- As the students give their answers, ask them to tell you the page number and the item that helped them decide.
  Look at the picture on page 66. What do you see? dolphins swimming in the water
- Show a video of the bottlenose dolphin, or play the sounds it makes.

Big Question
- Guide the students in reading the question.
  Where will you find the answer to this question? as the chapter is read

Q BACKGROUND

Chapter Photo
The chapter photo is a mother bottlenose dolphin and her baby dolphins. They are swimming in Dolphin Reef. Dolphin Reef is a horseshoe-shaped sea pen in the Gulf of Eilat. The gulf is in the northeast part of the Red Sea near the city of Eilat in southern Israel. The dolphins are not trained to perform but can interact with people.

Bottlenose Dolphins
Bottlenose dolphins send messages to one another in different ways, such as squeaking and squawking. A baby dolphin develops a special whistle soon after birth. A dolphin can be identified by its whistle, just like a person's name. A bottlenose dolphin is heavier than a piano, slower than a car, and shorter than a school bus. They eat fish, squid, and shrimp.

Visit TeacherToolsOnline.com for resources to enhance the lessons.

Animals

God created all things.

He created nonliving things.

The air and rocks are nonliving things.

God created living things.

Plants and animals are living things.

In Proverbs the Bible tells people to take care of animals.

We give glory to God when we obey Him.

67

⚠ HELPS

Class Pet

Consider having a class pet. Interaction with an animal provides many teaching opportunities. Relate concepts in the chapter to the pet. Provide times for the students to observe the animal's characteristics and behaviors.

LOOKING AHEAD

Science Materials

For Lesson 36 you will need one paper towel tube and a paper grocery bag for each student. See Lesson 32 for a letter to the parents and Lesson 36 for preparing the grocery bags.

Some preparation is required for the Investigation in Chapter 6 Lesson 41. See the **Helps** in Chapter 6 Lesson 39.

PREPARATION FOR READING

- Preview and pronounce the word *Proverbs*.
- Direct the students to read page 67 silently to find out how they can give glory to God.

TEACH FOR UNDERSTANDING

Who created all things? God

Who created the nonliving things? God

- Explain that God has not made *all* nonliving things, but He *has* made everything people need to make other nonliving things.
- Display Bible Verses and direct attention to Proverbs 12:10 on Activities page 59. As you read the verse aloud, ask the students to follow along on their pages.

What is a beast? It is an animal.

What does the word *regardeth* mean? Elicit that in this verse it means to "care for."

What is this verse in Proverbs telling us to do with animals? care for them

- Direct attention to Proverbs 27:23 and read the verse as the students follow along.

What does the word *diligent* mean? Elicit that it means to "be careful."

What groups of animals are called flocks? Possible answers: birds, chickens, sheep

What groups of animals are called herds? Possible answers: cows, sheep, goats, elephants

What is this verse in Proverbs telling us to do with animals? look out for them, know what they need, take care of them

What do both of these verses tell people to do? to take care of animals

How can you give God glory? by obeying Him

- Explain that to give God glory means to give God honor and praise. People worship God and show Him that they love Him when they obey Him.

- Preview and pronounce the vocabulary term *environment*.
- Preview and pronounce the word *imagine*.
- Direct the students to read pages 68–69 silently to find out what lives in an environment.

- Read page 68 aloud, pausing after each line so the students can find each of the things described in the picture on pages 68–69.
- What other things do you see in the picture? bridge, man sitting on a bench, swans, boy in the creek, girl standing on a rock

Nonliving Things May Move

The students may consider movement to be an indication of whether something is alive. Explain that though nonliving things may move, they cannot move on their own. Something else must make them move.

What is an environment? all the living and nonliving things around a plant or animal

- Explain that the picture is a park environment.

What is a living thing? something that needs food, water, air, and space to live and grow

Name some living things in the park environment. people, birds, bees, swans, grass, trees, flowers, rabbit

Put your finger on the rabbit. Because it is a living thing it needs food, water, air, and space to live and grow.

What can you find in the park environment that meets the rabbit's needs? Elicit that there is grass for food, a creek for water, air to breathe, and plenty of space to live and grow.

- Direct attention to the bees in the picture.

What can you find in the park environment that meets the bees' needs? Elicit that there are flowers for food, a creek for water, air to breathe, and plenty of space to live and grow.

What is a nonliving thing? something that does not need food, water, and air

Put your finger on the bridge. Does it need food? no Water? no Air? no

Is the bridge a living or nonliving thing? nonliving thing

Name some other nonliving things in the park environment. rocks, bridge, park bench, dirt, clothes, water, air

✓ **answer**

all the living and nonliving things around a plant or animal

PREPARATION FOR READING

- Direct the students to the green heading.
  What is the heading? "Needs of Animals"
- Direct attention to the two black headings.
  What are these headings? "Air, Water, and Food" and "Shelter and Space"
  What do you think you will be reading about on these two pages? the needs of animals; how air, water, food, shelter, and space are the needs of animals
- Direct the students to read pages 70–71 silently to find out what a shelter is.

TEACH FOR UNDERSTANDING

Who designed animals? God
- Discuss an animal's five main needs.
  What is a need? what a living thing must have to stay alive and grow
  Where can an animal's needs be met? in its environment
  What is one reason God designed an animal's environment? so that the animal can survive and grow
  What is the first need mentioned? air
  Why is air important for animals? They need air to breathe.
  What is the next basic need mentioned? water
  Do animals need water? yes
  Do all animals need the same amount of water? no
  Name some animals that would need more water than other animals. Possible answers: fish, dolphin, alligator, beaver
  Why does an animal need food? Food gives them energy and helps them grow.
  What is energy? what is needed to cause change
  How do animals get energy? by eating food
  Do all animals eat the same kind of food? no
- Direct attention to the picture.
  Where do you think the frog is? in a pond, in a lake
  Do you think the frog needs a little water or a lot of water to live? a lot
- Ask a volunteer to read the caption and answer the question.
  Where does the frog get its energy? from food, from the cricket
- Explain that God made the earth with many different kinds of places. He also made different animals to live in those places (Genesis 1:9–12, 20–25).
- Emphasize that God blessed the earth and people by giving us such a variety of places and animals (Genesis 1:28–31). We should praise Him for making the earth a wonderful place for us to live.

Needs of Animals

God designed animals.

An animal has five main needs to live.

Its needs can be met in its environment.

God designed an animal's environment so that the animal can survive and grow.

Air, Water, and Food

Three of an animal's needs are *air*, *water*, and *food*.

An animal breathes air and needs water to live.

Food gives energy and helps an animal grow.

Air, water, and food are found in an animal's environment.

70

ACTIVITIES

Build an Animal Shelter
Materials: an animal sticker, one per student; 20 toothpicks per student; 12 marshmallows per student (*Note:* For ease of distribution, place all the materials for each student in a plastic bag or small paper cup.)

Instruct the students to pretend that the animal on their stickers needs a shelter. With the toothpicks and marshmallows, quickly build something that will protect your animal.

Direct the students to keep the backing on their stickers as they work and put their animal stickers inside their shelters. Comment on the design and creativity of the structures.

Identify Shelters
Take a trip to a zoo or a living museum, or visit a virtual one. Encourage the students to identify the different shelters that living things use. Help them identify how the needs of the living things are being met. Discuss what is similar or different for each living thing.

Identify Main Needs
Materials: variety of pictures showing animals having their needs met

Display a picture. Encourage the students to identify which need is being met and how.

Shelter and Space

A *shelter* is a safe place to live.

A cow may have a barn.

A bug may use a plant for shelter.

A fox has a den.

All animals need shelter.

A fifth need of an animal is *space* to grow.

Animals need space to move and find food.

They need space to care for their young.

Shelter and space are found in an animal's environment.

God designed animals and their environments to work together so that the animals can survive and grow.

What five needs must animals have?

71

What is a shelter? a safe place to live

What is your shelter? my home

Do all animals have the same kind of shelter? no

What are some shelters that animals use? Possible answers: nests, dens, barns, rocks, plants

- Direct attention to the pictures.

What shelter does the fox have? a den, a hole in a cave

What shelter does the swan have? a pond, a nest

- Display the *Elephant*.

Does the elephant have enough space to grow? No; it does not have enough space to move around.

What other needs does the elephant not have inside the fence? water, enough food

Why do animals need space? to move and find food, to care for their young

Where can an animal's space be found? in its environment

Why do animals and their environments work together? so that the animals can survive and grow

- Direct attention to Genesis 1:29–30 on *Bible Verses*. Read the verses aloud as the students follow along on Activities page 59.

- Discuss how God provides the needs for every animal in its environment. People need to understand and use the earth in ways that allow the needs of most living things to continue to be met.

- Ask volunteers to share experiences of meeting the needs of a pet. Connect this to the truth that God wants us to care for the earth.

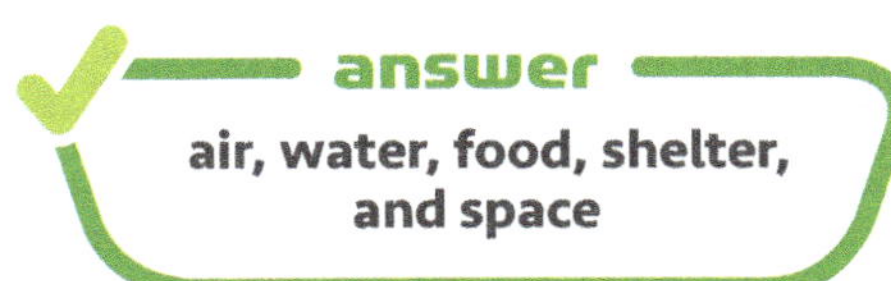

answer

air, water, food, shelter, and space

ACTIVITIES

Bible Verses, page 59
These Bible verses will be used in the teaching of this chapter.

Environments, page 61
This page reinforces the concepts taught in this lesson.

Animal Shelters in the Bible

Read Psalm 104:16–18 together. Allow volunteers to identify the shelters that are mentioned and the animals that live in each. Discuss God's wisdom and creativity in caring for His creation. Connect this worldview discussion to the truth that God is Creator. Challenge the students to think of other examples in the Bible of God's wisdom and care for animals.

Note: Some versions of the Bible list "conies" among the animals. Conies are commonly called hyraxes today. They are found in the Middle East and Africa.

Objectives
- Identify external characteristics of mammals, birds, and fish
- Classify animals as mammals, birds, and fish based on similar external characteristics
- Classify a zoologist as a scientist

Materials
- 7–8 magnets of different shapes, sizes, or colors
- clear jar
- cold, unfiltered tap water

> Fill the jar with the water and set it aside for use during the lesson. More bubbles will appear the longer the water sits.

Teacher Resources
- Visual 4.3: *Classify Animals*

Vocabulary
- fish
- bird
- mammal
- zoologist

INTRODUCTION

- Display the magnets in random order.

  What does the word *classify* mean? to group things that are alike

  What ways you can classify the magnets? by color, shape, or size

- Ask volunteers to classify the magnets and try different ways to classify them.

PREPARATION FOR READING

- Preview and pronounce the vocabulary terms *fish* and *bird*. Preview and pronounce the words *feathers* and *ostrich*.
- Direct attention to the green heading.

  What is the heading? "Classifying Animals"

  How did you classify the magnets? Answers will include the responses from the Introduction.

- Direct the students to read pages 72–73 silently to find one way scientists classify animals.

TEACH FOR UNDERSTANDING

Who named all the animals? Adam

Do we know what Adam named the animals? no

What is one way scientists classify animals? by the parts of the animals

- Display *Classify Animals* as the students look at the web on Activities page 63.

  What heading is in the center of the web? Animals

Lesson 26

Classifying Animals

Adam named all the animals.

We do not know the names he used.

Scientists have names for animals too.

These names help us classify animals.

Scientists look at the parts of animals.

They use these parts to classify animals.

Fish

A **fish** lives in water and has gills.

A fish spends all its life in water.

72

- Ask a volunteer to read the definition of fish, filling in the blanks as he reads. Allow time for the students to complete the definition.
- Direct attention to the fish on page 72.

  Where does a fish live? in water

  Does living in water make an animal a fish? no

  What are some other animals that live in water that are not fish? alligators, clams, crabs, seals, snails, and turtles

  What allows a fish to breathe underwater? gills

  Where are the gills on a fish? on the side of the head

  Can you breathe underwater as fish do? Answer should include that although people can hold their breath underwater, they cannot breathe underwater.

- Explain that people breathe with lungs and not gills.
- Display the jar of water. Allow the students to observe the air bubbles that have formed on the inside surfaces of the jar.
- Explain that all living things need air. People and most animals that are not fish breathe with lungs to get air. Water contains tiny bubbles of air, and a fish's gills are designed to take that air from the water.

  What do most fish have covering their bodies? scales

- Explain that a fish's scales are lightweight, strong, and overlap each other.

Birds

A **bird** is an animal with feathers and wings.

Feathers help a bird stay warm.

Feathers can also help a bird fly.

God gave birds two wings.

Some birds use their wings to fly.

The ostrich is the largest bird.

It has wings, but it does not fly.

macaw parrot

How is a bird different from a fish?

An ostrich runs fast.

73

What is a bird? an animal with feathers and wings

- Display the *Classify Animals* visual. Guide the students in completing the definition of a bird on Activities page 63.

What are two ways that feathers help a bird? They help it stay warm. Feathers also help some birds fly.

- Explain that an oil on feathers helps make the feathers waterproof so that water does not get to birds' bodies. When it rains, waterproof feathers help birds stay dry.

Some birds need waterproof feathers even when it is not raining. Name some of these kinds of birds. Answer should include any kind of water bird.

How is a bird's body different from most other animals' bodies? Birds have two legs and two wings instead of four legs.

How do most birds move from one place to another? They fly.

- Explain that God gave birds special bones so they can fly. Many of their bones are hollow. The hollow bones make birds lighter.

- Compare and contrast birds by appearance. Direct attention to the pictures.

Use your skill of observation to name some ways the birds are alike. They have feathers, beaks, wings, and two legs.

What are some ways the birds are different? They are different sizes and colors, have different beak shapes, and have different body shapes.

What is the largest bird? an ostrich

How does an ostrich move? It walks or runs very fast.

- Encourage volunteers to share descriptions of birds common to your area.

- Contrast how a bird breathes through lungs with how a fish breathes through gills.

Do birds live in the water like fish do? Elicit that they swim and dive in the water, but they do not stay in the water like fish do.

Can birds that live on the water and find food in the water breathe underwater like fish? No; they come to the top (surface) of the water for air.

Name some birds that live on the water and find food in the water. ducks, penguins, flamingos, sea gulls, swans, geese

answer

**A bird has feathers and wings.
Most birds fly.
A fish lives in water and has gills.
Fish swim.**

Oxygen

Although fish breathe by taking in water, they do not get oxygen from the water compound H_2O. Instead, they breathe oxygen gas dissolved in water.

Scales

Some fish, such as catfish, do not have scales. Other fish, such as sharks, have scales that are so tiny they are hardly noticeable.

Feathers

Feathers are made of keratin, the same material that claws, hooves, fur, human nails, and human hair are made of. There are different types of feathers. These feathers look different and have different purposes. Some are used for flying, some act like sensors, and some simply help the bird keep warm and dry. Birds can have anywhere from 1,000 to 25,000 feathers.

Bird Sounds

Discuss the sounds that birds in your area make. Allow students to demonstrate several common bird sounds, such as a crow's caw, a duck's quack, a dove's coo, a hawk's screech, or an owl's hoot. You may play audio recordings of bird songs.

PREPARATION FOR READING

- Preview and pronounce the vocabulary terms *mammal* and *zoologist*.
- Preview and pronounce the words *amounts, elephants, healthy,* and *aquariums*.
- Direct attention to the green heading on page 75.
  What are the headings? "STEM Careers" and "Zoologist"
- What is a career? a job
- Direct the students to read pages 74–75 silently to find out what a zoologist does.

TEACH FOR UNDERSTANDING

  What are mammals? animals with hair or fur
- Display the *Classify Animals* visual. Guide the students as they complete the web on Activities page 63 during the discussion.
- Discuss that some mammals have more hair than others.
  Name some mammals that have hair all over their bodies. dogs, lions, tigers, cows, horses, cats, giraffes
  Name some mammals that have only small amounts of hair on their bodies. elephants, whales, dolphins, walruses, pigs, hippopotamuses
- Direct attention to the pictures.
- Which picture shows a mammal with a lot of hair? the lion
- Which picture shows a mammal with a small amount of hair? the elephant
- Guide the students as they find hair on the elephant. The easiest place to see the hair is around the mouth, below the tusk.
- Name some places where mammals live. Possible answers: on farms, on mountains, in deserts, in wooded areas, in water
- Do mammals that live in the water breathe underwater like fish? No; they come to the top (surface) of the water to get air.
- Explain that some scientists classify people as mammals. People do have hair. But people are a special creation of God. Only people are made in God's image. Animals are not made in God's image.
- Direct attention to pages 44–45. As the students classify each of the animals, ask them why they chose to classify the animal that way.
  How would you classify the mouse? It is a mammal; it has fur or hair.
  How would you classify the owl? It is a bird; It has wings and feathers.

Mammals

Most pets are mammals.

And most farm animals are mammals.

A **mammal** is an animal with hair or fur.

Many mammals have hair or fur all over their bodies.

Dogs and lions have fur all over their bodies.

Some mammals only have small amounts of hair on their bodies.

Elephants and whales have small amounts of hair.

74

- Direct attention to page 66.
  How would you classify the dolphin? a mammal
- Explain that dolphins are born with hair around their snouts. Soon after birth the hair falls out. The little hole where the hair grew remains.
  Since dolphins are mammals, do they breathe with lungs or gills? lungs
  Dolphins can stay under the water for about 10 to 15 minutes.
- Direct attention to Number 1 on Activities page 64.
  Is the hamster a fish, bird, or mammal? mammal
  What makes the hamster a mammal? It has fur or hair.
  Is the goldfish in Number 2 a fish, bird, or mammal? fish
  Why did you decide the goldfish was a fish? It lives in water and has gills.
- Continue this procedure with each of the pictures.

STEM Careers

Zoologist

A **zoologist** is a scientist.

A zoologist learns how to care for animals.

He finds out about animal environments.

He learns what animals need to grow and survive.

Zoologists work in different places.

Some observe animals in their own environment.

Some work in aquariums or zoos.

Some teach in classrooms.

Zoologists enjoy keeping animals healthy and safe.

75

What is a zoologist? a scientist who learns how to care for animals

What did God tell the people to do with the earth in Genesis 1:26? to care for His earth

Note: If the students cannot remember, read the verse to them.

Are animals a part of God's world? yes

It pleases God when you and zoologists obey God and care for the earth and the animals.

What does a zoologist learn about animals? how to care for them, what their environments are, what animals need to survive and grow

Where are places that zoologists work? Possible answers: in the animal's environment, aquariums, zoos, classrooms

• Direct attention to the top picture.

Where is this zoologist working? Possible answers: an aquarium, a marine zoological park, a theme park

What is she doing? feeding an orca

Is an orca a fish, bird, or mammal? mammal

Where do you think the zoologist in the bottom picture works? a zoo, game park, national park

What is the zoologist doing? feeding a giraffe

Is a giraffe a fish, bird, or mammal? mammal

ACTIVITIES

Classifying Animals, pages 63–64
These pages were completed as part of the lesson.

Study Guide, pages 65–66
These pages review the concepts in Lessons 25–26.

Note: Show the students how to follow the two-column format. The page format is designed to give students practice in a standardized test format.

After the pages are corrected, direct the students to put the pages in their science notebooks to review for Test 4.

Classification of People

Scientific classification places humans as a kind of primate. Although human physical characteristics are similar to those of mammals, emphasize that people are a separate and special creation of God. All of God's creation brings glory to Him, but only people are made in God's image.

Mammal Species

Mammals do not have the largest number of species. Reptiles, birds, and fish each have more species than mammals do. However, mammals are typically the largest animals and are the most familiar to people.

Objective
- Relate the function of animal body parts to the survival and growth of animals

Materials
- globe
- picture of Arctic cod

INTRODUCTION

- Show the students where the Arctic Ocean is located on a globe. Explain that it is the coldest ocean in the world. It is often covered with ice.
  Who designed all the fish? God
- Display a picture of the Arctic cod. Explain that God designed the Arctic cod to live in these cold waters. He gave the fish a special substance in its blood to keep the blood from freezing.

PREPARATION FOR READING

- Preview and pronounce the word *lightweight*.
- Direct attention to the green heading.
  What is the heading? "Parts of Animals"
  What do you think this lesson will be about? animals' body parts, what the parts of animals do
- Direct the students to read pages 76–77 silently to find ways a bird gets food.

TEACH FOR UNDERSTANDING

What do most animals use to see? eyes
How does the sense of hearing protect animals? They can sense danger.
How did God design each animal's parts? to be able to survive and grow
- Discuss the parts of a fish.
  How do fish breathe under the water? with gills
  Why can you not breathe underwater? We cannot get our air from the water.
  Is a fish a living thing or nonliving thing? living
  What are the five needs of a living thing to survive and grow? air, water, food, shelter, and space
  Is air a need of a fish? Yes; air helps them survive.
  Why does a fish have fins? Fins help a fish swim.
  What are some reasons a fish needs to swim? to survive, to find food, to hide
  What do the scales of a fish do? help protect a fish
  Describe the scales. They are lightweight and strong.
  What does it mean to be lightweight? It does not weigh very much.

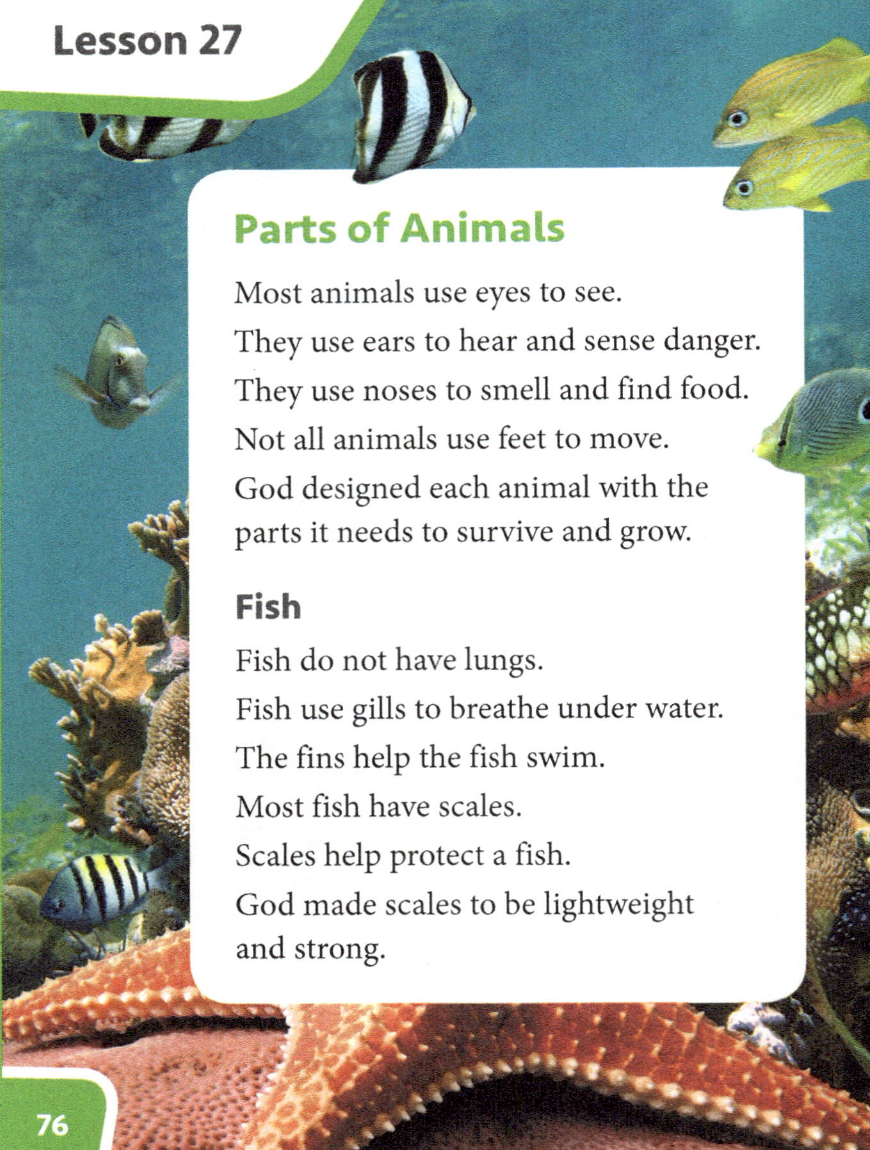

BACKGROUND

Gills and Lungs
Gills and lungs are similar. They both have small blood vessels that take in oxygen and release carbon dioxide. Gills are used by animals that get their oxygen from the water. Lungs are used by humans and land animals that get their oxygen from the air.

Sharks and Dolphins
A shark is a fish, but a dolphin is not. A dolphin is a mammal, a member of the whale order.

Largest Fish
The largest fish is the whale shark. Its average length is about 12 meters (39.4 ft). It eats plankton, tiny shrimp, and small bony fish. It is a filter feeder and catches food by swimming with its mouth open. Sharks are cartilaginous fish.

Birds

All birds breathe with lungs.

Most birds fly with their wings to find food.

Some birds use their beaks to get food.

Some birds use their feet to catch food.

Many birds carry twigs in their beaks.

They use twigs to make nests.

An eagle catches
its food with its feet.

This Vitelline masked weaver is
making a shelter for its babies.

✓ **What protects a fish?**

77

What is a bird? an animal with wings and feathers

How do you think feathers help a bird survive and
grow? They help it stay warm. The oil on feathers
helps make the feathers waterproof. Feathers also
help some birds fly.

What does a bird breathe with? lungs

What do most birds do with their wings? fly

How do wings help a bird to survive and grow?
help a bird fly to find food, help a bird fly away
from danger, help a bird fly to find water and
shelter

What are two ways birds can get food? use their
beaks, use their feet

What other way can birds use their beaks? to carry
twigs to build their nests

- Direct attention to the pictures.

What two parts of the body is the bald eagle using
to survive and grow? its feet to get food and its
wings to fly for food

What part of the body is the Vitelline masked
weaver using to survive and grow? its beak

How is the bird using its beak? to build a nest

Is a nest a need of the bird? Yes; the nest is the
bird's shelter.

✓ **answer**
scales

Arctic Cod

Arctic cod live in very cold waters. They have a protein molecule in their
blood that works the way antifreeze functions in a car. The molecule keeps
their blood from freezing even when the water temperature is below freezing.

Vitelline Masked Weaver

The Vitelline masked weaver can be found in parts of Africa's dry savanna
and arid scrubland. Its diet consists of seeds, insects, and nectar. It is also
called the southern masked weaver or the African masked weaver.

PREPARATION FOR READING

- Direct the students to read pages 78–79 silently to find ways God designed mammals to survive and grow.

TEACH FOR UNDERSTANDING

What is a mammal? an animal with fur or hair

What is a hoof? the hard part of a mammal's foot

Name some animals that have hooves. Possible answers: horse, cow, goat, pig, sheep, deer, bison, camel, giraffe

How do hooves help a mammal survive and grow? protect a mammal's feet

What do some mammals use their front legs for? hands; to hold things

- Direct attention to the picture of the horse.

Why is the horse a mammal? It has fur or hair.

What does the horse have to protect its feet? hooves

Can the horse use its tail as another arm or hand? no

Can the horse use its tail to get food? no

Can you think of how a horse uses its tail? swats away flies and other insects

- Read aloud *Fantastic Facts* as the students follow along.

What are ways an elephant uses its trunk? smelling, touching, getting food and water, hugging its baby, splashing each other

Mammals

Some mammals have hooves.

A hoof is the hard part of a mammal's foot.

Hooves help protect the animal's feet.

Some mammals use their front legs as hands.

Their front legs can hold things.

Fantastic Facts

The trunk on an elephant helps the elephant smell and touch. It also brings food and water to an elephant's mouth. A mother elephant hugs her baby with her trunk. Elephants use their trunks to splash each other at play.

78

Some mammals have long tails.

They use their tails like another arm or hand.

Front legs or tails can help them get food.

Some mammals use just their mouths for getting food.

Mammals have different kinds of shelters.

Monkeys have trees as their shelters.

God designed birds with the right body parts to survive.

He designed the fish and mammals to be able to get food.

God designed each animal with the parts it needs to survive and grow.

What can mammals use to help them get food?

79

What do some mammals use their tails for? as another arm or hand, to help them get food, to hang on to something

Name some parts of the body that a mammal uses to get food. tail, front legs, mouth

What is a shelter? a safe place to live

What are some kinds of shelters that mammals have? trees, dens, caves, meadows, forests, water, tunnels

• Direct attention to the pictures.

What does the monkey use its tail for? as a hand or arm, to help it get food, to hang on to the branch, as a help in swinging through the trees

What body parts is the woodpecker using? its beak to get food, its feet to hang on to the tree

• Discuss God's design for animals to survive and grow.

What does *survive* mean? to stay alive

How did God design the parts of animals? God designed them to survive and grow. He designed them to meet their needs.

answer

front legs, tails, mouths

ACTIVITIES

Parts of Animals, page 67

This page reinforces the concepts taught in the lesson.

Survive and Grow, page 68

This Enrichment activity gives each student an opportunity to draw an animal of his choice in its environment. Then the student writes a sentence telling how God designed the animal to survive and grow.

Objectives
- Identify and sequence the stages of the life cycle of an animal
- Name ways that animals care for their offspring
- Compare and contrast animals of the same kind
- Compare and contrast animals and their offspring
- Identify the Bible's explanation for animal death **BWS**

Materials
- pictures of a tricycle and a bicycle

Vocabulary
- offspring

INTRODUCTION

- Display the pictures of a tricycle and a bicycle.
- What is alike about these pictures? They both have wheels, seats, and handle bars.
- What is another word for wheel? cycle
  How many wheels or cycles does a bicycle have? 2
  How many cycles does a tricycle have? 3
- Put your finger on one of the wheels and trace the wheel as you say:
  It doesn't matter where I start on the cycle; it will always go in a circle.
- Repeat this, starting in a different place on the wheel each time.

PREPARATION FOR READING

- Direct attention to the green heading.
  What is the heading? "A Robin's Life Cycle"
  What do you think these pages will be about? the life cycle of a robin
- Do you know what a robin is? a bird
- Have you ever seen a robin?
  What is a cycle? a circle
- Direct the students to read pages 80–81 silently to find the steps of a robin's life cycle.

TEACH FOR UNDERSTANDING

Some animals lay eggs. What is another way that some animals give birth? to live babies
What are all the steps of the life of an animal called? a life cycle
What are the three steps of a simple life cycle? eggs, young, and adult
What is the first step in a robin's life cycle? eggs
How many eggs does a robin usually lay? three to five
Where does a robin lay its eggs? in a nest

Lesson 28

A Robin's Life Cycle

Some animals lay eggs.

Some animals give birth to live babies.

All the steps of the life of an animal are called a life cycle.

Most animals have a simple life cycle.

Their life cycle has three steps: eggs, young, and adult.

Eggs
A mother robin builds her nest.
She lays three to five eggs.
She sits on her eggs to keep them warm.

80

△ ACTIVITY

Observing Birds with Webcams
Many organizations provide webcams to observe various nesting birds, the eggs, and the young. To find some birds to observe, use the keywords "bird webcam."

Young

A robin's young are called chicks.
The parents feed the chicks fruit and worms.
The young will grow feathers and learn to fly.

Adult

The young will grow and change.
An adult can find its own food.
An adult will continue the life cycle.

✓ **What are the three steps in a robin's life cycle?**

81

- Explain that the number of eggs and the length of time for them to hatch is different for different kinds of birds.
- Ask volunteers to share experiences they have had observing birds' nests and eggs. Encourage them to compare and contrast the locations, sizes, and materials the nests were made of, as well as the sizes and colors of the eggs.
- Explain that if they find a nest, they should not touch it or the eggs. However, they can remove and keep an empty nest after they are certain the bird family has left.

What is the second step of a robin's life cycle? young

What kinds of foods do the robin's young eat? fruit and worms

What changes happen as the chicks grow? They grow feathers and learn to fly.

- Direct attention to the pictures for the young and the adult.
- Ask volunteers to compare and contrast how the birds look.

What is the third step of a robin's life cycle? adult

How do chicks continue to change as they become adults? They learn to find food on their own.

How do adult robins continue their life cycle? by laying more eggs

- Reinforce the concept of the life cycle by beginning with a different step and asking the students to name the steps.

Put your finger on the young robin. What steps come next? adult, eggs, and then young again

Put your finger on the adult robin. What steps come next? eggs, young, adult, and then eggs again

✓ **answer**
eggs, young, and adult

PREPARATION FOR READING

- Preview and pronounce the vocabulary term *offspring*.
- Direct attention to the green heading on page 82.
  What is the heading? "Parents and Their Offspring"
- Direct attention to the green heading on page 83.
  What is the heading? "How Parents and Offspring Are Alike"
  What do you think these pages will be about? parents and offspring
- Direct the students to read pages 82–83 silently to find out what an offspring is.

TEACH FOR UNDERSTANDING

What is an offspring? a baby
- Explain that they are the offspring of their parents.
  What are ways that animal parents care for their offspring? build shelters, keep them safe, keep them warm, feed them
  How do some offspring let their parents know they need something? cry or make a cheeping sound
  What are some ways that parents carry their offspring? in their mouths, on their backs, in pouches
- Direct attention to the pictures.
  How is the kangaroo carrying its offspring? in a pouch
  How is the lion carrying its offspring? in its mouth
  How does the sea turtle carry its offspring? It does not need to be carried.
- Ask a volunteer to read the caption for the baby sea turtle.
  Where is the baby sea turtle going? back to the sea
  Why would the sea turtle be going back to the sea? Possible answers: to find food, that is where it lives

Parents and Their Offspring

Many **offspring,** or babies, need care.

Parents build shelters to keep them safe.

Parents keep their offspring warm.

Offspring cry or make a cheeping sound.

Parents know to take offspring food.

Parents move their babies to keep them safe.

They carry offspring in different ways.

Some offspring do not need care after birth.

82

ACTIVITY

Parents and Offspring
Materials: pictures of animal parents and their offspring

Give each student a set of pictures of an animal and its offspring. Direct him to write one way the parents and offspring are similar and one way they are different.

Similarities and Differences
Show a video for students to observe similarities and differences between parent animals and their offspring.

How Parents and Offspring Are Alike

God said each animal will make the same kind of animal.

The offspring will be the same kind as its parents.

A dog will have puppies.

A dog does not have kittens.

The puppy and the parents are alike.

The puppy has the same parts as its parents.

What does God say about an animal's offspring?

- Read Genesis 1:24 aloud.
  "And God said, Let the earth bring forth the living creature after his kind, cattle, and creeping thing, and beast of the earth after his kind: and it was so."
- Explain that the word *kind* means to have the same offspring as the parent.
  What kind of offspring will a dog have? baby dogs, or puppies
  What kind of offspring will a kangaroo have? baby kangaroos, or joeys
  What kind of offspring will a cow have? baby cows, or calves
- Direct attention to the pictures.
  Does the puppy have all the same parts as its mother? yes
  Does the baby ring-tailed lemur have the same parts as its mother? yes

answer
The offspring will be the same kind as its parent.

83

PREPARATION FOR READING

- Direct attention to the green heading on page 84.
  What is the heading? "How Parents and Offspring Are Different"
- Direct attention to the green heading on page 85.
  What is the heading? "Result of Sin"
  What do you think these pages will be about? the differences between parents and their offspring; something happened because of sin
- Direct the students to read pages 84–85 silently to find out some differences between parents and their offspring.

TEACH FOR UNDERSTANDING

- Ask a volunteer to read the caption with the Dalmatian and her puppies.
  What other differences do you see between the Dalmatian mother and her puppies? The puppies are smaller than their mother. The puppies' eyes are closed.
- Explain that Dalmatian puppies start getting their spots when they are four to six weeks old.
- Ask a volunteer to read the caption with the zebra picture.
  What other differences do you see between the zebra and its offspring? The baby zebra is smaller. The stripes of the baby have more brown in them than the mother's. The baby zebra's tail is shorter than its mother's.

How Parents and Offspring Are Different

Offspring and their parents can be different.

A baby is not exactly like its parents.

The offspring is smaller than its parents.

A baby can be a different color than its parents.

A baby may have different spots than its parents.

A Dalmatian puppy does not have its spots yet.

A zebra has different stripes than its parents.

84

Result of Sin

God created the animals.

He created them perfect.

He created animals to eat plants.

When Adam sinned, animals changed.

Now some animals eat other animals.

This is called death.

Some animals also die because their needs are not met.

Some animals die because they are sick or old.

Death is now a part of the life cycle.

Death happened because of Adam's sin, or the Fall.

What caused animals to die?

85

Explaining the Gospel

One of the greatest desires of Christian teachers is to lead children to the Savior. God has called you to present the gospel to your students so they may repent and trust Christ, thereby being acceptable to God through Christ. Relying on the Holy Spirit, take advantage of the opportunities that arise during lessons for presenting the good news of Jesus Christ. You may find the *Explaining the Gospel* page in the Teacher Resources in the back of the Teacher Edition helpful.

Who created the animals? God

How did God create the animals? perfect

What did all the animals eat to survive and grow in the perfect world? plants

God created Adam and Eve. How did God create Adam and Eve? He created them perfect.

- Explain that God told Adam and Eve that they could eat the fruit of any tree in the garden except of the tree of the knowledge of good and evil.

What did Adam do? He ate from the tree of the knowledge of good and evil.

- Explain that when Adam disobeyed God, Adam sinned against God (Romans 5:12). Because Adam sinned, God changed the whole world (Romans 8:19–22).

- Direct attention to the pictures on page 53 to remind the students how plants changed when Adam sinned.

How did plants change when Adam sinned? There are now weeds, thorns, and thistles.

What do some animals eat now? They eat other animals.

What do we call this? death

- Explain that there was not supposed to be death. What happens to animals when their needs are not met? They die.

What happens to some animals when they are sick and old? They die.

Why did death begin to happen? because Adam sinned

What is Adam's sin called? the Fall

- Explain that every time you see a dead animal it will remind you of your sin. The Bible tells us that everyone has sinned (Romans 3:23).

One day everything will work together as it should. Jesus will rule the earth. Everyone who turns away from sin and trusts in Jesus will live with God forever (John 3:16).

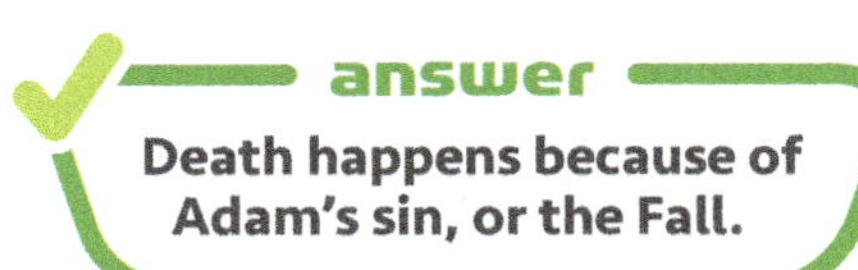

answer

Death happens because of Adam's sin, or the Fall.

ACTIVITIES

Study Guide, pages 69–70

These pages review the concepts in Lessons 27–28. Remind the students how to follow the two-column format. After the pages are corrected, ask the students to put them in their science notebooks to review for Test 4.

Objectives
- Identify a real-life human problem
- Design a solution to a human problem by using biomimicry
- Draw and label the design
- Explain how the design solves the problem

Materials
- used piece of tape or used sticky note
- picture of a gecko
- picture of a gecko's feet
- index card

Teacher Resources
- Visual 2.6: *STEM The Engineering Design Process*

INTRODUCTION

- Show the students the used piece of tape or the used sticky note.
- Why does this piece of tape (or sticky note) no longer stick to the wall? Elicit that dirt, hair, and other small things have stuck to the stickiness of the tape and the stickiness no longer works.
- Display the pictures of the gecko and the gecko's feet.
- What is this animal? a gecko
  Have you ever seen a gecko?
- Explain that a gecko is a kind of lizard that lives in warm areas all over the world. Scientists and other people have been interested in the gecko's feet.
- What do you think is interesting about a gecko's feet? Accept any reasonable answer.
- Point out that God created the gecko with feet that can climb walls and walk across ceilings. The gecko's feet stay clean; dirt, hair, and other tiny things do not stick to its feet.
  Two scientists had been studying the gecko's feet for over twenty years. Working together, they figured out how to use their knowledge of the gecko's feet to invent Geckskin™.
- Show the index card. Explain that a piece of Geckskin the size of an index card can hold up to 700 pounds on a smooth area. When the Geckskin™ is taken off, there is no dirt left behind and the Geckskin will work again.
  These scientists copied God's design in creation. You will have an opportunity to copy God's design today.

PREPARATION FOR READING

- Preview and pronounce the word *solve*.
  What does it mean to solve a problem? to find an answer for it

Lesson 29
Copying God's Design

Plants have parts to help them survive and grow.

Plants have roots, stems, leaves, and flowers.

Animals have parts to help them survive and grow.

Most animals have eyes, a nose, ears, and legs.

Some animals have long necks and hard shells.

86

Biomimicry
Point out that many of the ideas that scientists have come from things that God designed in nature. The scientists seek to copy and use what they see. This technology is called *biomimicry*.

Geckos and Geckskin™
Biologist Duncan Irschick, polymer scientist Alfred Crosby, and their teams at University of Massachusetts Amherst have been studying gecko's feet. Geckos have the ability to adhere to smooth surfaces. On the gecko's feet are millions of very tiny hairs called setae. The setae act as a soft substance that conforms to smooth surfaces.

In 2009 the teams began working together. They hypothesized and discovered that geckos have stiff tendons attached to their toepads. The teams began work on stiff fabric adhesives. Geckskin was formally revealed to the world in February of 2012.

People get ideas from these parts.

They use the ideas to design something that will help solve a problem.

How can an animal body part help solve a human problem?

How can a plant part help solve a human problem?

87

- Instruct the students to remove pages 71–72 from their Activities books.
- Direct them to read pages 86–87 and Activities pages 71–72 silently before beginning the STEM activity.

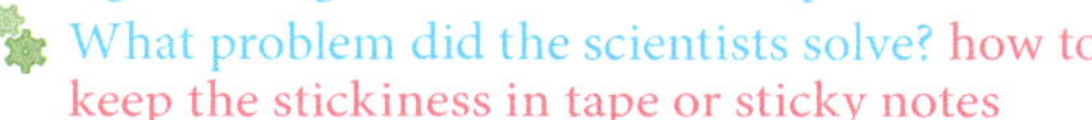

TEACH FOR UNDERSTANDING

- Display *STEM The Engineering Design Process.* Explain that for this STEM activity the students will *Imagine, Plan,* and *Share* a design to find an answer to the problem. Point out that they will not *Make* or *Make Better* their designs.

  What are some of the parts of a plant that help it survive and grow? roots, stems, leaves, and flowers

  What are some of the parts of an animal that help it survive and grow? eyes, nose, ears, mouth and legs

  What do people do with the ideas they have from plants or animals parts? They design something that will help solve a human problem.

  Remember how scientists used the way God designed the gecko's feet to solve a problem?

  What problem did the scientists solve? how to keep the stickiness in tape or sticky notes

- Direct the students to study the pictures on pages 86–87. Explain that they should look at the animals and plants to see how using one of their outside parts helps them survive and grow. Point out that they can mimic or copy God's design to help solve a human problem.
- Each student may design his own solution to a human problem, or the students may work in science groups to collaborate as they design a solution.
- Invite the students to share their designs with a partner or another science group.

ACTIVITIES

STEM Activity: Copying God's Design, pages 71–72

In this STEM activity, the students will design a way to copy God's design of an animal or plant part to solve a human problem.

ASSESSMENT

Rubric

Use the prepared rubric or design a rubric to include your chosen criteria.

⚠ HELPS

Students Who Are Stuck
When students have difficulty getting started, ask leading questions such as the following.
Which plant (or animal) part would help you to . . .?
Which plant (or animal) part reminds you of this . . .?
What can the elephant use its trunk for . . .?
What does the cactus use its spikes for . . .?

Mimicking Ideas
Here are some ideas for each of the plants and animals pictured.

Burr: anything that uses Velcro®; remind students that a burr sticks to their socks and pants.

Elephant: trunk for a flexible robotic arm; ears for some kind of hearing device; trunk for reaching something up high

Turtle: shell for a helmet or some type of head gear; shell for armor

Giraffe: long neck for reaching things up high or things down low. Giraffes can eat thorns with their leathery mouths. A student that is aware of this design might use it to solve a human problem.

Cactus: spines for collecting and storing water; spines for needles or pins; spines for protection or safekeeping of an item

Objective
- Recall terms and concepts from Chapter 4

REVIEW

- Material for Test 4 will come from the Study Guides on Activities pages 65–66 and 69–70. You may review any or all of the material during the lesson.
- You may choose to review Chapter 4 by playing the *Animal Mix-Up* review game or another game from the Game Bank (Teacher Resources).

🏆 REVIEW GAME

Animal Mix-Up
Materials: *Animal Cards*, copied on cardstock (Visual 4.4)

Arrange the cards face-down on an empty surface. Divide the students into two teams. Ask each team questions. If the team answers correctly, a member of the team chooses a card. If he can identify the animal group the animal belongs to, the team keeps the card. If not, he places the card face-down again.

The goal is to get at least one card for each animal group (fish, bird, mammal).

If a student correctly identifies a card but his team already has that kind of animal, you may choose to let him replace the card and choose another. The student must correctly identify the new card to keep it, and he may not choose a third card.

NOTES

Objective
- Recall and apply terms and concepts from Chapter 4

ASSESSMENT

- Administer Test 4.

NOTES

UNIT INTRODUCTION

- Direct attention to pages 88 and 89.

  What is the title of Unit 3? "Let's Learn About Our Bodies"

  What do you see in the picture? hands, a running heart

  What are the hands showing? the bones

  Do you know what kind of picture lets you see inside your hands? x-ray

- Invite students to share an experience with having an x-ray.

  Is this the kind of heart that you see during Valentine's Day? no

  What kind of heart is it? the organ; the kind that's in my body

- Allow students time to look through Unit 3, Chapters 5 and 6, pages 90–123 to decide what the unit is about.

  What do you think we are going to be talking about in Unit 3? Possible answers: bones, muscles, inside my body, health, safety, fire

- As the students give their answers, ask them to identify the page number and the item that helped them decide.

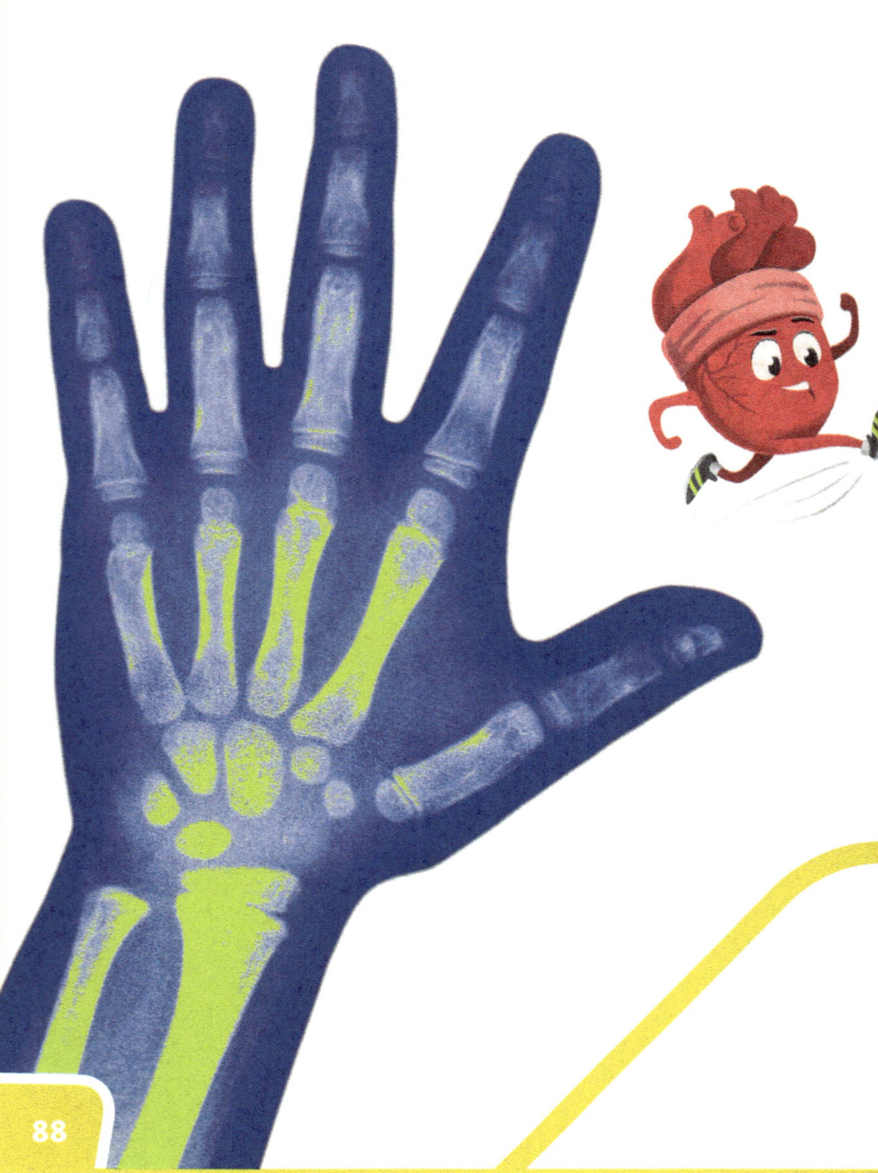

88

Science Materials

Some preparation is required for the Investigation in Chapter 6 Lesson 41. See the Helps in Chapter 6 Lesson 39.

Visit TeacherToolsOnline.com for resources to enhance the lessons.

Let's Learn About

3

Our Bodies

89

Unit Photo
The photo on pages 88–89 is an x-ray of a child's hands with a color overlay.
The little jogger represents exercising for the health of your heart.

NOTES

Chapter Objectives
- Compare and contrast the needs of living things
- Defend the biblical truth that people are different from animals **BWS**
- Describe some internal and external human body parts
- Explain how God's design of human body parts helps people survive and grow **BWS**
- Apply knowledge of a human body part to give praise to God **BWS**

Lesson Objectives
- Infer the topic of the unit and the chapter based on the pictures and headings
- Compare and contrast the needs of animals to the needs of people
- Explain how God created the first man and woman **BWS**
- Evaluate the statement that people are no different from animals **BWS**

Teacher Resources
- Instructional Aid 5.1: *Parent Letter*, one per student (See Helps.)
- Visuals 5.1–5.3: *Bible Verses A*; *Bible Verses B*; *Animals and People Venn Diagram*

CHAPTER INTRODUCTION

- Ask a volunteer to read the title of Chapter 5 on page 91.
  - What is a human? another word for man, woman, person, girl, or boy
- Direct attention to the picture.
  - What are the children doing? blowing bubbles
  - What parts of their bodies are they using to blow bubbles? hands to hold and dip the wand; mouths to blow into the wand
  - Have you blown bubbles?
- Allow time for students to share their experiences with blowing bubbles.

Big Question
- Guide the students in reading the question.
  - When will you find the answer to this question? as the chapter is read

⚠ **HELPS**

Parent Letter
Paper grocery bags will be needed in Lesson 36. Prepare the *Parent Letter* (Instructional Aid 5.1) with the day and date for your lesson and sign the letter before copying for each student.

The Human Body

Plants and animals are living things.

People are alive too.

All living things have needs.

Living things meet their needs in their environments.

God designed living things and their environments.

He designed them to work together so that living things can survive and grow.

91

- Direct the students to read page 91 silently to find out how God designed living things to work.

Name some living things. plants, animals, and people

Where do living things have their needs met? in their environments

Who designed living things and their environments? God

How did He design them to work? God designed them to work together so that living things can survive and grow.

PREPARATION FOR READING

- Direct the students to the yellow heading.

  What is the heading? "Animals and People"

  What are the two headings in black? "How We Are the Same" and "How We Are Different"

  What do you think these two pages will be about? how animals and people are the same and how they are different

- Direct the students to read pages 92–93 silently to find out if animals were created in God's image.

TEACH FOR UNDERSTANDING

- Display *Animals and People Venn Diagram*. Direct attention to Activities page 75.

  What does each circle picture? One circle pictures how animals are different from plants, and the other circle pictures how plants are different from animals.

  Look at the area where the circles overlap, or cross each other. What does this area picture? how animals and plants are alike

  Are people living things, like animals? yes

  Do people have the same needs as animals? yes

  What are the needs of people and animals? air, water, food, shelter, and space

  Do animals depend on their environment to meet their needs for survival? yes

  Do people depend on their environment to meet their needs for survival? yes

  How did God design people, animals, and their environments? to work together so they can survive and grow

- Direct attention to *Animals and People Venn Diagram*. Guide the students as they complete the alike section of the Venn diagram on Activities page 75.

Lesson 32

Animals and People

How We Are the Same

Animals are living things.

People are alive too.

Both have the same needs.

Both need air, water, food, shelter, and space.

All their needs are met in their environments.

God designed people, animals, and their environments to work together so they can survive and grow.

92

How We Are Different

The Bible tells us that God created people different from animals.

God created the first man, Adam, from the dust of the ground.

God formed the first woman from Adam's rib.

He created all people in His image.

God did not create animals in His image.

Ribs

Most adults have twelve pairs of ribs, but the number can vary in both men and women.

Adam's Rib

The Bible does not state the number of ribs that Adam or Eve had. However, the Bible does tell us that God removed one of Adam's ribs and used it to form Eve (Genesis 2:21–22). Removing the rib did not affect Adam's genetic make-up, so its removal would not have affected how many ribs his children had.

How do you know that God created people different from animals? The Bible tells us.

- Display *Bible Verses A* and direct attention to Activities page 73. Read aloud Genesis 2:7 as the students follow along.

  Who was the first man? Adam

  What did God create Adam from? the dust of the ground

  What are nostrils? openings of the nose

  What did God breathe into Adam's nostrils? the breath of life

- Display *Bible Verses B* and direct attention to Activities page 74. Read Genesis 2:21–22 aloud as the students follow along.

  God decided that Adam needed someone to help him. Who did He create to be Adam's help mate? the first woman, or Eve

  What did God form Eve from? Adam's rib

- Show the students where the ribs are located on their bodies.

- Display *Bible Verses A* and direct attention to Activities page 73. Read aloud Genesis 1:26–27 as the students follow along.

  How did God create all people? in His image

  What does *His image* mean? His likeness

- Read Genesis 2:19 aloud while the students listen to find out if God created animals in His image.

  "And out of the ground the LORD God formed every beast of the field, and every fowl of the air; and brought them unto Adam to see what he would call them: and whatsoever Adam called every living creature, that was the name thereof" (Genesis 2:19).

  Did God create the animals in His image? no

  Is this a way that animals and people are alike or different? different

- Direct attention to *Animals and People Venn Diagram*. Guide the students as they complete the first difference on the Venn diagram on Activities page 75.

**Both are living things.
Both have the same needs.**

PREPARATION FOR READING

- Preview and pronounce the words *special, psalmist, fearfully,* and *wonderfully.*
- Direct the students to read page 94 silently to find other ways that people and animals are different.

TEACH FOR UNDERSTANDING

Whose image are people created in? God's
Can people talk with each other? yes
Can you talk to God? yes, through prayer
Can God talk to you? yes, through the Bible; Elicit that God talked directly to several people in the Bible.

- Explain that Moses saw a burning bush before he went to lead the people of Israel out of Egypt. Read Exodus 3:4 aloud, asking the students to listen for what God said to Moses.

What did God call to Moses? "Moses, Moses"; his name

- Read Exodus 3:5 aloud, asking the students to listen for what God said.

What did God tell Moses to do? to take off his shoes because he was standing on holy ground

Do animals talk to God? no

Does God talk to the animals? no

Who can people love? God and each other

Does God love people? yes

Will people live forever? Yes, they will live somewhere forever.

Do animals live forever? no

All people were created in God's image. Does this mean that all people are the same? No, God created each one of us special.

- Display *Bible Verses B* and direct attention to Activities page 74. Read aloud Psalm 139:14 as the students follow along.

How does the Bible say you are made? fearfully and wonderfully

- Explain that to be "fearfully and wonderfully made" means that people have been designed by God in a wonderful way.

Think about taking a hike in the woods. Imagine turning the corner and suddenly seeing a beautiful waterfall for the first time. You might breathe in loudly and whisper, "Wow!"

God made us so wonderfully that we should whisper, "Wow!"

- Explain that because of God's great design, we praise, or thank, Him for who He is. People are the most important part of God's creation, and we respect Him for making us the way He planned.

People are created in God's image.
Therefore we can talk with God and each other.
God can talk with people through the Bible.
People can love God and each other.
All people will live somewhere forever.

Each person God created is different.
Each person is special to God.
We can say with the psalmist, "I will praise thee; for I am fearfully and wonderfully made" (Psalm 139:14).

How are people different from animals?

94

What should you do because you are fearfully and wonderfully made? praise God

- Direct attention to *Animals and People Venn Diagram*. Guide the students as they complete the remaining differences on the Venn diagram on Activities page 75.

answer

People were created in God's image.
People talk with God and each other.
God talks with people.
People can love God and each other.
People will live somewhere forever.
Animals do not do any of these things.

ACTIVITIES

Bible Verses, pages 73–74
These Bible verses will be used in the teaching of this chapter.

Animals and People, page 75
This page was completed during the lesson.

EXPLORATION

Lesson 33

My Head

God created you just the way He wanted you to be.

We should praise God for making each of us special.

To **praise** means to thank God for who He is and what He does.

In this Exploration, you will study your head.

You will name the parts of your head.

You will tell God's purpose for each part.

How are your five senses used?

How will you praise God with a part?

95

Objectives
- Observe the human head
- Identify body parts found on the head
- Identify purposes for why God designed the body parts located on the head **BWS**
- Associate each of four senses with the correct body part
- Apply knowledge of a human body part to give praise to God **BWS**

Materials
- Mr. Potato Head™ and his accessories

Teacher Resources
- Visual 5.2: *Bible Verses B*

Vocabulary
- praise

INTRODUCTION

- Display Mr. Potato Head. Place the eyes, ears, nose, and mouth where the students can see them. Include one or two other pieces as distractors.
- Ask a volunteer to choose a part that is the same as a part he has himself.

Chapter 5: The Human Body

- Direct him to attach it to Mr. Potato Head.
- Follow the same procedure until the eyes, ears, nose, and mouth have been attached.
  Some of the parts of your face use the senses. What are the senses? sight, hearing, smell, taste, touch
  Today you will study your head and choose a way to praise God.

PREPARATION FOR READING

- Preview and pronounce the vocabulary term *praise*.
- Direct the students to read page 95 and Activities pages 77–78 silently before beginning.

TEACH FOR UNDERSTANDING

- Direct attention to and discuss the **Process Skills**, using information from Chapter 2.
- Ask a volunteer to read the **Purpose** aloud.
  What does the word *praise* mean? to thank God for who He is and what He does
  What are the three things you are going to do? Name the parts of my head, write one of God's purposes for each part, and tell how I can praise God with one of the parts of my head.
- Ask a volunteer to read Steps 1–2 of the **Procedure**.
- Direct the students to complete Activities pages 77–78 independently.
- When the students have completed the pages, allow several students to share the purposes they wrote for the parts of the head.
- Discuss the **Conclusions**. Allow several students to share how they can praise God with one of the parts of their heads.
- Display *Bible Verses B* and read Psalm 139:14 aloud.
  How does the Bible say you are made? fearfully and wonderfully
  What parts of your head can you say are fearfully and wonderfully made? all of them
  What should you do because you are fearfully and wonderfully made? praise God

ACTIVITIES

Exploration: My Head, pages 77–78

ASSESSMENT

Rubric
Use the prepared rubric, or design a rubric to include your chosen criteria.

Objectives
- Recall and describe the body parts of the head
- Describe the head, arm, and leg
- Label the head, arm, and leg
- Explain ways that God's design of the human outside body parts helps people survive and grow (Psalm 139:14) **BWS**

Materials
- rope for tug-of-war
- masking tape to mark a center line on the floor for a game of tug-of-war

Teacher Resources
- Visual 5.2: *Bible Verses B*

Vocabulary
- head
- arm
- leg

INTRODUCTION

- Choose three volunteers for each team to demonstrate a game of tug-of-war. Ask the students to play the game.
- What job do the students on each team have? to pull the other team across the center mark
- How is each team doing that? by working together to pull the rope
- Direct one of the team members to stop pulling.
- Discuss what happened when he did not do his job and pull with his team.
  The team had to work together to do their jobs. The parts of the body must work together too.

PREPARATION FOR READING

- Preview and pronounce the vocabulary term *head*.
- Ask the students to turn to the glossary in their books. Explain that the definitions to all the bold vocabulary words can be found in the glossary.
  What letter does the word *head* begin with? *h*
- Explain that the words in the glossary are listed in alphabetical order.
- Ask the students to say the alphabet together as they see the letters in the circles on the glossary pages, stopping when they get to the letter *h*.
- Ask a volunteer to read the definition of *head*.
- Direct attention to the diagram on page 96. Tell the students to read the captions and to look at the diagram as they read.
- Direct the students to read pages 96–97 silently to find out what the parts of the head do. They do not need to read *Creation Corner* at this time.

Lesson 34

Outside Your Body

Your Head

God designed your head.

He placed it at the top of your body.

The **head** holds the brain.

Your head has its own parts to help you survive and grow.

God gave you *eyes* to see His creation.

You see the food you eat.

You see dangers around you.

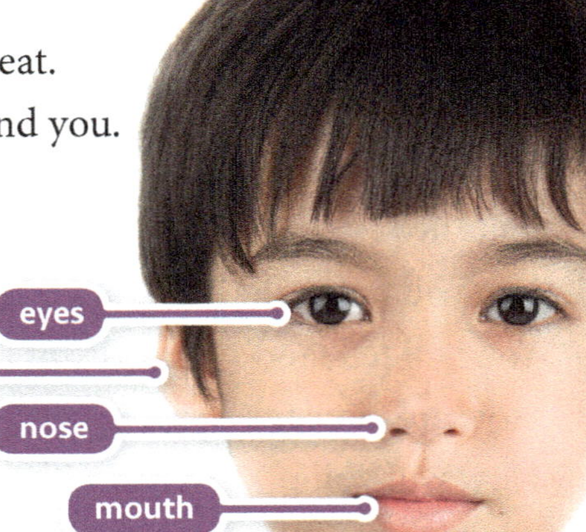

96

TEACH FOR UNDERSTANDING

Who designed your head? God
Where did God place your head? at the top of my body
What does your head hold? my brain

- Direct attention to the Outside Your Body diagram on page 100. Ask the students to put their fingers on the head.
- Explain that the brain controls everything in the body. *Note:* Discussion about the brain is in Lesson 35.
- Direct attention to the Parts of the Head diagram on page 96 and ask the students to locate the parts as each one is discussed.
  What body part did God create for you to see? eyes
  What are some of the things you can see? food you eat, dangers around you, your friends, where you are walking

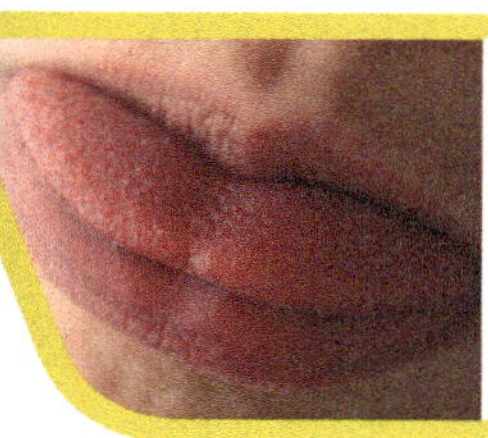

Creation Corner

God created you with a way to taste foods. Is your food sweet or salty? What helps you decide? Your tongue has many taste buds. Taste buds help you taste your food. Taste buds are too small to see, but they have a huge job.

God gave you a *mouth* to speak, sing, and eat.

You talk to family and friends.

You sing and pray to God.

God gave you *ears* to hear.

You hear your mother calling you

to come and eat.

You hear the birds sing.

✔ **What did God create at the top of your body?**

97

What body part did God give you to speak, sing, and eat? mouth

What are some things you do with your mouth? talk to family and friends, sing and pray to God, eat food, whistle

- Read the *Creation Corner* box aloud as the students follow along.

What did God create so that you can taste foods? taste buds

Where are the taste buds found? on your tongue

What size are your taste buds? too small to see

- Explain that the tongue has about ten thousand taste buds.

What did God give you to hear? ears

What are some things that you can hear with your ears? your mother calling you, birds singing, music playing, cars roaring by, ambulance sirens

- Direct attention to the picture.

What parts of the head is the choir using? mouths to sing, eyes to see the music, ears to hear the music

Taste Buds

Taste buds are special clusters of cells that react to chemicals in food. Each taste bud has sensory cells in it. Sensory cells gather information and send it to the brain, which tells you how your food tastes. Taste buds work only when food is wet. That is one way saliva helps you taste. Saliva also helps spread the taste of the foods around in your mouth. Scientists recognize five basic tastes: sweet, sour, salty, bitter, and umami.

Scientist: Kikunae Ikeda

The fifth taste was identified in 1908 by Japanese professor Kikunae Ikeda. One day he was thinking about the tastes of foods. Tomatoes, cheeses, and meats all seemed to have a common taste, but it was not one of the four basic tastes. Ikeda discovered this taste was strong in a broth made from a type of seaweed. He named the taste *umami*, which means "savory." Later he discovered that monosodium glutamate (MSG) had a strong *umami* taste. Now MSG is often sold as a food seasoning.

PREPARATION FOR READING

- Preview and pronounce the vocabulary terms *arm* and *leg*.
- Direct the students to the glossary to find the definition for each term. Ask volunteers to read the definitions aloud.
- Preview and pronounce the words *delicious, thumb,* and *balance.*
- Direct the students to read pages 98–99 silently to find out what the legs do.

TEACH FOR UNDERSTANDING

What did God give you to smell? my nose
What can you smell with your nose? delicious food, flowers, perfumes, smoke

- Direct attention to the picture.

What is the boy using his eyes for? to see a cross-walk sign
What is the boy using his ears for? listening to the sound of clapping
What is he using his nose for? to smell the rose
What is he using his mouth for? to eat an apple
What senses does your head use? sight, taste, hearing, smell, touch (tongue; skin)
What are some ways your head helps you survive and grow? Possible answers: My eyes, ears, and nose can sense danger and help protect me. My eyes see food that I can eat to help me grow. My mouth helps me eat and taste food to help me grow.

God gave you a *nose* to smell.
You smell delicious food.
You sniff roses and lilies.
God designed each of these parts so that you can survive and grow.

98

Your Arms

God designed your arms.

The **arms** let you reach for things.

Your arms bend and move in different ways.

At the end of each arm is a hand.

Your hands have fingers and a thumb.

These help you grab and hold things.

Your Legs

God designed your legs.

The **legs** help you stand, walk, and run.

At the end of each leg is a foot.

Each foot has toes.

Your feet and toes help you keep your balance.

They also help you walk and run.

What are your arms for?

99

What did God design to let you reach for things? my arms

Where is your hand found? at the end of my arm

Where are the fingers and thumb found? on my hand

What helps you grab and hold things? fingers and a thumb; a hand

- Demonstrate the movement of the arms.

  Hold your arms straight up in the air.

  Bend your hand toward me.

  Wiggle your fingers.

  Hold your arms straight out to your sides.

  Bend your arm and touch your head with your fingers.

  Grab and hold your book with your fingers and thumb.

- Direct attention to the diagram Outside Your Body on page 100 and ask the students to put their fingers on an arm.

  What did God design to help you stand, walk, and run? my legs

  What is on the end of each leg? my foot

  What is on your feet? my toes

  What helps you keep your balance? feet and toes

  What does it mean to keep your balance? to keep me from falling over

- Invite the students to stand by their desks so they can demonstrate the movement of the legs.

  Walk two steps forward.

  Walk two steps backward.

  Run in place by your desk.

  Squat down on your legs.

  Kneel on your legs.

- Direct attention to the diagram Outside Your Body on page 100 and ask the students to put their fingers on a leg.

- Direct attention to the picture at the top of page 99.

  What outside body parts is the boy using? head to see, arms to swing the jump rope, legs to jump

- Direct attention to the picture at the bottom of the page.

  What outside body parts is the boy using? head to see, arm to pull the wagon, legs to walk

answer

to let you reach for things

PREPARATION FOR READING

- Direct the students to read page 100 silently to find out what the outside body parts do.

TEACH FOR UNDERSTANDING

- Direct attention to the diagram.
  Name the outside parts of your body. head, arm, and leg
  Which outside body parts is the boy using? head for seeing, an arm for drawing on the sidewalk, an arm for holding him up, and legs for kneeling on the sidewalk
  Who designed your outside body parts? God
  Why did God design your outside body parts? to help me survive and grow
  How do the outside body parts help you survive and grow? get food and water, keep me safe, exercise to become strong and grow, take care of my body (bathing, brushing teeth, brushing hair), play with others
- Display *Bible Verses B* and direct attention to Psalm 139:14 on Activities page 74. Direct the students to read the verse aloud with you.
  What are we to praise God for? for fearfully and wonderfully making me

ACTIVITIES

Study Guide, pages 79–80
These pages review the concepts taught in Lessons 32–34. After completion, direct the students to keep them for review for Test 5.

answer

to reach for things; to help you stand, walk, and run; to see, smell, sing, talk, eat, hear; to help you survive and grow

All these parts are on the outside of your body.

These parts help you get food and water.

These parts help you keep safe.

God designed these parts to work together.

He designed them to help you survive and grow.

Let's praise Him.

We are "fearfully and wonderfully made" (Psalm 139:14)!

What do the parts on the outside of your body help you do?

100

Lesson 35

Inside Your Body

Your Bones

God designed the inside of your body too.

He designed the bones in your body to work together.

The **bones** are hard parts of the body.

Your bones are alive.

They grow and change.

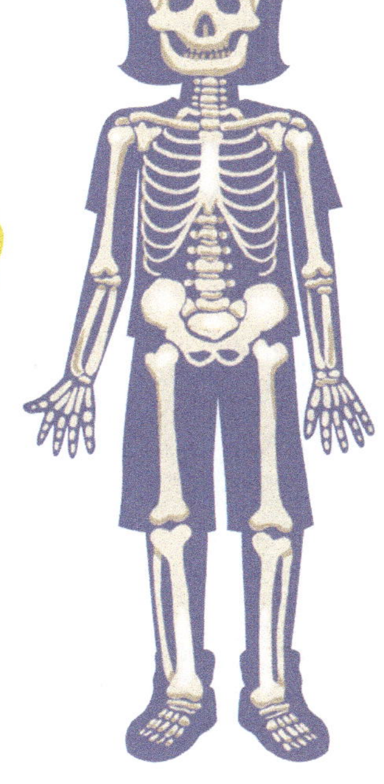

How did God design your bones?

101

Objectives

- Describe the function of the brain, lungs, heart, stomach, bones, and muscles
- Label the brain, lungs, heart, stomach, bones, and muscles on a diagram
- Explain ways that God's design of the human body parts helps people survive and grow **BWS**

Materials

- 2 similar hand puppets

Teacher Resources

- Visuals 5.1–5.2, 5.4–5.5: *Bible Verses A*; *Bible Verses B*; *Path of Food*; *Path of Food Key*

Vocabulary

- bones
- muscles
- heart
- lungs
- brain
- stomach

INTRODUCTION

- Give two students each a puppet. Ask one student to hold his puppet up without putting his hand inside. Ask the other student to put his hand

---

inside the puppet and make the puppet move.

How are the puppets different? Possible answers: They are different in how they look: one is floppy, and the other moves.

What does having a hand inside one puppet allow it to do? Possible answers: move, pretend to talk

- Explain that part of the fun of the puppet is seeing it move and pretend to talk. It can do that only when someone's hand is inside.

Today we will be learning about what helps our hands and the rest of our bodies move.

PREPARATION FOR READING

- Preview and pronounce the vocabulary term *bones*.
- Direct the students to the yellow and black headings.

What are the headings? "Inside Your Body" and "Your Bones"

Where do you think your bones are? inside your body

- Direct the students to read page 101 silently to find out if your bones are alive.

TEACH FOR UNDERSTANDING

Who made the inside of your body? God
How did God design your bones? to work together
What are your bones? the hard parts of the body
Are your bones alive or dead? alive

- Explain that the old, dry bones in a museum are dead, but not the bones inside the students' bodies.

Will your bones always be the same size? No, they grow and change.

- Direct attention to the picture.

Put your finger on the bones in the head.

What do you call the bone that protects your brain? the skull

Put your finger on the bones in your chest.

What do you call these bones? the ribs

- Explain that these bones protect the heart, lungs, and stomach.

What did God do with one of Adam's ribs? made the first woman; made Eve

- Ask the students to feel their ribs.
- Direct the students to put their fingers on the bone that seems to be coming out of the middle of their ribs.
- Direct the students to reach around and feel the bones in the middle of their backs. Explain that these bones are called the spine.

answer

to work together

PREPARATION FOR READING

- Preview and pronounce the vocabulary term *muscles*.
- Direct the students to the glossary to find the definition. Ask a volunteer to read it aloud.
- Direct the students to read pages 102–3 silently to find out what the jobs of the bones are.

TEACH FOR UNDERSTANDING

God gave your bones three important jobs to do.

- Direct a volunteer to read aloud the sentence that tells the first job of the bones.

What is the first job? to give my body shape

- Refer to the Introduction activity.

Which puppet showed the job of the bones—the puppet with the hand inside it or the puppet without the hand? the puppet with the hand inside it

- Direct attention to the top picture on the page.

What is this a picture of? a frame of a house or building

How are the frame of a house and the bones of your body alike? They both give shape. The frame of the house gives the house its shape. The bones of the body give my body shape.

- Direct a volunteer to read aloud the sentence that tells the second job of the bones.

What is the second job? to protect the soft parts inside your body

What does the word *protect* mean? to keep safe

Name some soft parts inside your body. brain, heart, lungs, stomach

- Direct attention to the silhouette of the girl. Explain that the orange parts are the lungs, which we breathe with.

What is protecting the lungs? bones; the ribs

- Direct a volunteer to read aloud the sentence that tells the third job of the bones.

What is the third job? to help your body move

How do the bones help you to survive and grow? give my body shape, protect the soft parts, help me to move, to get food, and to keep me from danger

- Direct attention to the picture with the doctor.

What is wrong with the boy? He has hurt his leg or the bones in his leg.

Why is he in the wheelchair? because he cannot put weight on the bones in his leg; to give his leg time to heal; because he cannot walk on his leg

What do you think the doctor is telling him? Possible answers: what he can or cannot do; how long to stay off the leg; when he can get up and walk again

God gave your bones some important jobs.

One job is to give your body shape.

They also protect the soft parts inside your body.

A third job of bones is to help your body move.

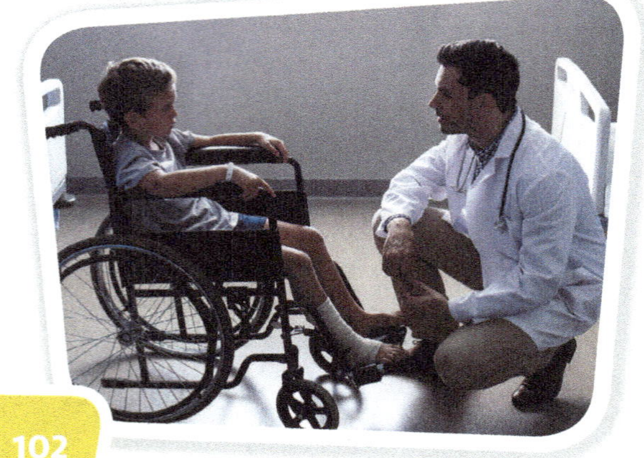

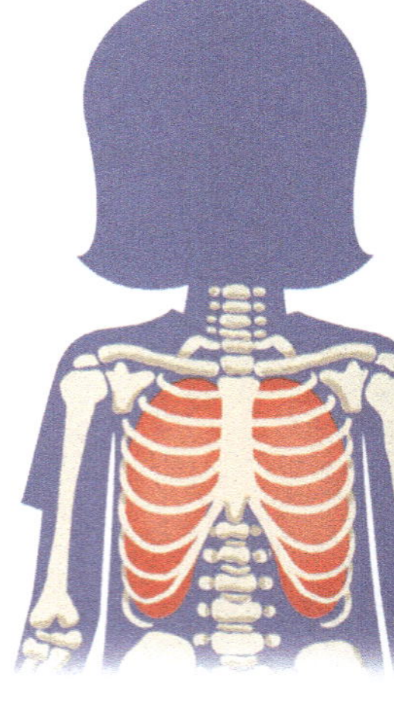

🔍 BACKGROUND

The Spine
The spine is made of a total of 33 vertebrae (7 cervical, 12 thoracic, 5 lumbar, 5 fused sacral, and 4 fused coccygeal).

The Skull
The skull is actually made of 22 bones, not just one. Eight of the bones protect the brain, and the others form the facial features.

⚠ ACTIVITY

Show How Bones Protect the Body
Materials: 2 marshmallows, paper cup, hard plastic cup, heavy books
Note: For a fair test, use books that are the same weight and cups that are about the same size.

Place a marshmallow in each cup.

Which of these cups do you think will protect the marshmallow better? the hard plastic cup

Demonstrate the support of each cup by placing an equal amount of heavy books on each until the paper cup collapses.

Your bones are like the hard plastic cup. They give support and protect the soft parts in your body.

Your Muscles

God designed the muscles to work with the bones.

The **muscles** make the bones and other body parts move.

Many of your muscles connect to bones.

When the muscles move, the bones move.

Some muscles move without you thinking about them.

God made muscles that work even while you sleep.

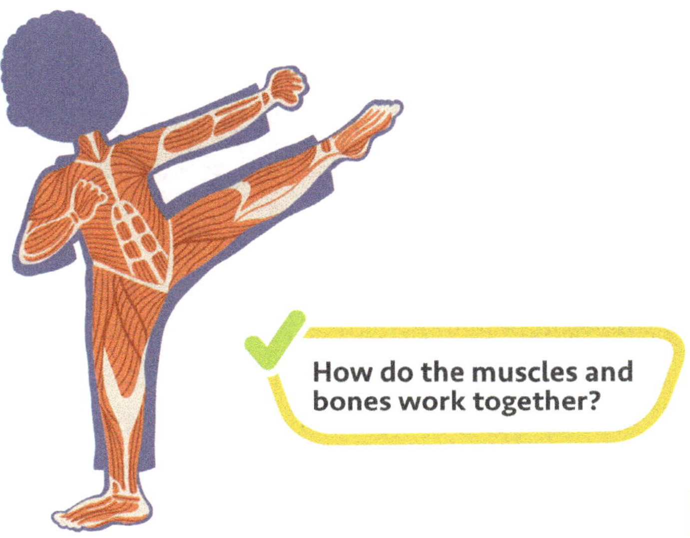

How do the muscles and bones work together?

103

What did God design to work with the bones? muscles

What do the muscles do? make the bones and other body parts move

What do many of your muscles connect to? bones

- Explain that some muscles are very small. Direct the students to move their little fingers up and down to see a small muscle in action.
- Point out that the muscles in the top of the leg are very large muscles.

What can the large muscles in the top of the leg help you to do? walk, run, stand

When the muscles connected to bones move, what else moves? bones

How do the muscles attached to bones know to move? You control these muscles by thinking about moving them.

- Direct the students to raise their right hands. Ask them to lift their left feet.

How did your right hand get in the air? You thought about moving it.

How did your left foot get in the air? You thought about moving it.

- Explain that some muscles move without you thinking about them. Some even move while you are sleeping.
- Direct the students to put their hands on their chests and sit very still.

What did you feel? my heart beating

Did you tell your heart to beat? No; it works all the time whether I think about it or not.

How do your muscles help you to survive and grow? They help me to move to get food and water; to move away from danger; to breathe; to grow; to see food; to see danger.

answer
When the muscles move, the bones move.

Play "Simon Says"

Explain that the students should do the activity you tell them to only if you say "Simon says" before saying the activity. Give the students many instructions that involve specific muscles.

Spread the fingers of your left hand.	Touch your toes.
Stretch your right foot forward.	Shake your head no.
Lean your head back.	March in place.
Blink your eyes.	Pucker your lips.
Yawn.	Turn around in a circle.
Wave.	Snap your fingers.

Include a few things that the students cannot do and discuss why they cannot.

Make your heart stop beating.
Keep your eyes from blinking.

- Preview and pronounce the vocabulary terms *heart, lungs,* and *brain.*
- Preview and pronounce the words *powerful, breathe, minute,* and *breath.*
- Direct attention to the three black headings.
  What parts of the body will you be reading about? heart, lungs, and brain
- Remind the students to be looking for the definitions of the bold words.
- Direct the students to read pages 104–5 silently to find out how Adam became a living person.

What pumps blood through the body? the heart
What is the heart? a powerful muscle
Where can you find your heart? in my chest
What bones protect the heart? the ribs
- Refer to the picture on page 102 as needed.
  How big is your heart? about the size of my fist
- Direct the students to make a fist and to place their fists against their chests to demonstrate the size of their hearts.
  When does this powerful muscle rest? in between beats
- Ask the students to put their fingers on the heart in the diagram.
  What are on either side of the heart? the lungs
  How many lungs do you have? two
  What need do the lungs get for your body? air
  What do the lungs do? fill with air when you breathe in

Your Heart

The **heart** pumps blood through the body.

It is a powerful muscle.

God placed your heart in your chest.

Your heart is about the size of your fist.

Each time it beats, it pumps blood through the body.

It beats all the time.

The only rest your heart gets is between beats.

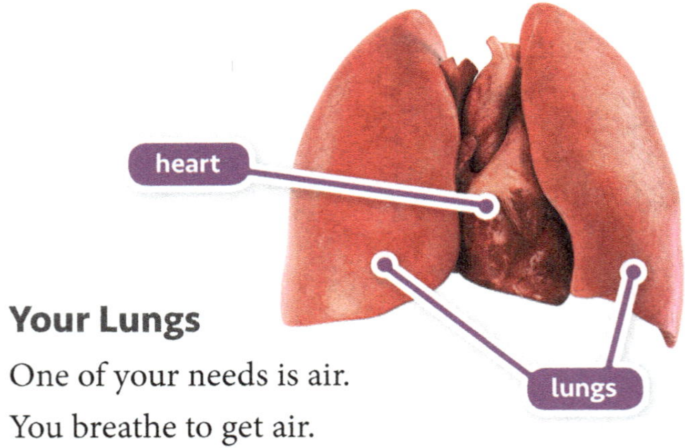

Your Lungs

One of your needs is air.

You breathe to get air.

God placed two lungs inside your chest.

The **lungs** fill with air when you breathe in.

Find Your Heart Rate
Materials: stopwatch or a watch with a second hand
Help each student find his heart rate while in a resting (sitting) position. Explain that the heart rate lets you know how fast your heart is beating.
Demonstrate how to locate a pulse at the wrist.
Turn one hand over so your palm is facing up. Place two fingers of your other hand on the thumb side of your wrist. Press your wrist lightly until you find your pulse. If you press too hard, you will not be able to feel it.
Direct the students to count the beats as you watch the time for 30 seconds. Explain that the number they counted is half the number of times the heart beats in a minute. Help the students calculate their heart rates by writing the count twice and adding.

Count Breaths
Materials: stopwatch or a watch with a second hand
Divide the class into pairs. Direct one student to count the number of breaths his partner takes in one minute while breathing normally. Tell the student when to start counting and when to stop, and have him keep time for one minute. Write the count for display. Repeat, having the partners switch tasks.

They empty when you breathe out.

You breathe about 20 times every minute.

Genesis 2:7 tells us that God created the first man.

God breathed into him the breath of life.

With this breath, Adam became a living person.

Breathing is a gift from God.

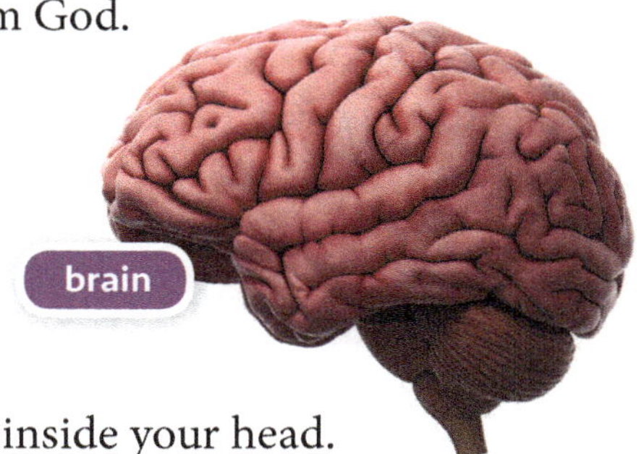

Your Brain

God designed a brain inside your head.

The **brain** controls everything in the body.

It helps you blink, breathe, and see.

It helps you walk, learn, and grow.

The brain never stops working.

What two body parts never stop working?

What happens when you breathe out? The lungs empty.

About how many times a minute do you breathe? 20

- Direct attention to Genesis 2:7 on *Bible Verses A* and Activities page 73. Ask a volunteer to read the verse aloud.

  Who gave Adam the breath of life? God

  How did God give Adam the breath of life? God breathed into Adam's nose or nostrils.

  What did Adam become? a living person

- Direct attention to Psalm 139:14 on *Bible Verses B* and Activities page 74. Read the verse together.

  Breathing is a gift from God. What should you do to thank God for this gift? praise God

  How can you show praise to God? praying to God, worshiping Him, singing songs to Him

- Direct the students to put their hands on their heads.

  What are the bones of the head called? the skull

  What can you find inside the head or the skull? my brain

- Direct attention to the picture.

  What does the brain control? everything in the body

  Name some things the brain helps you do. blink, breathe, see, walk, learn, grow, eat, drink, sing

  When does the brain rest? It doesn't. It never stops working.

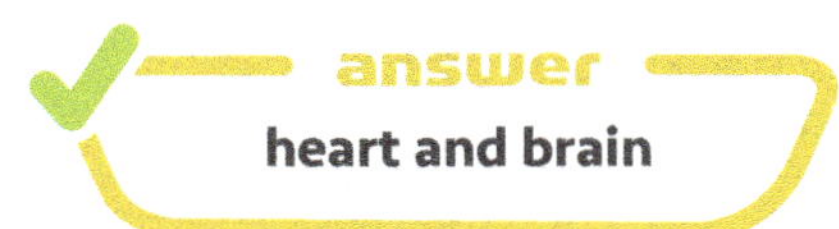

Heart Rates

An elementary child's heart beats about 90–120 times per minute. An adult's heart usually beats about 60–100 times per minute. The more physically fit a person is, the slower his heart rate usually is.

Asthma

Asthma is a common respiratory disorder that causes the small bronchial tubes in the lungs to become narrow from time to time, making it harder for the person to breathe. Asthma can begin in childhood or adulthood. Some symptoms of asthma are difficulty breathing, rapid breath, shortness of breath, tightness in the chest, and coughing. Most people with asthma take medication to help control the symptoms and prevent asthma attacks.

PREPARATION FOR READING

- Preview and pronounce the vocabulary term *stomach*. Preview and pronounce the word *swallow*.
- Direct attention to the black heading.
  What part of the body will you be reading about? stomach
- Direct the students to read page 106 silently to find out how the stomach helps you to survive and grow.

TEACH FOR UNDERSTANDING

- Display *Path of Food*. Ask the students to read each of the steps silently. Number the steps in order as the students describe the path of food.
  What is the first thing that happens to food when you put it in your mouth? Your teeth chew the food.
  What do your teeth do to the food? break it up
- Ask the students to put their fingers on the mouth on the diagram.
  What happens after you have broken up the food? You swallow it.
  Where does the food go next? down a tube
- Explain that this tube is called the esophagus. Direct the students to move their fingers down the tube, or esophagus, like the food would travel.
  Where does the food reach after it travels down the esophagus? the stomach
  What does the stomach do with the food? It breaks down the swallowed food.
  What does your body do with the food from your stomach? It uses the food to survive and grow. It uses the food to get energy.
- Use the labeled parts on the diagram for the students to tell how each body part helps you to survive and grow.
  brain: controls everything to help me survive and grow; helps me breathe to live and grow; helps me run from danger; helps me to see and get food
  lungs: help me breathe to stay alive and to grow
  heart: helps pump the blood through my body; my body needs blood to live
  stomach: breaks down the food to give me energy to survive and grow

✓ **answer**
to break down the swallowed food

ACTIVITIES

Study Guide, pages 81–82
These pages review the concepts taught in Lesson 35. After completion, direct the students to keep them for review for Test 5.

Your Stomach

You need food to survive and grow.

Your teeth chew food.

Then you swallow the broken-up food.

The food travels down a tube to your stomach.

God designed your stomach.

The **stomach** breaks down the swallowed food.

Your body uses the food from your stomach to survive and grow.

Your body gets energy from the food.

⚠ **HELPS**

Directions for Paper Vest (Lesson 36)
- Cut out both sides of the paper bag.
- Cut a slit up the back of the paper bag for ease in putting the vest on.
- Cut a hole for the student's head in the bottom of the paper bag.

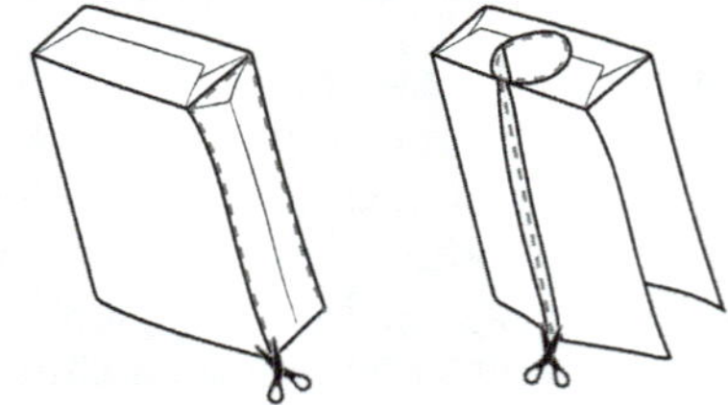

Sandwich Bags
It will be easier for the students to tape the sandwich bags to the straws and paper tubes if the bags are not resealable.

Stomach
If you choose for the students to model how the stomach works, each student will need a third sandwich bag and a cracker that will fit down the paper towel tube. If you choose not to model the stomach, you may want the students to mark through those steps on the Activities pages.

EXPLORATION

Process Skills
Observe
Communicate

Lesson 36

How My Lungs Work

Close your eyes.

Think about talking, working, or playing.

Did you think about breathing at the same time?

Most of the time, you do not think about breathing.

God created us to breathe without thinking.

How wonderful is God's creation!

In this Exploration, you will show how the lungs work.

You will see how God's design of the lungs helps people survive and grow.

107

Objectives
- Assemble internal body parts to show location
- Construct a model that shows how the lungs work
- Explain ways that God's design of the lungs helps people survive and grow **BWS**

Materials
- See Activities page 83 (See Helps on previous page.)
- completed paper-bag vest as an example
- clothespin
- 2 short pieces of chenille wires for butterfly antennae
- sandwich bag, one per student (optional)
- small cracker, one per student (optional)

Teacher Resource
- Visual 5.6: *Model of Butterfly*, for display (Make according to the directions.)

INTRODUCTION

- Direct the students to think about a butterfly that they have seen.
  If you made a butterfly, what would you use for the wings? paper, fabric

What would you use for the body of the butterfly? chenille wires, pompoms, stick, clothes pin
What would you use for its antennae? chenille wires, paper clips
- Display the butterfly model.
  Is this a real butterfly? no
- Explain that the butterfly that you made is called a *model*.
  What does the model show? the parts of a butterfly
  How are the model of the butterfly and the real butterfly the same? They have the same parts that are similar in shape.
  How are they different? The size is different. The materials are different from the real butterfly. The model is not alive.
- Explain that scientists make models to show the parts of something or to show how something works.
  Today you will make a model to show how your lungs work.

PREPARATION FOR READING

- Direct the students to read page 107 and Activities pages 83–84 silently before beginning.

TEACH FOR UNDERSTANDING

- Direct attention to and discuss the **Process Skills**, using information from Chapter 2.
- Choose a student to read the **Purpose** aloud.
- Distribute the **Materials**.
- Instruct the students to follow the **Procedure** steps to make their models. Remind them to check off each step as it is completed.
- The model has several pieces. Remind the students to be careful with the parts as they cut them out. You may want to tell the students to write their names on the back of each piece as it is cut out.
- Direct the students to work with a science partner to communicate the answers to the **Conclusions** questions.

ACTIVITIES

Exploration: How My Lungs Work, pages 83–89
The students will make a model to show how the lungs work.

ASSESSMENT

Rubric
Use the prepared rubric, or design a rubric to include your chosen criteria.

Objective

- Recall terms and concepts from Chapter 5

REVIEW

- Material for Test 5 will come from the Study Guides on Activities pages 79–82. You may review any or all of the material during the lesson.
- You may choose to review Chapter 5 by playing the *My Body* review game or another game from the Game Bank (Teacher Resources).

🏆 REVIEW GAME

My Body

Display a stick figure with two blank spaces for the teams to duplicate the stick figure as they answer questions.

Divide the students into two teams. For each correct answer to a review question, the student draws a part of the stick figure for his team. The team with the most parts drawn wins.

NOTES

Objective
- Recall and apply terms and concepts from Chapter 5

ASSESSMENT

- Administer Test 5.

NOTES

Chapter Objectives
- Explain why people should be treated with kindness and respect **BWS**
- Formulate a plan to show how you can biblically treat others **BWS**
- Explain why healthy and safe habits are important
- Apply the scientific method to investigate the effect that hand washing has on germs
- Explain fire prevention and safety responses
- Apply steps of the engineering design process to solve a real-life problem

Lesson Objectives
- Identify kind and respectful behavior
- Explain why we should treat other people with kindness and respect **BWS**
- Formulate a plan to show how to treat another person with love, care, and respect **BWS**
- Identify healthy habits for a strong body

Materials
- apple
- knife
- paper bag, lunch size

Teacher Resources
- Visuals 6.1–6.2: *Bible Verses A*; *Bible Verses B*

Vocabulary
- healthy
- habit

CHAPTER INTRODUCTION

- Begin the *Apple Decay* activity to illustrate tooth decay, but do not discuss tooth decay during this lesson.

  What is the title of Chapter 6? "Care for the Human Body"
- Direct the students to look through the chapter.

  What are we going to be learning about in this chapter? Possible answers: being kind; how to be healthy; how to be safe
- Direct attention to the picture.

  What is the boy doing? rollerblading

  This boy is being healthy and safe. In this chapter you will learn about being healthy and safe.

Big Question
- Direct attention to the Big Question. Ask a volunteer to read the question aloud.
- Explain that the answer to the Big Question will be found as they read Chapter 6.

> Visit TeacherToolsOnline.com for resources to enhance the lessons.

ACTIVITY

Apple Decay
Materials: apple, knife, paper bag

Show the students an unbruised apple. Use a knife to make a small, inch-deep hole in the apple. Show the hole. Explain that they will observe the apple each day. Place the apple in a paper bag and set the bag aside. *Note:* Placing the apple in the bag will hasten decay. You may take photos of the apple as a visual record of the apple's condition from start to finish. Observations will be made through Lesson 43.

HELPS

Lesson 41 Gelatin Cups
Prepare the gelatin cups ahead of time for the "Clean Hands" Investigation in Lesson 41. You may use commercially prepared nutrient agar and petri dishes. Liquid agar can be bought from science supply companies and then poured into sterile petri dishes. Most science supply companies also sell petri dishes already filled with agar.

If using commercially prepared agar, be sure to buy nutrient agar. Many varieties of agar exist, but nutrient agar is safe for elementary students.

Agar provides faster results (usually in one or two days) than the homemade version, which may take four or five days.

Care for the Human Body

God made people in His image.

God made our bodies.

He gave us the job of caring for them.

We glorify God when we take care of our bodies.

We show that we care for other people when we are kind.

109

- Direct the students to read page 109 silently to find out how we can glorify God.

Whose image are people made in? God's
Who made our bodies? God
How can we glorify God? when we take care of our bodies

- Explain that to glorify God means to give God honor and praise. We worship God and show Him that we love Him when we take care of our bodies.
How do we show that we care for other people? when we are kind

Gelatin Cup Recipe

Materials: 4 envelopes plain, unflavored gelatin; 4 cups cold water; 8 tsp sugar; 4 tsp beef or chicken bouillon granules; foil muffin cups, 2 per student; muffin pans; plastic wrap; resealable sandwich bags, 2 per student

Mix the ingredients in a saucepan. Bring to a boil, stirring constantly. Cool slightly.

Place foil muffin cups in muffin pans and fill each cup half full with the hot gelatin mixture. Cover with plastic wrap and refrigerate until solid. Place each cup in a resealable sandwich bag and refrigerate. Use within 2–3 days. Makes about 40 muffin cups.

Note: Sterile petri dishes can be substituted for the muffin cups and pans.

Both the commercial agar and the homemade gelatin provided in this lesson need to be refrigerated until used. No matter which substance is used, it is important to keep it sterile until the students add the germs. Be careful not to touch or breathe on it. Each student needs two petri dishes or two gelatin cups, one in a resealable sandwich bag labeled "Unwashed" and one in a resealable bag labeled "Washed."

PREPARATION FOR READING

- Introduce and pronounce the word *value*.
- Direct the students to read pages 110–11 silently to find out what are some kind actions.

TEACH FOR UNDERSTANDING

Why are people the most important part of God's creation? People are made in God's image.

- Direct attention to Activities page 91. Display *Bible Verses A* and read Genesis 1:26–27 aloud.

What does it mean that people are made in God's image? People are made to be like God. Nothing else in creation is made in His image or like Him.

What does it mean to be made in God's likeness? Like God, people can also think, love, talk, and tell the difference between right and wrong.

- Remind the students that being made in God's image makes people the most important part of God's creation.

Why does it please God when you are kind? Being kind shows that I care about other people.

- Direct attention to Activities page 92. Display *Bible Verses B* and read Ephesians 4:32 aloud. Ask a volunteer to read the first part of the verse aloud, stopping with "another."

What does the Bible say about how you should treat other people? I should be kind.

According to the Bible, why should your words be kind? Kind words are good for other people to hear.

- Direct attention to Activities page 91. Display *Bible Verses A* and read Proverbs 16:24 aloud.

What does honey taste like? It is sweet.

Is "sweet" a good taste or a bad taste? Elicit good.

- Point out that the Bible tells us that pleasant words are good words. When we say kind things to another person, we are speaking in a way that will be good for that person.

What are some kind words you can say? Possible answers: thank you, please, hello

Lesson 39

Be Kind

People are the most important part of God's creation.

You are made in God's image.

It pleases God when you are kind to others.

Being kind shows that you care about other people.

In Words

Your words should be kind.

The Bible tells us that kind words are good for other people to hear.

You are kind when you say "thank you."

You are kind when you say "please."

You are kind when you say "hello" to a friend.

110

🔍 BACKGROUND

The Image of God

People are unique in God's creation because they are made in God's image. The image of God in people means that we have a nature that is like God's nature in some ways. How humans share in God's nature cannot be reduced to a specific list, but there are some evidences of God's image in man. People have a conscience and a self-consciousness. We have the ability to think rationally and intelligently, to feel emotions, and to communicate with God in thought and word. People have an eternal soul/spirit that will live forever. Those who put their faith and trust in Jesus Christ alone for their salvation will live in God's presence for all eternity. Those who do not trust Christ as the sole means of salvation will live in punishment, separated from God for all eternity.

Explaining the Gospel

One of the greatest desires of Christian teachers is to lead children to the Savior. God has called you to present the gospel to your students so that they may repent and trust Christ, thereby being acceptable to God through Christ. Relying on the Holy Spirit, you should take advantage of the opportunities that arise during lessons for presenting the good news of Jesus Christ. You may find the *Explaining the Gospel* page in the Teacher Resources.

In Actions

Your actions should be kind.
You are kind when you smile.
You are kind when you share.
You are kind when you obey.
Kind words and actions show that
you love and value other people.

111

Not only should your words be kind, but what else
should be kind? my actions

What are actions? the way I act or behave; the
things I do

- Direct attention to Activities page 92. Display *Bible
Verses B* and read Matthew 7:12 aloud.

According to Matthew 7:12, how should you treat
other people? I should treat other people the way I
want them to treat me.

How many of you like to have people treat you
with kindness? Ask for a show of hands.

What are kind actions? smiling, sharing, and
obeying

- Direct attention to Activities page 92. Display *Bible
Verses B* and read Mark 12:31 aloud.

Why are people the most important part of God's
creation? People are made in God's image; people
can think, love, and talk.

- Explain that the Bible tells us that we should love
other people the same way we love ourselves. Love
moves us, or drives us, to be kind.

- Direct attention to the picture on pages 110–11.

How do you know the children in the picture are
being kind? They are saying "hello"; they are offer-
ing to share what they have; they are smiling.

answer

**People are made in God's image.
It pleases God when we are kind
to others.
Being kind shows that you care
about other people.**

Honey

Honey is smooth and sweet tasting. However, honey is more than a sweeten-
er. Honey is also an effective medicine. Raw honey contains natural antibi-
otics, and it can be applied topically to a cut to prevent infection and speed
up healing. For this reason, honey has also been used as a first-aid treatment
for burns. The vitamins and minerals in honey help wounds to heal, often
with little or no scarring. Honey can aid the healing of ulcers and has a pos-
tive effect on people with allergies. When mixed with lemon juice, honey
can soothe a sore throat and calm a cough. Honey is "sweet to the soul, and
health to the bones" (Proverbs 16:24).

Read-Aloud Book for Chapter 7, Lesson 47

A book on the Big Bang will be used in Lesson 47. A small portion of the
book will be read aloud to the students, after which they will biblically
evaluate the truth of the Big Bang. Select an age-appropriate book such as *A
Cartoon History of the Earth: The Birth of the Earth No. 1* by Jacqui Bailey.
Public libraries are a source for such a book.

PREPARATION FOR READING

- Preview and pronounce the vocabulary terms *healthy* and *habit*.
- Direct the students to read page 112 silently to find out what a healthy habit helps you to do.

TEACH FOR UNDERSTANDING

What is a healthy person? someone who is not sick or hurt

What do you call something that you do over and over again? a habit

- Discuss good habits that the students have observed or have made. Such habits may be praying before a meal, praying before bedtime, going to bed early, exercising regularly, and saying "please" and "thank you."

What does a healthy habit help you to do? care for my body; keep me well

What are some healthy habits that will keep your body strong? eating food that is good for me; running and playing outside; getting the sleep I need.

Are you healthy all the time? no

- Explain that having healthy habits does not mean you will never get sick. Point out that these are ways to take care of your body and help it to be healthy and strong.
- Direct attention to the picture in the upper-right corner.

What is the boy eating? an apple

Are apples a healthy food? yes

What other healthy foods can you eat? fruit, vegetables, milk, eggs, cheese, meat, whole grain bread, and cereal

- Direct attention to the picture on the bottom of the page.

What are the children doing that is healthy? They are running and playing.

- Direct attention to the picture on page 108.

What is the child doing that is a healthy habit? rollerblading; playing outside

- Allow time for the students to share some of their healthy habits. Be aware of students with disabilities and be sure to include healthy habits they have developed.

answer

eating food that is good for you, running and playing outside, getting the sleep you need

Healthy Habits

A **healthy** person is not sick or hurt.

A **habit** is something you do over and over again.

Healthy habits help you care for your body.

For a Strong Body

You should eat food that is good for you.

You should run and play outside.

You should get the sleep you need.

These are healthy habits that will keep your body strong.

What are some healthy habits to keep your body strong?

112

ACTIVITIES

Bible Verses, pages 91–92
These Bible verses will be used in the teaching of this chapter.

Good Behavior, pages 93–94
These pages reinforce the concepts taught in Lesson 39.

Lesson 40

To Keep Germs Away

Some habits help keep germs away.

Germs are things that can make you sick.

You cannot see germs.

You can get rid of some germs by using soap and water to wash your hands.

You should wash your hands before you eat.

You should use soap and water to keep cuts on your skin clean.

You should take baths and wash your hair to keep clean.

113

LESSON 40

Objectives
- Identify ways to prevent the spread of germs
- Identify healthy habits for strong teeth
- Explain the importance of developing healthy habits
- Practice healthy habits

Materials
apple in paper bag (from Lesson 39)

Teacher Resources
- Instructional Aid 6.1: *Hand Washing Poster*, for display and one per student
- Visuals 6.3–6.4: *Hand Washing Song*; *Tooth Decay*

Vocabulary
- germs
- dentist
- cavity

INTRODUCTION

Have you ever been told to wash your hands? Today you will learn why it is important for you to wash your hands.

PREPARATION FOR READING

- Preview and pronounce the vocabulary term *germs*.
- Direct the students to read page 113 silently to find out how you get rid of germs.

TEACH FOR UNDERSTANDING

Healthy habits keep your body strong. What else can some healthy habits do? help keep germs away

What are germs? things that can make you sick

Can you see germs? No; they are too small to be seen.

- Explain that germs are so small that they can be seen only with a science tool called a microscope. Point out that germs can be found just about everywhere and that there are many types of germs that can make people sick.

How can you get rid of some germs? by using soap and water to wash my hands

- Display the *Hand Washing Poster*. Distribute one per student and discuss each step.
- Display the *Hand Washing Song*. Direct the students to pretend they are at a sink. Instruct them to follow the hand-washing steps.
- Direct them to sing the song to the tune of "Row, Row, Row Your Boat" while they pretend to scrub their hands for 20 seconds.
- Explain that they could also sing "Happy Birthday" twice while scrubbing their hands for the required 20 seconds.

When should you wash your hands with soap and water? Possible answers: before eating, drinking, cooking, or touching food; after playing outside or with animals; after visiting a sick person; after using the bathroom; after handling garbage; whenever they look or feel dirty

Why should you wash your hands with soap and water before eating? to help keep germs off my food; to help keep germs from getting into my mouth and making me sick

What is another reason to use soap and water? to keep cuts on the skin clean

What is another healthy habit to keep you clean? taking baths and washing my hair

- Direct the students to complete Activities page 95. When they have finished, discuss the correct hand-washing sequence with the students.

- Introduce and pronounce the words *cough, sneeze,* and *tissue.*
- Direct the students to read pages 114–15 silently to find out why brushing your teeth is a healthy habit.

What do some healthy habits help to do? keep germs away from other people

What should you do when you cough or sneeze? cover my mouth and nose

- Explain that covering a cough or sneeze prevents germs from traveling through the air. Point out that germs in the air can be breathed in by another person, which may cause the other person to get sick. You may use the *How Far Can Germs Spread?* activity to demonstrate the importance of covering a sneeze.

What should you use when you sneeze? a tissue

What should you do with a used tissue? throw it in the trash

Why is throwing a dirty tissue in the trash a kind action and a healthy habit? Putting the tissue with the germs in the trash will keep other people from getting sick; I am treating others the way I would want them to treat me (Matthew 7:12).

What should you do after throwing away a dirty tissue? wash my hands

Why should you wash your hands? to get rid of germs; to keep from spreading germs to other people

- You may use the *Germs and Hand Washing* activity to demonstrate that hand washing correctly helps to stop the spread of germs.
- Direct attention to the picture of the girl.

What is the girl doing? covering her mouth while sneezing or coughing

What is the girl using to cover her mouth? her arm or elbow

- Point out that when the students do not have a tissue to cover their cough or sneeze, it is best to "catch" the cough or sneeze in the crook of their arms.

Why do you think the girl is not using her hand to cover her mouth? The germs would go onto her hand and then onto whatever she touches. She will spread fewer germs by coughing or sneezing into her elbow or her arm.

Why is covering your mouth when you cough or sneeze a kind action? I am trying to keep others from getting sick.

To Keep Germs Away from Others

Some healthy habits help to keep germs away from other people.

You should cover your mouth when you cough or sneeze.

You should use a tissue when you sneeze.

You should throw your used tissue in the trash.

You should wash your hands.

All these habits will help keep others from getting sick.

How Far Can Germs Spread?

Materials: spray bottle, water, paper towels

Cover an area of the floor with paper towels. Fill a spray bottle with water and hold it upright near the mouth and nose.

Explain that when you squeeze the spray bottle, it will represent a sneeze or a cough. Squeeze the spray bottle and have the students observe where the drops fall on the paper towel. Direct the students to count how many steps away the farthest drop is from where the bottle was sprayed. Explain that germs from an uncovered sneeze travel about 100 mph and can land more than 10 feet away from the person who sneezed.

Germs and Hand Washing

Materials: hand lotion or cooking spray, glitter

Pretend to cough or sneeze into your hand. Spray your hand with cooking spray or spread hand lotion on your hand. Sprinkle some glitter on the palm of your hand. Pretend that the pieces of glitter are germs. Show your hand. Shake a student's hand and have that student shake another student's hand. Choose one student to wash his hands using soap and water, scrubbing his hands for 20 seconds. Direct the other student to use a tissue or paper towel to wipe his hands. Compare the students' hands. Explain that hand washing

For Strong Teeth

Brushing your teeth is a healthy habit.

You use your teeth to cut and chew food.

God made your teeth strong.

Brushing helps your teeth stay healthy and strong.

You need to take care of your teeth.

What are three ways to keep germs away from others?

115

- Explain that if the students make an effort to practice healthy habits, eventually they will do them automatically, or out of habit.
- Demonstrate the proper way to cough or sneeze into a tissue, to dispose of the tissue, and to wash one's hands.
- Demonstrate how to cough or sneeze into the elbow.
- Allow time for the students to practice these healthy habits to keep germs away from others.
- Direct attention to Activities page 96. Read the statements at the bottom of the page aloud and instruct the students to write their name on the line if they agree with each statement. Explain that they are now a "Germbuster Team Member."

Teaching for page 115 begins here.

What is a healthy habit for your teeth? brushing
What do you use your teeth to do? to cut and chew food

- Direct attention to the girl and boy at the bottom of the page.

What are the girl and boy doing? using their teeth to eat a banana and bite a carrot
Would eating be easier or harder without teeth? harder
Who made your teeth strong? God
Why is brushing your teeth a healthy habit? Brushing helps my teeth stay healthy and strong; I take care of my teeth when I brush them.

- Direct attention to the Big Question on page 108.

Why are healthy habits important? Healthy habits help you care for your body; they will help keep germs away from you and from others; they will keep your teeth healthy and strong.

answer

**cover your mouth when you cough or sneeze,
throw used tissues in the trash,
wash your hands**

with soap and water helps to get rid of more germs (and glitter) than wiping hands on a paper towel will. Point out that clean hands help to stop the spread of germs. Explain that washing hands correctly does not wash away all the germs, but it does get rid of many of them.

Variation: After sprinkling glitter "germs" on your hand, touch various objects around the room and direct attention to the "germs" left behind on the surfaces. Have students touch the surfaces with the "germs." Observe the "germs" on their hands. Remove the "germs" from the surfaces by cleaning them with sanitizing wipes and instruct the students to wash their hands with soap and water for 20 seconds. Observe the surfaces and the washed hands.

PREPARATION FOR READING

- Preview and pronounce the vocabulary terms *dentist* and *cavity*.
- Direct the students to read page 116 silently to find out what a *cavity* is.

TEACH FOR UNDERSTANDING

What is a STEM Career? a job that uses science, technology, engineering, and math

What is a doctor who takes care of teeth called? a dentist

What does a dentist look for? cavities

What is a cavity? a hole in a tooth

- Display the apple. Remind the students of the little hole.

What happened to the little hole? It got bigger; it is more noticeable; it has turned brown.

- Display *Tooth Decay.* Explain that the word *decay* means "to rot."
- Explain that the hole in the apple is like a cavity in a tooth. Point out that an acid in our mouths can attack the outside of our teeth and make a hole.

What does a dentist do to a cavity? He fills it.

- Explain that the dentist fills the cavity to keep the hole from getting bigger. Mention that if the dentist does not fill a cavity, the cavity will keep getting bigger until the entire tooth decays.

What do you think will happen to the apple because of the hole? Accept any reasonable answer at this time.

- Return the apple to the bag. Explain that they will observe the apple three more times to see what happens to it as a result of the hole.

What does a dentist teach you to do? how to take care of my teeth

What does brushing your teeth do? takes away food and germs

What should you use to brush your teeth? a toothbrush and toothpaste

How many times a day should you brush your teeth? at least twice a day

- Point out that after breakfast and before bed are two good times for the students to brush their teeth.

Why is it good to go to the dentist and brush your teeth? Doing these things will keep my teeth healthy and strong.

STEM Careers

Dentist

A **dentist** is a doctor who takes care of teeth.

He looks for cavities.

A **cavity** is a hole in your tooth.

A dentist will fill the cavity in your tooth.

He uses special tools to fill the cavity.

A dentist teaches you how to care for your teeth.

Brushing your teeth helps take away food and germs.

Brush your teeth with a toothbrush and toothpaste.

Brush your teeth at least twice a day.

Doing these things helps keep your teeth healthy and strong.

116

- Instruct the students to remove pages 97–98 from their Activities books. Direct attention to the tooth-brushing steps on page 97. After reading each step aloud, allow the students time to simulate brushing their teeth.
- Direct attention to Activities page 98. Explain that this chart will be used at home to help the students develop a healthy habit for strong teeth. Point out that each day they are to brush their teeth at least two times. Each time they brush their teeth, they are to put a check mark on the chart. Explain that when they brush after breakfast, they will put a check in the first half of the box. When they brush before bedtime, they will put a check in the second half of the box.

ACTIVITIES

Hand Washing, page 95; Germbuster, page 96; Tooth Brushing, page 97; Tooth Care Chart, page 98
These pages were completed or assigned during the lesson.

Study Guide, pages 99–100
These pages review Lessons 39–40. After completion, direct the students to keep them for review for Test 6.

INVESTIGATION

Lesson 41

Clean Hands

Are your hands clean?

You wash your hands with soap and water.

You are told that this will help get rid of germs.

In this Investigation you will see if washing your hands does help get rid of germs.

Process Skills
• Observe
• Infer

117

Objectives
• Formulate a hypothesis to determine the effect that washing hands has on germs
• Record observations
• Draw conclusions from data collected

Materials
• apple in paper bag (from Lesson 40)
• See Activities page 101. *Note:* Prepare the gelatin cups ahead of time. See suggestions in Lesson 39 Helps.

INTRODUCTION

• Display the apple in the paper bag. Observe and discuss any noticeable changes. Return the apple to the bag.

Have you ever been told to wash your hands when you thought they were already clean?

Today you will do an Investigation to find out whether your hands are as clean as you think they are.

PREPARATION FOR READING

• Direct the students to remove pages 101–2 from their Activities books. Direct them to read page 117 and Activities pages 101–2 silently before beginning.

TEACH FOR UNDERSTANDING

• Direct attention to and discuss the **Process Skills**, using information from Chapter 2.
• Choose a student to read the **Problem** aloud.
• Direct the students to complete the **Hypothesis** to answer the problem.
• Distribute the **Materials**. Instruct the students to handle the cotton swabs as little as possible so that they remain somewhat sterile, or germ free.
• Instruct the students to follow along together as **Procedure** Steps 1–5 are completed.
• Read Step 1. Explain that "unwashed" refers to their hands, not to the cup itself. Direct the students to use the entire hand to touch various surfaces (desk, doorknob, floor, shoe, etc.). *Note:* Touching the surfaces and seeing the resulting bacterial growth later will help the students realize that germs from other surfaces can move onto their hands.
• Direct the students to place Activities pages 101–2 in their science notebooks, or you may choose to collect the pages and return them after four or five days.
• Instruct the students to make **Observations** of the gelatin cups daily. Do not allow the students to open the bags when checking their cups. After four or five days, instruct the students to record what they observe by drawing a picture of each cup on Activities page 102. *Note:* Some germs do not grow in this medium. The growth observed on the gelatin cups reflects only some of the germs present on the students' hands.
• When the observations are complete, throw away all the cups and bags without opening them.
• Direct the students to draw **Conclusions** based on their observations. They may do this individually or with their science group. Discuss the conclusions the students arrived at.

ACTIVITIES

Investigation: Clean Hands, pages 101–2
This Investigation will be started during this lesson and completed in four to five days.

ASSESSMENT

Rubric
Use the prepared rubric or design a rubric to include your chosen criteria.

Objectives
- Identify safe habits when at play and in the car
- Explain the importance of safe habits

Materials
- apple in paper bag (from Lesson 41)
- bicycle helmet

INTRODUCTION

- Display the apple in the paper bag. Observe and discuss any changes. Return the apple to the bag.
- Invite students to tell about times they have been hurt while playing.

  Today you will learn about ways to keep from getting hurt while at play and in the car.

PREPARATION FOR READING

- Preview and pronounce the vocabulary term *safe*. Introduce and pronounce the word *distract*.
- Direct the students to read pages 118–19 silently to find out what you should wear when riding a bike or a scooter.

TEACH FOR UNDERSTANDING

What does it mean to be *safe*? to stay away from things that could hurt you

Why should you be safe? to take care of my body

Who should you always tell where you are going when you are playing? an adult

- Explain that before going anywhere to play, the students should get permission from a parent or another trusted adult. Point out that adults know where it is safe to play.

  What should you wear when riding a bike or a scooter? a helmet

- Why should you wear a helmet? Possible answers: to be safe; to protect my head
- Direct attention to the picture on pages 118–19.

  What are the children doing who are wearing helmets? The girl is riding a scooter and the boy is riding a bike.

- Direct attention to the picture on page 108.
- How is this boy being safe while rollerblading? Possible answer: He is wearing a helmet, knee pads, and elbow pads.
- Display the bicycle helmet and use a volunteer to show the proper way to wear it.
- Why should you wear sunglasses, sunscreen, and a hat when it is sunny? to protect my eyes and skin from the sun
- Mention that even on cloudy days, sunscreen protects the skin from the harmful rays of the sun.

Fitting a Bicycle Helmet
A helmet should fit snugly on the head and be level, not tilted to the front or back. The helmet should sit one or two finger widths above eyebrows, and the chin strap should be snug.

⚠️ **ACTIVITY**

Why It Is Important to Wear a Helmet
Materials: 2 eggs, 2 resealable plastic sandwich bags, cardboard box, foam packing peanuts, bubble wrap

Place each egg in a resealable sandwich bag. Fill the box with packing peanuts and bubble wrap several inches deep. Direct a student to hold one of the eggs above the floor with his arm stretched out straight. Ask the students what they think will happen to the egg when it is dropped. Direct the student to drop the egg on the floor and observe the results. Explain that bike or scooter accidents may seem small but can be serious. Point out that the egg was not protected when it had the accident. Put the second egg inside the box, protected with the packing materials. Explain that these materials are similar to the material inside a helmet. Point out that the packing materials will be a protection for the egg, just like a helmet protects the head. Direct the student to drop the box

with the second egg. Observe the results. The egg should not crack or break. If it does crack, discuss that helmets do not prevent all injuries, but they do lessen the severity of injuries.

Booster Seats
When a child is no longer required to be in a car seat, most states require a booster seat for children seven and younger, depending on height and weight.

What are some playground rules? Climb up the ladder, not the slide; do not stand on swings; do not push or shove; do not throw dirt, rocks, or sticks.

Why should you obey playground rules? so I am safe; so I do not get hurt; so I do not hurt other people

Teaching for page 119 begins here.

- Direct attention to the picture of the car.

What do you think the family is getting ready to do? Possible answer: take a trip in the car

- Choose a student to read the first sentence on page 119 aloud.

- Explain that when the driver of a car is distracted, he is not paying attention to his driving, because he is now paying attention to something else.

What are some ways the driver of a car could be distracted by a child? Possible answers: yelling or fighting, kicking the driver's seat, not wearing a seat belt, not staying seated in the booster seat

How do you know the child in the car is not going to distract the driver? The child has a book to read to help him sit quietly.

What should you sit in, and what should you put on when you get in the car? I should sit in a booster seat and put on my seat belt. Why? They will help keep me safe in the car.

- Explain that car seats and booster seats are designed to keep children safe when riding in the car. A booster seat should always be used until the student is big enough for the adult seat belt to fit properly.

- Direct the students to page 103 in their Activities books. Instruct them to complete the page. Discuss each picture and ask the students to explain how the child is being safe or not being safe.

answer

At play: tell an adult where you are going; wear a helmet when riding a bike or scooter; wear sunglasses, sunscreen, and a hat; obey playground rules

In the car: do not distract the driver; sit in a booster seat; put on a seat belt

At Play and In the Car, page 103
This page was completed during the lesson.

Objectives
- Identify safe habits at home and in the community
- Identify fire hazards
- Explain the proper response in an emergency
- Identify trustworthy adults to go to in a dangerous situation

Teacher Resources
- Visuals 6.4–6.5: *Tooth Decay*; *Safety Song*
- Instructional Aid 6.2: *Fire Safety Cards*, one for display and one per student

Materials
- apple in paper bag (from Lesson 42)

INTRODUCTION

Tooth Decay activity ends here.

- Display the apple in the paper bag. Remind the students of the little hole.
- What does the hole in the apple remind us of? a cavity
- What happened to the little hole in the apple? It got bigger.
- What happened to the apple because of the hole that got bigger? The whole apple went bad; it rotted.
- Display the *Tooth Decay* visual.
- How are these pictures like what happened to the apple? The cavity keeps getting bigger and bigger, just like the hole in the apple.
- Point out that teeth that are not brushed can decay. Remind the students that an acid in our mouths can attack the outside of the teeth and make a cavity. Explain that the cavity will keep getting bigger until the entire tooth decays.
- What does a dentist do to stop a cavity from getting bigger? He fills it.
- How can you keep from getting cavities? brush teeth twice a day; floss; eat foods that are good for your teeth

Introduction for Lesson 43 begins here.

- Invite students to identify hot objects that they have been warned not to touch. Possible answers: a stove, an iron, a pot or pan
- Why did the person warn you? to keep you from getting burned; so you would not get hurt
  Today you will learn about ways to be safe at home and in your community.

Lesson 43

At Home

It is good to have safe habits at home.

Always obey your parents.

Keep your house doors locked.

Do not open your doors to strangers.

Know how to escape if there is a fire.

Know what to do if someone gets sick or hurt.

Fire Safety
1. Be sure your house has a smoke detector.
2. Leave your house quickly if you hear the alarm.
3. Crawl low if there is smoke in a room.
4. Stop, drop, and roll if your hair or clothes catch on fire.
5. Never run back into a burning house.
6. Call 911 when you are safely outside.

120

PREPARATION FOR READING

- Introduce and pronounce the words *escape*, *smoke detector*, and *community*.
- Direct the students to read pages 120–21 silently to find out what to do before crossing the street. Explain that they do not need to read the *Fire Safety* box at this time.

TEACH FOR UNDERSTANDING

- Ask a volunteer to read the first two sentences on page 120 aloud.
  What should you always do? obey my parents
- Point out that the Bible tells children to obey their parents (Ephesians 6:1; Colossians 3:20). Explain that when they obey their parents, they are obeying God. Point out that parents want to keep their children safe and that obeying their parents will help to keep them safe.
  What are some habits to keep you safe in your house? Keep your house doors locked; do not open your doors to strangers.
- Why should you keep your house doors locked? to keep strangers out; to keep the people and things inside the house safe

In Your Community

It is good to have safe habits when you are away from home.

Stop, look, and listen before you cross a street.

Always have a buddy with you.

Do not talk to strangers.

Do not go with someone you do not know.

Police officers and firefighters are your friends.

Call 911 if there is danger.

What are some safe habits at home and in your community?

121

- Remind the students that to be safe at home, they should never open the door to strangers.

What should you know in case there is a fire in your home? how to escape

Direct attention to the picture.

What do you think is happening in this picture? Elicit that the family's house is on fire and the boy is using a ladder to escape from his bedroom window.

- Direct attention to the *Fire Safety* box. Read and discuss each step. Explain that it is important to have smoke detectors in the home and to know what to do if the alarm sounds. Remind the students that they should never hide during a fire, even though it is scary. Explain that it is good to practice how to escape a fire quickly. Point out that smoke is often the most dangerous part of a fire because it gets thick and dark and makes it hard to breathe. Allow time for the students to practice crawling low and to practice how to "stop, drop, and roll."

During a fire emergency, when should you call 911? when I am safely outside

- Display and distribute *Fire Safety Cards*. Read and discuss the statement on each card. Point out that the students may color each picture and cut out the cards to remind themselves of some safe habits at home.

What should you do if someone gets sick or hurt? get a parent or another adult; call 911; use any medical training I have to help the person

---

- While some students may have some medical training, such as how to perform the Heimlich maneuver, many do not. Encourage the students to seek adult help immediately or to call 911 if adult help is not available. Emphasize that 911 is only for a true emergency.

Teaching for page 121 begins here.

You have learned that it is good to have safe habits at play, in the car, and at home. Where else is it good to have safe habits? in my community; when I am away from home

- Explain that *community* means the people living and working together in an area. Point out that their neighborhood is part of their community.

What should you always do before you cross a street? Stop, look, and listen.

Why should you stop, look, and listen? to see if any cars are coming; to be safe

- Explain that students should *stop* before stepping out into the street. They should *look* both ways for traffic.

Why should you *listen*? I may hear a car or truck that I cannot see.

- Direct attention to the picture of the two boys.

What are the boys doing? Possible answers: waiting to cross the street; looking both ways before crossing the street; listening for traffic they cannot see

Where else should you stop, look, and listen? Possible answers: parking lots, anywhere cars can be

- Display the *Safety Song*. Review what the students should do before crossing the street by having them sing the song together.

What is a *buddy*? a friend

Why is it a safe habit to always have a buddy with you? If you get hurt, there is someone who can go for help.

- Explain that as a buddy, the students are showing kindness by caring about and being helpful to their friends.

Who should you not talk to? strangers

Who is a stranger? someone I do not know

Should you go anywhere with a stranger? No; it is not safe.

Who are your friends in the community? police officers and firefighters

- Remind the students that police officers and firefighters are there to help them and keep them safe. They are strangers the students can trust.

- Explain the difference between a stranger who is a bad person who wants to hurt the students, and a stranger who they can trust. Point out that they should never talk to strangers who offer them a ride or a gift, who want their help to find a lost pet, or who want to tell them a secret. Explain that it is always good to yell "No!" and leave the area immediately when a stranger approaches them or when they do not feel safe. Point out that if they are alone and need to approach a stranger for help, they should look for a police officer, a firefighter, or another trusted adult.

Who are some other adults you can trust? Possible answers: pastors, teachers, librarians, security guards, store clerks

What phone number should you call if there is danger? 911

answer

**obey your parents;
keep doors locked;
do not open doors to strangers;
know how to escape a fire;
stop, look, and listen;
have a buddy with you;
do not talk to strangers;
do not go with a stranger;
find a police officer or firefighter,
or call 911, if there is danger**

Teaching for page 122 begins here.

PREPARATION FOR READING

- Preview and pronounce the words *life jacket*.
- Direct the students to read page 122 silently to find out how you should treat your body.

TEACH FOR UNDERSTANDING

Who gave you your body? God

How should you treat your body? I should take care of it.

Why should you take care of your body? God gave it to me; I glorify God when I take care of it.

- Direct attention to the Big Question on page 108.

Why are healthy and safe habits important? They help me take care of my body.

- Direct attention to the *Be Safe Around Water* box. Read and discuss each statement.

God gave you your body.
You should take care of it.
Healthy habits and safe habits are good.
They help you to take care of your body.

Be Safe Around Water

1. Learn to swim.
2. Do not swim alone.
3. Wear a life jacket when on a boat.

122

- Point out that if a person falls into a body of water, a life jacket helps the person to float. Life jackets help prevent drowning.
- Direct attention to the picture.

How are the boy and his father being safe? Possible answer: They are each wearing a life jacket.

- Instruct the students to complete Activities page 104. Discuss each picture and ask the students to explain how the child is being safe or not being safe.

ACTIVITIES

At Home and In Your Community, page 104
This page was completed during the lesson.

Study Guide, pages 105–6
These pages review Lessons 42–43. After completion, direct the students to keep them for review for Test 6.

STEM

Lesson 44

Safe Shoes

Getting a new pair of shoes can be fun.

Some new shoes have slick soles.

Slick soles can be dangerous.

New shoes can cause a person to slip or fall.

In this STEM activity you will find an answer to slick shoes.

123

Fire Safety Presentation
Invite a firefighter from a local fire department to give a safety presentation. Most fire stations provide educational fire safety lessons and materials for students. You may also plan a field trip to a fire station.

Fearful Students
Be aware of any students who may be frightened or concerned about any of the topics discussed in this chapter. Remind the students that God is in control of all things and He is a "very present help in time of trouble" (Psalm 46:1).

Objectives
- Propose a possible solution to the real-life problem of slick-soled shoes
- Construct a design to solve the problem
- Communicate to others how the design solves the problem

Teacher Resources
- Visual 2.5: *STEM The Engineering Design Process*

INTRODUCTION

What do you like about getting a new pair of shoes? Possible answers: They fit better than the old shoes; they look nicer than the old shoes.

Sometimes, getting a new pair of shoes is fun, but new shoes can also be unsafe. Today you will think of ways to make new shoes safe.

PREPARATION FOR READING

- Preview and pronounce the words *dangerous* and *slick*.
- Direct the students to remove pages 107–8 from their Activities books.
- Direct the students to read page 123 and Activities pages 107–8 silently before beginning the STEM activity.

TEACH FOR UNDERSTANDING

- Display *STEM The Engineering Design Process*. Explain that for this STEM activity the students will *Imagine*, *Plan*, and *Share* a design to find an answer to the problem. Point out that they will not *Make* or *Make Better* their designs.
- For this activity, each student may design his own solution to the problem of slick or slippery soles on shoes, or the students may work in science groups to collaborate as they design a solution.
- Invite the students to share their designs with a partner or another science group.

ACTIVITIES

STEM Activity: Safe Shoes, pages 107–8
In this STEM activity, the students will design a way to make shoes with slick soles safe.

ASSESSMENT

Rubric
Use the prepared rubric or design a rubric to include your chosen criteria.

Objective
- Recall terms and concepts from Chapter 6

REVIEW

- Material for Test 6 will come from the Study Guides on Activities pages 99–100 and 105–6. You may review any or all of the material during the lesson.
- You may choose to review Chapter 6 by playing *Be Kind, Be Healthy, Be Safe* review game or another game from the Game Bank (Teacher Resources).

🏆 REVIEW GAME

Be Kind, Be Healthy, Be Safe
Prepare an equal number of pictures showing people being kind and engaging in healthy habits and safe habits. Display three columns. Head each column with one phrase: "Be Kind"; "Be Healthy"; or "Be Safe."

Divide the students into three teams, each representing one of the phrases. Ask review questions. For each correct answer, give the student a picture and have the student place the picture in his team's column. The team with the most pictures at the end of the game wins.

NOTES

Objective
- Recall and apply terms and concepts from Chapter 6

ASSESSMENT

- Administer Test 6.

NOTES

Materials
- photo of the *Apollo 11* rocket in flight
- photo of the *Apollo 11* command module on the moon

UNIT INTRODUCTION

- Direct attention to pages 124–25.

  What is the title of Unit 4? "Let's Learn About Earth and Space"

  What do you see in the picture? the moon, a rocket

  Where do you look to see the moon? the sky

  When can you see the moon? at night

- Direct attention to the rocket. Help the students understand that this is a cartoon image of a real rocket.

- Explain that some men have traveled to the moon in a rocket. Display a picture of the *Apollo 11* rocket. Point out that this was the first rocket to take Americans to the moon.

- Discuss ways the rockets are similar and ways they are different.

- Discuss the Background information and display a picture of the *Apollo 11* command module that orbited the moon. The term *astronaut* will be introduced in Lesson 51.

- Direct the students to look through Chapters 7–9, pages 124–79. Guide the students through several pages, stopping to look at headings and pictures.

  What do you think we are going to be learning about in Unit 4? the earth, sun, moon, stars, how the earth moves, seasons, weather

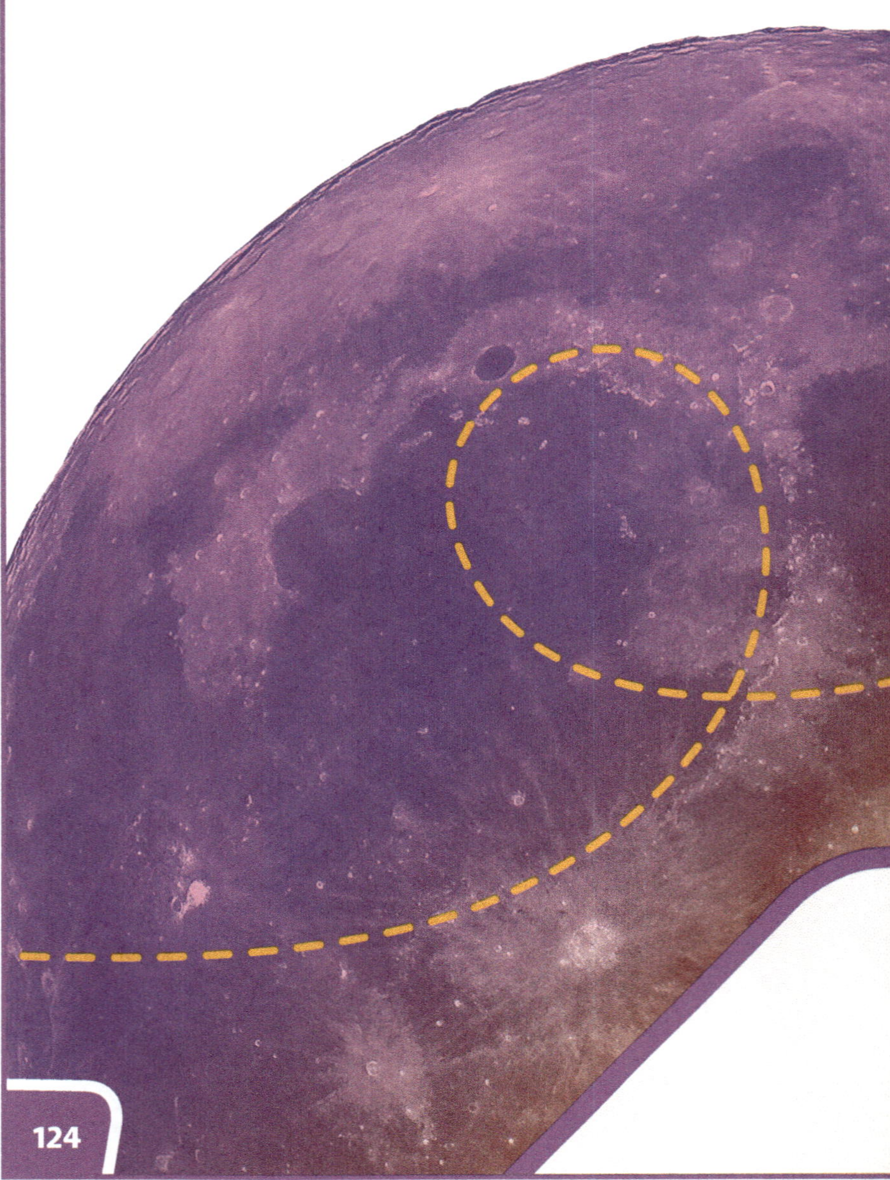

124

🔍 BACKGROUND

The photo of the moon has a color overlay. On July 20, 1969, Neil Armstrong and Edwin "Buzz" Aldrin Jr. became the first men to walk on the moon. The *Apollo 11* command module was piloted by Michael Collins. Men had flown in space before and had orbited the moon, but these men were the first ones on the moon.

NOTES

Chapter Objectives
- Evaluate from the Bible an opposing view of how the earth, sun, moon, and stars formed **BWS**
- Identify the characteristics of the created lights in the sky
- Analyze patterns of the earth, sun, moon, and stars
- Apply science process skills of observing, predicting, and inferring
- Identify the telescope as a tool to magnify things that are far away
- Apply our knowledge of the earth, sun, moon, and stars to praising God for His greatness and goodness **BWS**

Lesson Objectives
- Infer topics by previewing the unit and chapter
- Explain from Genesis 1 how the earth, sun, moon, and stars were formed **BWS**
- Evaluate from the Bible an opposing view of how the earth, sun, moon, and stars formed **BWS**

Materials
- read-aloud library book on the Big Bang

Teacher Resources
- Visual 7.1: *Bible Verses*

CHAPTER INTRODUCTION

- Direct attention to pages 126–27.
- Direct attention to the picture.
-  What is this a picture of? the earth, sun, moon, and stars in space
- Instruct the students to look at the title of this chapter on page 127.

   What is the title of Chapter 7? "The Earth and Its Lights"

   What lights are referred to in the chapter title? sun, moon, and stars

   In this chapter you will learn about the earth and its lights—the sun, moon, and stars.

Big Question
- Direct attention to the Big Question. Read the question aloud.

   In this chapter we will learn how the sky changes each day.

⚠ **HELPS**

Big Bang Read-Aloud Library Book
A book on the Big Bang will be used in this lesson. A small portion of the book will be read aloud to the students, after which they will biblically evaluate the truth of the Big Bang. Select an age-appropriate book such as *A Cartoon History of the Earth: The Birth of the Earth No. 1* by Jacqui Bailey. This book begins with the words, "Once upon a time." For the purposes of this lesson, only the first two pages would need to be read aloud to the students.

Creation/Evolution Discussions
Answers in Genesis and the Institute for Creation Research offer many resources to support biblical creation.

Visit TeacherToolsOnline.com for resources to enhance the lessons.

The Earth and Its Lights

The earth was made by God.

When God first made the earth, there was only darkness.

Then God said, "Let there be light."

He called the light *day*.

He called the darkness *night*.

Later God created the sun, moon, and stars.

God wisely designed the sun, moon, and stars to be lights for the earth.

127

PREPARATION FOR READING

- Direct the students to read page 127 silently to find out what God designed the sun, moon, and stars to be for the earth.

TEACH FOR UNDERSTANDING

Who made the earth? God

- Direct attention to the Bible verses on page 109 in the Activities book. Display *Bible Verses*. Ask a volunteer to read Genesis 1:1 aloud. Mention that *only* God could make the heaven and the earth.

What was it like when God first made the earth? There was only darkness.

- Poll the students to find out how many do and do not like to be in the dark.

What did God say that tells us He did not want His creation to be in darkness all the time? "Let there be light" (Genesis 1:3).

What did God call the light? day

What did God call the darkness? night

After God made day and night, what did He create? the sun, moon, and stars

What did God design the sun, moon, and stars to be for the earth? lights

- Display *Bible Verses* and read Genesis 1:16 aloud.

What is the greater light? the sun

What is the lesser light? the moon

- Explain that to "rule" means that the sun would be a very important object in the sky during the day and the moon would be a very important object during the night.

- Point out that these lights, as well as the stars, were created on Day Four of the Creation week. You may point out that the light created in Genesis 1:3 is not the light of the sun, moon, or stars. Only God knows what light He made to divide day from night before the creation of the sun, moon, and stars.

PREPARATION FOR READING

- Preview and pronounce the word *all-powerful.*
- Direct the students to read pages 128–29 silently to find out why God made the earth.

TEACH FOR UNDERSTANDING

Who is all-powerful? God

What does the word *all-powerful* tell you about God? There is no one who has more power than God has.

How did God create the earth? He spoke.

How did God make the sun, moon, and stars? He spoke.

- Explain that only God can speak and make something when there is nothing. Emphasize that God is great and He is all-powerful.

Lesson 47

What the Bible Says

God is great.

God is all-powerful.

The Bible says that God made the earth, sun, moon, and stars.

Genesis 1 tells us that God spoke and the earth was created.

God spoke again and the sun, moon, and stars were made.

128

Why did God make the earth? to be our home
Why did God make the sun, moon, and stars? to be lights in our sky

- Direct attention to the picture on pages 128–29. Explain that in this picture the artist is showing some of what God created.

Put your finger on the sun.
Put your finger on the moon.
Put your finger on one star.

- You may point out other parts of God's creation: the dry land and sea, the vegetation, various sea creatures, birds, and land animals, as well as Adam and Eve in the water in the foreground. You may explain that dinosaurs are pictured with Adam and Eve because dinosaurs were created by God on the same day that Adam was created.

answer
God spoke.

PREPARATION FOR READING

- Direct the students to read page 130 silently to find out what book is always right.

TEACH FOR UNDERSTANDING

What do some books say about how the earth, sun, moon, and stars were made? Some books say that God did not make them.

If these books do not say that God made the earth, sun, moon, and stars, do these books agree with the Bible? no

- Read aloud a short portion of a book on the Big Bang. Select an age-appropriate book such as *A Cartoon History of the Earth: The Birth of the Earth No. 1* by Jacqui Bailey. This book begins with the words, "Once upon a time." For the purposes of this lesson, only the first two pages would need to be read aloud to the students.

- Guide the students as they biblically evaluate the truth of the Big Bang.

According to the pages from the book that was just read aloud, how do some scientists think the earth, sun, moon, and stars were made? They were made by the Big Bang that happened a long, long, long time ago.

According to the Bible, how do we know the earth, sun, moon, and stars were not made by the Big Bang? God told us in the Bible that He made them (Genesis 1:1; Genesis 1:16); God spoke and they were made.

- Display *Bible Verses* and direct attention to Activities page 109. Review Genesis 1:1 and Genesis 1:16.

If a book does not agree with the Bible, which book is right? the Bible

If a book says that the earth, sun, moon, and stars were made by a Big Bang, and not by God, is that book right? No; when a book says something that disagrees with the Bible, the Bible is right and the other book is wrong.

- Explain that when science and the Bible disagree, the Bible is right and science is wrong. God's Word, the Bible, is always true and always trustworthy.

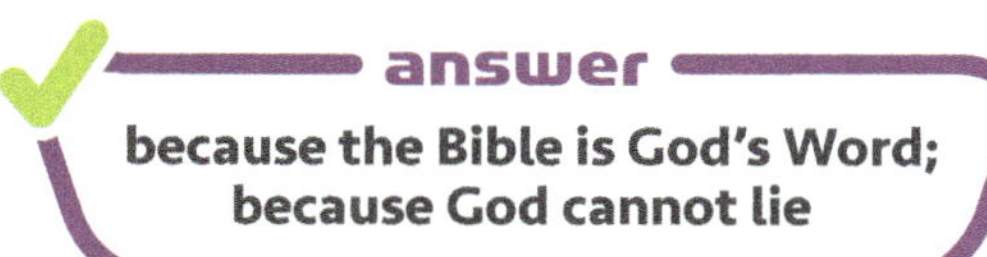

answer
because the Bible is God's Word; because God cannot lie

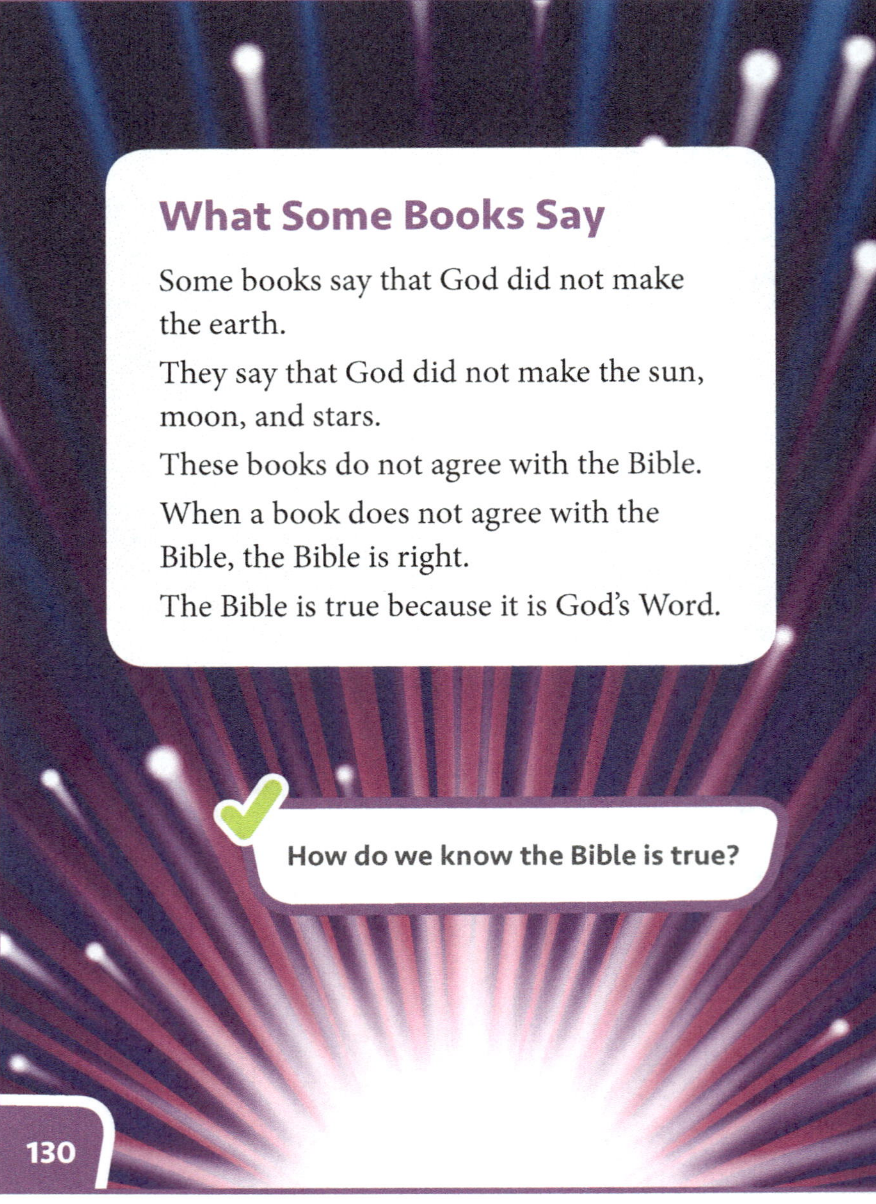

ACTIVITIES

Bible Verses, page 109
These Bible verses will be used in the teaching of this chapter.

In the Beginning, page 111
This page reinforces the concepts in the lesson.

BACKGROUND

The Big Bang
The Big Bang is a belief that teaches that the universe began billions of years ago by natural means. The Bible teaches that everything was created supernaturally by God in six days. The Big Bang will also be discussed in Chapter 10.

Lesson 48

The Earth

The earth **rotates**, or spins.

The earth rotates once every day.

A **day** is the time that the earth takes to rotate one time.

We see different lights in the sky because the earth rotates.

What does the earth do once every day?

131

LESSON 48

Objectives
- Describe the earth's daily motion
- Identify the sun as a star
- Identify the beneficial properties of the sun
- Explain from Genesis 1 why God made the sun **BWS**
- Describe and predict the sun's pattern across the sky

Materials
- flashlight
- 54.5" diameter paper circle
- 110 mini marshmallows

Vocabulary
- rotate
- day
- sun
- star
- light
- sunset
- sunrise

INTRODUCTION

- Direct attention to the picture.
  What do you see in the picture? the earth, sun, and stars
- Point out North America, the clouds, and the oceans in the picture of the earth.
  Today you will learn about the earth and the sun.

PREPARATION FOR READING

- Preview and pronounce the vocabulary terms *rotate* and *day*.
- Direct the students to read page 131 silently to find out what a day is.

TEACH FOR UNDERSTANDING

What does the earth do when it rotates? It spins.
How many times does the earth rotate in a day? once
What is a day? the time that the earth takes to rotate one time
What do we see in the sky because the earth rotates? different lights

- Explain that because the earth rotates, or spins, we see different lights in one day.
- Direct the students to stand beside their desks. Instruct them to rotate, or spin, one time, keeping their eyes open.
  What did you see as you turned? Answers should include the entire room and all the people and things in it.
- Point out that the students did not see just one part of the room as they rotated; they saw the whole room. Explain that because the earth rotates, we do not see just one light; we see different lights.
  What are some lights in the sky? sun, moon, and stars

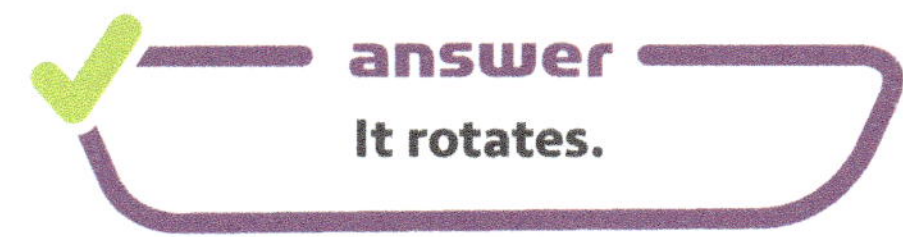

PREPARATION FOR READING

- Preview and pronounce the vocabulary terms *sun*, *star*, and *light*.
- Direct the students to read pages 132–33 silently to find out which star is closest to the earth.

TEACH FOR UNDERSTANDING

Who made the sun? God
When do you see the sun in the sky? during the daytime

- Direct attention to the diagram on pages 132–33. Explain that the earth is rotating in the direction of the arrows. Direct attention to the light and dark sides of the earth. Explain that the sun is shining on the light side of the earth, and it is not shining on the dark side.

When is it daytime for part of the earth? when that part of the earth faces the sun
When is it nighttime for part of the earth? when that part of the earth does not face the sun

- Invite two volunteers to demonstrate daytime and nighttime. Instruct the students to face each other. Identify one student as the earth and the other as the sun. Give the "sun" a flashlight and instruct the student to stand still and shine the light on the "earth." Dim the lights. Instruct the "earth" to rotate slowly one time.

- Explain that when the students are facing each other, the side of the "earth" facing the sun is in daytime and the "earth's" back is in nighttime. When the "earth" is facing away from the light of the "sun," the "earth's" face is in nighttime and the "earth's" back is in daytime.

- Instruct the "earth" to rotate slowly and direct the other students to identify when the "earth's" face is in daytime and when it is in nighttime.

How big is the sun? It is very big; it is bigger than the earth.

- Conduct the *How Big Is the Sun?* activity to compare the sizes of the sun and the earth.

⚗ ACTIVITY

How Big Is the Sun?

Materials: 110 mini marshmallows, 54.5″ diameter paper circle

Note: Use any small counting objects, such as round cereals, that are a half inch in diameter. The paper circle needs to have a diameter equivalent to 54.5″ when using half-inch objects for counters. Use bulletin board paper or newsprint. Draw a straight line across the diameter of the circle. The circle is used again in Lessons 51–52.

Place the circle on the floor and explain that it represents the sun. Explain that the sun is many times bigger than the earth. Show one mini marshmallow. Direct the students to pretend that the earth is about as big as the marshmallow. Ask the students how many marshmallows they think they would have to line up to make a line as wide as the sun.

Lead in counting the other marshmallows as you place them side by side on the straight line. Continue until 109 marshmallows have been put down.

Explain that the sun is about 109 times bigger than the earth. If the earth was the size of one marshmallow, the sun would be the size of the 109 marshmallows side by side.

Remove all but one marshmallow from the circle to compare the sizes of the earth and the sun.

The **sun** is a star.

The sun is the closest star to the earth.

A **star** is an object in the sky that makes its own light.

The sun is bright.

The sun is hot.

It warms the earth.

Genesis 1 tells us that God made the sun to give the earth light during the day.

The sun gives light to the earth.

Light is used to see objects around us.

Plants need light to make food.

How does the sun help the earth?

133

What is the sun? a star

Which star is the closest to the earth? the sun

- Point out that even though the sun is the closest star, it is still very far away. It is millions of miles away. Explain that if a person could drive to the sun, it would take them more than 163 years to get there.

What is a star? an object in the sky that makes its own light

What does it mean that the sun is bright? It gives off a lot of light.

- Point out that the sun seems to be the brightest star because it is the star that is closest to the earth.

- Explain that because the sun is so bright, the students should never look directly at it, even when wearing sunglasses. Point out that looking directly at the sun, or any very bright light, can damage their eyes.

How is the sun different from the stars that you see at night? It is closer; it is bigger; it is brighter; I see the sun during the day, not at night.

- Explain that the sun is a medium-sized star. There are stars that are bigger than it is, and there are stars that are smaller.

What kind of temperature does the sun have? It is hot.

Because the sun is hot, what does it do for the earth? It warms the earth.

- Explain that God placed the sun in exactly the right place in the sky for us to live on the earth. If the sun were closer, the earth would be too hot for us to live. If the sun were farther away, it would be too cold on the earth for us to live.

According to Genesis 1, why did God make the sun? to give the earth light during the day

What do you use light from the sun to do? to see objects around us

What do plants use sunlight to do? to make food

- Remind the students that one of the needs of plants is light.

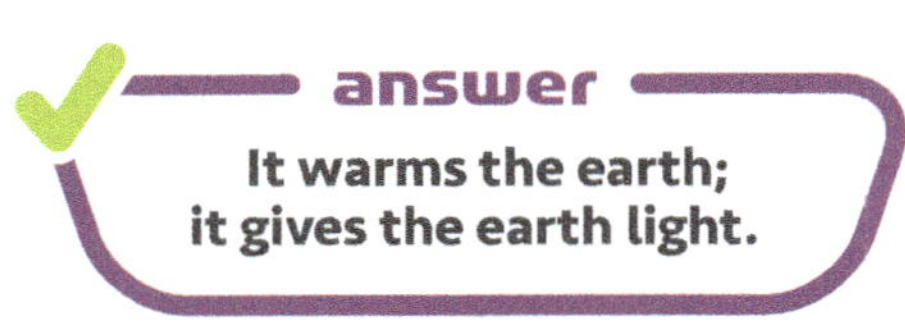

answer
**It warms the earth;
it gives the earth light.**

Distance to the Sun

The distance between the sun and the earth is approximately 150 million kilometers (93 million miles). A distance that large is hard for first graders to comprehend. The statistic given in the discussion is based on driving 65 mph nonstop.

Light from the Sun

Without light, we cannot see. We can see the objects around us only because light rays reflect off them. Light makes things visible and gives objects color. Light also enables plants to make food through photosynthesis. Herbivores eat the plants and obtain some of the food energy from the plants. Other animals and people eat both plants and animals, and the food energy is transferred. Light is important for life.

Light will be discussed in more detail in Chapter 10.

PREPARATION FOR READING

- Preview and pronounce the vocabulary terms *sunset* and *sunrise*.
- Direct the students to read pages 134–35 silently to find out when the sun is overhead.

TEACH FOR UNDERSTANDING

What does it look like the sun is doing in the sky? moving across it

Is the sun moving across the sky? no

What does the earth do that makes the sun look like it is following a path? The earth rotates.

- Point out that the sun is not moving across the sky; it is the earth that is moving. Remind the students that because the earth rotates, they will see different lights in the sky, and the sun is one of those lights.
- Direct attention to the diagram of the sun on pages 134–35.

What is this diagram showing? the path of the sun across the sky

- Explain that you will follow the direction of the arrows as you discuss the path of the sun.
- Direct attention to the compass rose on page 135. Explain that a compass rose is used on maps and diagrams to show direction. Point out the letters.

What letters are around the compass rose? *N, E, S,* and *W*

- Explain that the *N* stands for north and the *S* for south.

 What does the *E* stand for? east

What does the *W* stand for? west

Put your finger on the *E* on the compass. The sun at the bottom of page 135 is in the east.

Where is the sun in the morning? in the east

What does the sun seem to do in the morning? rise

What do we call the time when we can first see the sun? sunrise

⚠ HELPS

Direction and the Compass Rose

Explain that a direction is a point toward which something faces. When the student faces the front of the class, that is the direction in which he is facing. Maps use a compass rose so that directions can be standard for everyone looking at the map. Top, bottom, left, and right can vary depending on where each person is in relationship to the map.

- Direct the students to use their index fingers to trace the arrow on page 135.
  What direction did the arrow take you on the page? up
  Where is the sun about noon (12:00 PM)? overhead
- Explain that the sun will be at its highest point in the sky around the middle of the day.
- Direct the students to use their index fingers to trace the arrow on page 134.
  What direction did the arrow take you? down
  Where is the sun in the evening? in the west
  What do we call it when the sun begins to disappear? sunset
- Direct the students to place their index fingers on the sun in the east. Choose a student to read the caption aloud.
- Direct the students to trace the arrows on pages 135 and 134 and to place their fingers on the sun in the west when they have finished tracing the arrows. Choose a student to read the caption aloud.

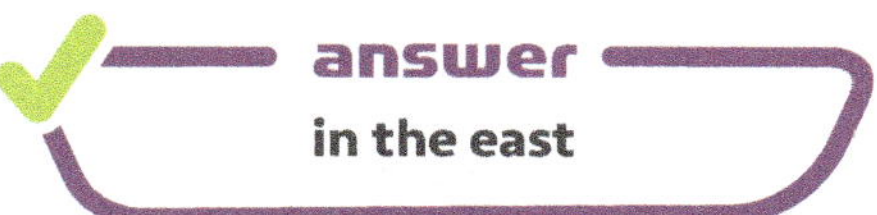

ACTIVITIES

Study Guide, pages 113–14
These pages review the concepts taught in Lessons 47–48.

BACKGROUND

Where Is the Sun at Noon?

Noon is not just a time of day (12:00 PM); it is also a scientific term. Solar noon describes the time when the sun has reached its highest elevation, or zenith; the time when the sun is no longer rising. Depending on latitude and time of year, the sun may be overhead at noon, or 12:00 PM. For example, at the equator, the sun is overhead at noon, or 12:00 PM, on the equinoxes.

Objectives
- Formulate a hypothesis for why it is hard to see stars during the daytime
- Observe simulated stars in various lighting
- Infer why it is hard to see stars, other than our sun, during the daytime

Materials
- See Activities page 115

INTRODUCTION

Have you ever watched for the first star at night? What do you think happens to the stars during the day? Accept any reasonable answer.

Is there a star that you can see during the day? yes; the sun

In this Investigation, you will find out why it is hard to see stars, other than the sun, during the day.

PREPARATION FOR READING

- Direct the students to remove pages 115–16 from their Activities books. Direct them to read page 136 and Activities pages 115–16 silently before beginning.

TEACH FOR UNDERSTANDING

- Direct attention to and discuss the **Process Skills**. If necessary, refer to Chapter 2.
- Choose a student to read the **Problem** aloud.
- Direct the students to complete the **Hypothesis** to answer the problem.
- Distribute the materials.
- Instruct the students to follow along together as **Procedure** Steps 1–8 are completed.
- Read Step 1 and demonstrate how to attach the foil to the flashlight.
- Read Step 2 and remind the students to make small holes.
- Keep the room brightly lit for Steps 3 and 4. Instruct the students to find the box labeled "Bright Room" under **Observations** and to draw what they observed.
- For Step 5, make the room dim and for Step 6, brighten the room.
- For Step 7 make the room dark and for Step 8, brighten the room.
- Remind the students of the importance of accurately recording their observations.
- Direct the students to draw **Conclusions** based on their observations. The conclusions may be done individually or with their science group. Guide a discussion of their conclusions.

INVESTIGATION

Lesson 49

Stars in the Day

The sky is full of stars.

Sometimes it is easy to see stars.

Sometimes we cannot see any stars.

In this Investigation you will find out why it is hard to see stars during the daytime.

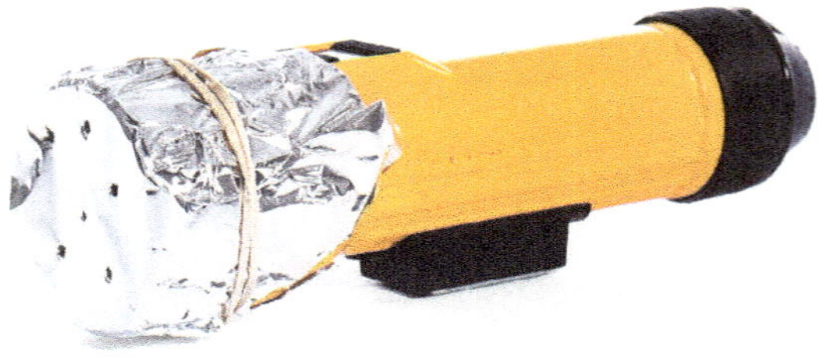

136

ACTIVITIES

Investigation: Stars in the Day, pages 115–16
This Investigation will help to explain why it is hard to see stars during the daytime.

ASSESSMENT

Rubric
Use the prepared rubric or design a rubric to include your chosen criteria.

⚠ HELPS

Don't Be Foiled by the Aluminum Foil
It is easier to punch small holes in the foil if the flashlight has a small raised ridge around the lens. You may want to slide the foil a little away from the lens before the student pokes his holes.

The holes should be small. Folding the foil to a double thickness or using a heavy duty aluminum foil will lessen the chance of holes that get too large. Keep extra foil on hand in case a student tears the foil while poking holes.

Lesson 50

The Stars

God made many, many stars.

The sun is a star that you see in the daytime.

We see the other stars at night.

The stars are far away from the earth.

They look like points of light.

Like our sun, stars are always shining.

We can see many stars just by looking at the nighttime sky.

137

Objectives
- Identify the characteristics of stars, other than the sun
- Identify the telescope as a magnifying tool to observe stars, other than the sun
- Identify the groups of stars called the Big Dipper and the Little Dipper
- Identify the North Star

Materials
- map, showing streets and roads
- flashlight
- ladle

Vocabulary
- telescope

INTRODUCTION

- Display the map.
 Why do people use a map like this one? Possible answer: to find out how to get from one place to another; to keep from getting lost

People use maps like this to know how to get to different places. There is something in the sky that can also help people find their way.

What do you think that is? stars

- Explain that for many years people have enjoyed looking at stars and studying them. Sailors and travelers from long ago knew the locations and names of many stars. They used the stars to help them know which direction they were going in. Today you will learn more about stars.

PREPARATION FOR READING

- Direct the students to read page 137 silently to find out when stars are shining.

TEACH FOR UNDERSTANDING

What is a star? an object in the sky that makes its own light

What star can you see during the daytime? the sun

When do you see the other stars? at night

What do stars look like in the night sky? points of light

Why do stars look like points of light? They look like points of light because they are very, very far away from the earth.

Why does the sun look bigger in the sky than the other stars? It looks bigger because it is the closest star to the earth.

- Review the size of the sun and its distance from the earth. Remind the students that our sun is a medium-size star. Explain that some stars are smaller and some are bigger than our sun, even though they all look like tiny points of light from the earth.

When are stars shining? Stars are always shining.

Why do you only see stars, other than our sun, by looking at the nighttime sky? The sun is so bright during the daytime, the other stars cannot be seen.

What does the earth do once every day that allows us to see the stars at night? It rotates.

- Use a flashlight to repeat the rotation-of-the-earth demonstration used in Lesson 48. Remind the students that when it is nighttime for them, the sun is shining on the other side of the earth where it is daytime. Because the earth rotates and there is nighttime, they see different lights in the sky.

- Preview and pronounce the vocabulary term *telescope*.
- Preview and pronounce the words *Big Dipper* and *Little Dipper*.
- Direct the students to read pages 138–39 silently to find out where the North Star can be found. Explain that they do not need to read the *Science and History* box at this time.

What science tool helps you see things that are far away? a telescope

What does a telescope do? It makes far away objects look closer and bigger.

If you look through a telescope, will you see more stars or fewer stars? more

- Explain that there are some stars that we cannot see because they are so far away. Point out that with a telescope we can see more stars because the telescope makes the stars look closer and bigger.

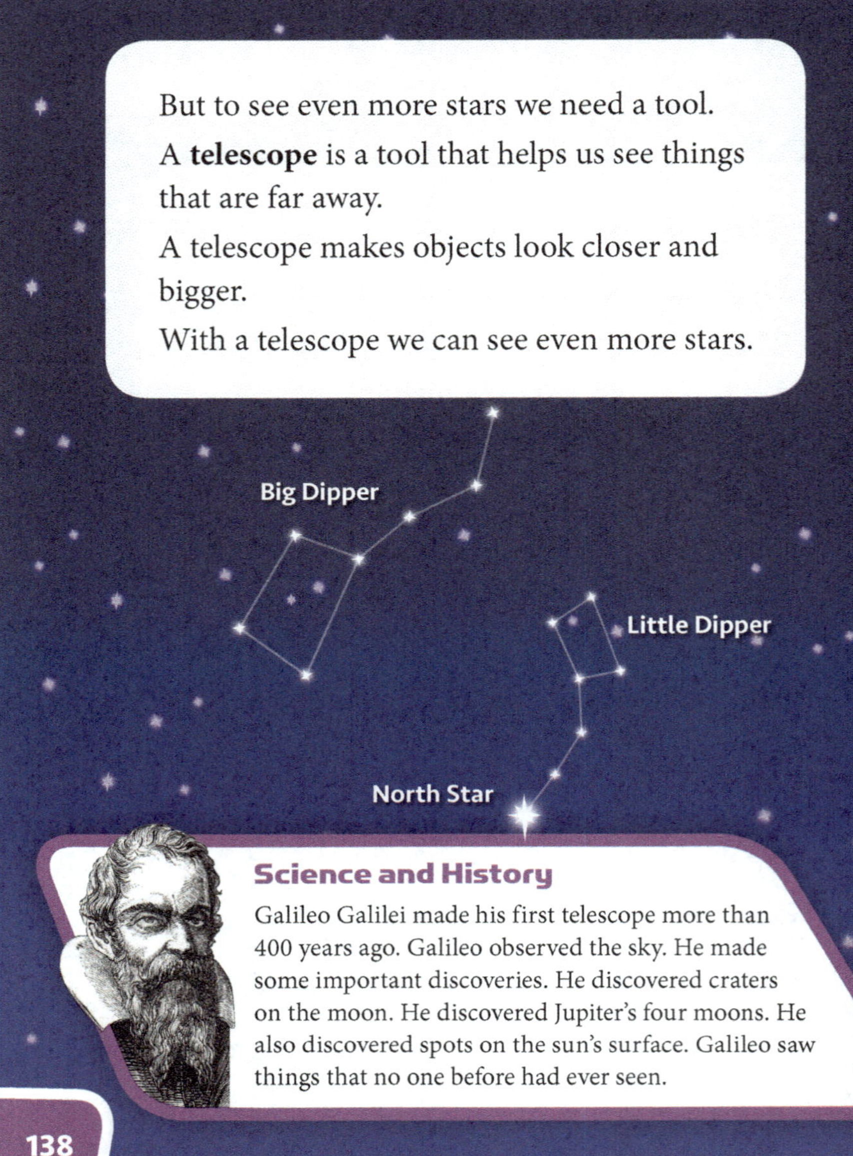

Telescopes

No one knows for sure who invented the telescope, but in 1608 Hans Lippershey, a Dutch spectacle maker, was the first to apply for a patent. He is usually credited with being the inventor.

In 1609, Galileo Galilei made his first telescope. Galileo was the first person to make scientific discoveries by studying the sky with a telescope. He is said to have "reinvented" the telescope when his discoveries made it popular.

Constellations

Groups of stars that make pictures are called constellations. Down through history, not all countries and cultures used the same names for the constellations. Today, there are 88 official constellations chosen by the International Astronomical Union (IAU).

The Big Dipper and Little Dipper are groups of stars in patterns in the night sky. The Big Dipper forms the back and tail of the constellation named Ursa Major, or the Great Bear. The Little Dipper forms the constellation Ursa Minor, or the Little Bear.

The North Star

When observing the North Star, you are facing true north. At the North Pole, the North Star is almost directly overhead.

What is the name of the one star that is bright and easy to see in the Little Dipper? the North Star

Where is the North Star in the Little Dipper? at the end of the "handle" of the "dipper"

Would it be easier to see stars when you are around other lights or when it is very dark? Elicit that it is easier to see stars when you are away from bright lights; it is easier to see stars in the country than in the city.

- Direct attention to the *Science and History* box. Read aloud about Galileo's telescope and his discoveries as the students follow along.

What did Galileo use his telescope to do? observe the sky

What important discoveries did Galileo make by using his telescope? the moon has craters, Jupiter has four moons, there are spots on the sun's surface

- Point out that it is not safe to look directly at the sun. Explain that a sun filter is needed for safe viewing through a telescope. Point out that without a sun filter, permanent damage to the eyes can occur, including blindness.

ACTIVITIES

Star Pictures, page 117
This page reinforces the concepts taught in Lesson 50.

- Direct attention to the picture.

What is the boy doing? He is looking at stars through a telescope.

What do some groups of stars seem to make? pictures

What is one group of stars called? the Big Dipper

- Direct attention to the Big Dipper on page 138.
Note: This is the position of the Big Dipper and Little Dipper in June, as seen from North America.

How many stars make up the Big Dipper? 7

- Display the ladle. Explain that a ladle is sometimes called a "dipper."

What could you dip with this ladle? Possible answers: water, soup, gravy

- Discuss the similarities between the ladle and the Big Dipper. You may hold the ladle in the same position as the Big Dipper as you discuss their similarities.

Near the Big Dipper is another group of stars that makes a picture. What is the name of that star picture? Little Dipper

How many stars make up the Little Dipper? 7

How do the Little Dipper and Big Dipper look alike? Possible answer: they both have what looks like a "dipper" made of four stars with a "handle" made of three stars.

Objectives

- Identify the characteristics of the moon
- Identify what an astronaut does
- Identify the changes in the shape of the moon over the course of a month
- Predict the phases of the moon over the course of a month
- Explain from Genesis 1 why God made the moon **BWS**
- Explain how the sky changes each day

Materials

- flashlight
- mirror
- circle from Lesson 48
- one ½" diameter mini marshmallow
- one ⅛" diameter button, or another ⅛" round object
- lamp, without a shade
- ball

Teacher Resources

- Visual 7.1: *Bible Verses*

Vocabulary

- moon
- reflect
- astronaut
- phases of the moon

> Two days have been alloted for this lesson. You may divide the lesson where it works best for your instruction time.

INTRODUCTION

- Direct attention to the Bible verses on page 109 in the Activities book. Display *Bible Verses.* Ask a volunteer to read Genesis 1:16 aloud.

  What did God make to give light during the day? the sun

  What did God make to give light at night? the moon and the stars

  Today we will learn more about the moon.

PREPARATION FOR READING

- Preview and pronounce the vocabulary terms *moon*, *reflect*, and *astronaut*.
- Direct the students to read pages 140–41 silently to find out what the moon is made of. Explain that they do not need to read the *Science and History* box at this time.

TEACH FOR UNDERSTANDING

Why did God make the moon? to give the earth light at night

What is the moon made of? rock

Lessons 51–52

The Moon

God made the moon to give the earth light at night.

The **moon** is a big ball of rock that moves around the earth.

It does not make its own light.

The moon reflects the light of the sun.

To **reflect** means to bounce off.

Light from the sun bounces off the moon.

The moon is closer to the earth than the sun is.

The moon is smaller than the sun or the earth.

140

- Explain that for a long time people did not know what the moon was made of. Many people made up stories about it. We now know it is made of rock.

  Does the moon make its own light? no

  Where does the moon get its light from? the sun

  What does it mean that the sun's light *reflects* off the moon? Light from the sun bounces off the moon.

- Explain that the moon reflects light from the sun much like a mirror reflects light. Demonstrate this concept with the *Reflecting Light* activity.

  Does the mirror make its own light? No; the mirror reflects the light from the flashlight.

- Explain that the sun shines on the moon, and we see the light of the sun reflecting, or bouncing off, the moon at nighttime.
- Direct attention to the picture.

  Why can you see the moon? The sun's light is reflecting off the moon.

  What is the moon's light reflecting off? water

  Which is closer to the earth—the sun or the moon? the moon

  Which is smaller—the sun or the moon? the moon

  Is the moon bigger or smaller than the earth? smaller

154

Some people have walked on the moon.

These people are called astronauts.

An **astronaut** is a person who travels into space.

> **Why did God make the moon?**

Science and History

Alan Shepard was the first American astronaut in space. Shepard went 116 miles above the earth in 1961. His trip into space lasted about 15 minutes. He later led a flight to the moon. He was the fifth man to walk on the moon.

Commander Alan Shepard on the moon

141

ACTIVITIES

Reflecting Light

Materials: flashlight, mirror

Turn on the flashlight and shine it at a wall. Dim or darken the room as needed. Put the mirror in the path of light and show how the direction of the light changes.

Size of the Moon

Materials: paper circle from Lesson 48, one ½" mini marshmallow, one ⅛" button (e.g., a man's shirt-collar button)

Show the paper circle and mini marshmallow. Discuss what each represents. Show the ⅛" button. Explain that the button is representing the moon. Point out that the moon is about one-fourth the size of the earth.

BACKGROUND

Men on the Moon

Between July 1969 and December 1972, several American astronauts walked on the moon. Alan Shepard commanded *Apollo 14* and walked on the moon in February 1971.

What does the moon move around? the earth

- Use the *Size of the Moon* activity to compare the sizes of the sun, earth, and moon.
- Direct the students to Activities page 119. Complete the graphic organizer about the moon with the students.

> Teaching for page 141 begins here.

Have any people walked on the moon? yes; astronauts

What is an astronaut? a person who travels into space

- Direct attention to the *Science and History* box. Read aloud about Alan Shepard as the students follow along.

Who was the first American astronaut in space? Alan Shepard

How many miles above the earth did he travel? 116

- Direct attention to the picture. Ask a volunteer to read the caption.

Where is Alan Shepard? on the moon

- Explain that Alan Shepard walked on the moon about 10 years after the first American flight into space.
- Point out the astronaut's space suit.

Why do you think Alan Shepard is wearing a space suit? There is no air to breathe on the moon like there is on earth. The space suit gives the astronaut air to breathe. It helps to keep the astronaut safe.

How do we know that the moon is made of rock? The astronauts brought rocks and soil back from the moon and told people about the moon.

- Point out the American flag. Explain that the American astronauts were representing *all* Americans when they visited the moon. Explain that they left the flag on the moon when they returned to earth. You may point out that there is no wind on the moon and that a pole across the top is used to hold the flag straight out.

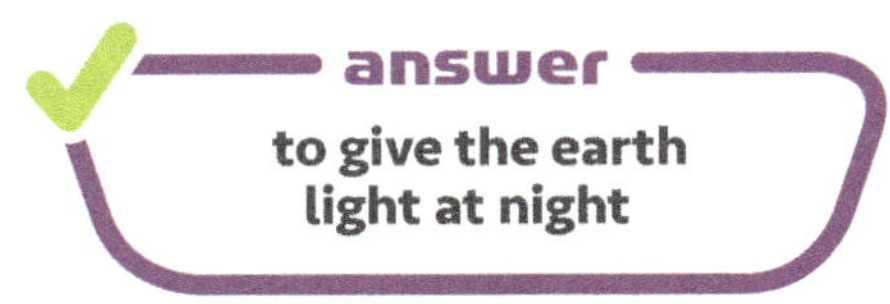

> **answer**
>
> **to give the earth light at night**

- Preview and pronounce the vocabulary term *phases of the moon*.
- Direct the students to read pages 142–43 silently to find out where the moon is in the sky in the evening.

When does the moon look like it is moving across the sky? during the nighttime

Why does the moon look like it is moving? The earth rotates, and the moon follows a path.

What object in the sky looks like it is following a path during the daytime? the sun

Where is the moon in the evening? in the east

What does the moon seem to do in the evening? rise

- Direct the students to use the compass as their guide. Instruct them to put their fingers on the moon in the east. Direct them to trace the purple arrow in the east.

In what direction did your finger trace the arrow? up

- Explain that around the middle of the night the moon is at its highest point in the sky.
- Instruct the students to put their fingers on the moon in the middle of the picture. Direct them to trace the purple arrow in the west.

In what direction did your finger trace the arrow? down

Where is the moon in the morning? in the west

What does the moon seem to do in the morning? set

What does the moon move around? the earth
When you look at the moon, does the moon look the same every night? no
What have you seen the moon look like? Possible answers: a fingernail, a banana, a cradle, a ball, a circle
What does the moon seem to do every month? change shape
When you look at the moon, what part of it do you see? only the part that is reflecting the sun's light
Does the moon change a lot or a little each night? a little

- Explain that the shape of the moon changes a little each night because the moon moves around the earth. As the moon moves, we usually cannot see all of the area that is reflecting the sun's light.

What do we call the different shapes of the moon? the phases of the moon

- Direct attention to the diagram. Point out the position of the sun, the earth, and the phases of the moon. Explain that the four moons show where the moon is located as it goes around the earth.

- Instruct the students to put their fingers on the new moon and trace the red arrows that show the moon's path around the earth. Read together the name of the moon's phase as the students follow the arrows.

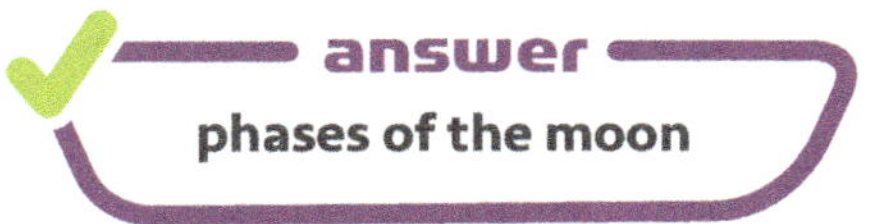

Phases of the Moon

Not all of the moon's phases are taught in SCIENCE 1. The moon's phases are the new moon, waxing crescent, first quarter, waxing gibbous, full moon, waning gibbous, third quarter, and waning crescent. *Waxing* means that the amount of surface that is lighted is getting larger. *Waning* means that the lighted surface is getting smaller. In the gibbous phases, the moon is more than half-lit but not completely lit.

PREPARATION FOR READING

- Preview and pronounce the words *new moon*, *quarter moon*, and *full moon*.
- Direct the students to read page 144 silently to find out what the phase of the moon is called when we see half the moon.

TEACH FOR UNDERSTANDING

What is the phase of the moon called when none of the moon is bright? new moon

- Direct attention to the pictures. Instruct the students to find the new moon.

Why is a new moon all black? We cannot see the moon's surface; it is not reflecting the sun's light.

- Explain that the moon cannot be seen in the sky when it is a new moon.

What is the phase called when you see half the moon? quarter moon

- Direct attention to the pictures. Instruct the students to find the quarter moons.

How many quarter moons do you see? 2

- Ask a volunteer to read the names of the quarter moons.

What is the phase of the moon called when the moon looks like a big, bright ball? full moon

- Direct the students to point to the full moon.

How many days does it take to see all the phases of the moon? about 29

- Explain that the moon goes around the earth one time in about 29 days, and this movement causes the phases of the moon.
- Direct the *Modeling the Phases of the Moon* activity as the students refer to the diagram on page 143.
- Instruct the students to complete the diagram of the phases of the moon on Activities page 120.

How does the nighttime sky change each day? The phase of the moon changes a little each day.

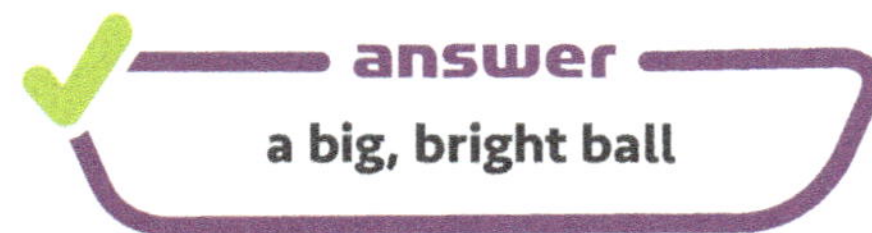
answer
a big, bright ball

ACTIVITIES

Bible Verses, page 109
The Moon, pages 119–20
These pages were completed during the lessons.

Study Guide, pages 121–22
These pages review Lessons 50–52. After completion, direct the students to keep them for review for Test 7.

Sometimes none of the moon is bright.

This is called a *new moon*.

Sometimes we see half the moon.

This is called a *quarter moon*.

Sometimes the moon looks like a big, bright ball.

This is called a *full moon*.

It takes about 29 days to see all the different phases of the moon.

New moon   **First quarter**   **Third quarter**   **Full moon**

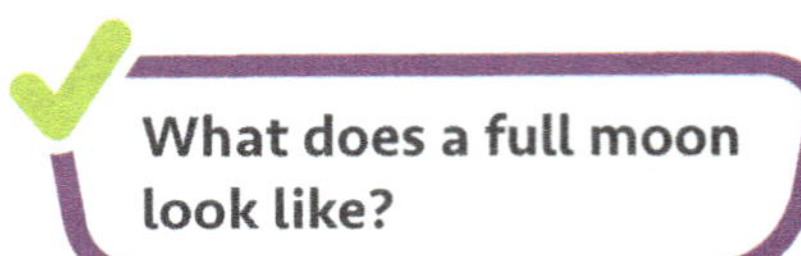

What does a full moon look like?

BACKGROUND

Quarter Moons
The name of the phase of the moon refers not to how much of the moon is lit but to how far the moon has traveled in its path around the earth. When the moon is half lit, it is either one-fourth or three-fourths of the way around the earth. The full moon is the second quarter, or halfway around the earth. The term *revolution* will be taught in Chapter 8.

ACTIVITY

Modeling the Phases of the Moon
Materials: lamp without a shade, ball
Place the lamp on a table. Explain that the sun lights the surface of the moon similarly to the way the lamp lights the surface of the ball. Ask a volunteer to help you model the phases of the moon. *New moon:* direct the student to stand about 2' from the lamp, facing the lamp. Instruct him to hold the ball about 10" in front of his eyes and to look at the ball. He should not see any light on the ball's surface. *Quarter moon:* direct the student to turn so his side faces the lamp, still holding the ball away from his eyes. He should see half of the ball's surface lit. *Full moon:* direct the student to turn his back to the lamp, still holding the ball in front and slightly to the left. He should see all of the ball's surface lit.

EXPLORATION

Process Skills
Observe
Predict

Lesson 53

Changes in the Sky

Our sky changes every day.

Sometimes we see the sun.

Sometimes we see other stars.

Sometimes we see the moon.

In this Exploration you will observe the nighttime sky and the daytime sky.

You will predict what the moon will look like in about one month.

145

Objectives
- Compare and contrast the nighttime sky with the daytime sky
- Predict the moon's phase
- Infer the cause for the changes in the sky each day
- Apply our knowledge of the earth, sun, moon, and stars to praising God for His greatness and goodness **BWS**

Materials
- clear nighttime sky
- clear daytime sky

INTRODUCTION

- Invite the students to tell about times they observed the sun, moon, and stars.
  In this Exploration, you will observe the nighttime sky and the daytime sky. You will observe how the sky changes.

PREPARATION FOR READING

- Direct the students to read page 145 and Activities pages 123–25 silently before beginning.
- Explain that this Exploration will be completed when the sky is clear, on two different days.

TEACH FOR UNDERSTANDING

- Direct the students to the **Process Skills**.
  What science process skills will you be using in this Exploration? Observe and Predict
- Review the process skills *observe* and *predict*.
- Direct attention to the **Purpose**.
  What is the purpose of this Exploration? to observe the changes in the sky and to predict how the sky changes
- Direct attention to the list of **Procedures**. Remind the students to follow each step. Encourage the students to check off each step as it is completed.
- Point out that this Exploration will be completed in two days. Day 1 will be completed at night and at home. Point out that they may have help from their parents.
- Day 2 will be completed on the next clear day. *Note:* Remind the students that it is never safe to look directly at the sun. They may draw the sun in the sky without looking directly at it.
- Discuss the **Conclusions**.
- Direct the students to work alone or collaborate in science groups to complete Activities page 127.

ACTIVITIES

Exploration: Changes in the Sky, pages 123–25
The students will explore the nighttime and daytime skies.

Thank You, God, page 127
This page was completed during the lesson.

ASSESSMENT

Rubric
Use the prepared rubric or design a rubric to include your chosen criteria.

Objective
- Recall terms and concepts from Chapter 7

REVIEW

- Material for Test 7 will come from the Study Guides on Activities pages 113–14 and 121–22. You may review any or all of the material during the lesson.
- You may choose to review Chapter 7 by playing *Night Sky Five in a Row* review game or another game from the Game Bank (Teacher Resources).

🏆 REVIEW GAME

Night Sky Five in a Row
Prepare several moon shapes and star shapes to use as markers. Display a grid, five squares high by five squares wide.

Divide the students into two teams, assigning one team to be stars and the other to be moons. Ask review questions and allow each team to put its shape in one of the squares when a question is answered correctly.

The first team to get five of its shapes in a row wins.

NOTES

Objective
- Recall and apply terms and concepts from Chapter 7

ASSESSMENT
- Administer Test 7.

NOTES

Chapter Objectives
- Identify the causes of the seasons
- Sequence the season cycle
- Relate temperature changes using a thermometer
- Apply the process skills of measuring, inferring, and communicating
- Identify characteristics of winter, spring, summer, and fall
- Defend, using Scripture, that seasonal patterns exist by God's design **BWS**

Lesson Objectives
- Recall that the earth rotates once each day
- Identify that the earth revolves around the sun
- Identify that one complete revolution around the sun is equal to one year
- Identify the two things that cause the seasons
- Sequence the cycle of the seasons

Materials
- desk globe

Vocabulary
- revolve
- year
- tilt
- season

CHAPTER INTRODUCTION

- Direct attention to the chapter title on page 147.
  What is the title of this chapter? "Seasons"
  What do the four trees in the picture show? different kinds of trees; trees in different seasons
- Observe the students as you ask them to put their finger on the tree you describe. (*Note:* Some students may not be able to recognize the season for each tree.)
  Put your finger on the snow-covered tree. Which season does this tree show? winter
  Put your finger on the pink blossom tree. Which season does this tree show? spring
  Put your finger on the tree with green leaves. Which season does this tree show? summer
  Put your finger on the tree with orange leaves. Which season does this tree show? fall
- Invite students to identify the tree that is their favorite.
- Ask volunteers to tell about any of these trees that are in their yard.

Big Question
- Direct attention to the Big Question on page 146. Read the question aloud.
  What causes the seasons?

BACKGROUND

Chapter Photo
The photo on page 146 is of four different kinds of trees, one in each season.

Snow-covered tree: deciduous tree in the winter

Pink blossom tree: a spring pink cherry blossom in spring

Green-leafed tree: a maple tree in the summer

Orange-leafed tree: golden forest tree near Mount Fitz Roy in Argentina, South America

Visit TeacherToolsOnline.com for resources to enhance the lessons.

Seasons

God designed the earth to move.

The earth moves in different ways.

As it moves there are changes.

It may become colder or hotter.

There may be more daylight or less daylight.

The time of year may change, but God never changes.

147

- Explain that the answer to the Big Question will be found as they read Chapter 8.

In this chapter you will learn more about the seasons and the changes each one may bring.

PREPARATION FOR READING

- Direct the students to read page 147 silently to find what changes there are because the earth moves.

TEACH FOR UNDERSTANDING

In Chapter 7 you learned about one way the earth moves. What way did you learn the earth moves? It rotates.

- Explain that in this chapter the students will learn about the second way that the earth moves.

How do you know that the earth moves? God designed it to move.

What changes are there because the earth moves? Answers should include that it may get colder or hotter; there may be more daylight or less daylight; the time of year may change.

Does God ever change? No; God never changes.

⚠ HELPS

Different Seasons
Throughout this chapter, students will learn about the characteristics of each season based on North America's seasonal characteristics. An Exploration at the end of this chapter provides an opportunity for students to infer and communicate what the seasons look like where they live. This activity considers the fact that the entire world does not experience the seasons at the same time and that some parts of the world may not experience distinct seasons at all.

LOOKING AHEAD

Fall Leaves
Bring in some fall leaves to display during Lesson 60. Artificial fall leaves will work in the lesson.

Chapter 9 Science Materials
An outdoor thermometer, pinwheel, and a small flag will be used in Chapter 9 to determine air temperature and how much wind is blowing.

PREPARATION FOR READING

- Preview and pronounce vocabulary terms *revolve* and *year*.

  What does the word *rotate* mean? to spin

  What does the earth do when it rotates? It spins.

  How long does it take the earth to rotate one time? one day

  Today, you will learn about a second way the earth moves.

- Direct the students to read pages 148–49 silently to find out how long it takes for the earth to move around the sun. (*Note:* The students do not need to read *Meet the Scientist* yet.)

TEACH FOR UNDERSTANDING

  What are some objects that spin? a toy top, a ball, the earth

  Why is it important for the earth to rotate? Answers should include that we get day and night; we see the sunrise and the sunset; we see the sun, moon, and stars.

- Encourage the students to follow as you read aloud *Meet the Scientist*.

  Who was Nicolaus Copernicus? a scientist

  What did he observe? the heavens; that the earth moves

  What did he prove? The earth goes around the sun.

Lesson 56

The Earth Moves

It Rotates

A toy top spins.

A ball spins.

The earth spins.

Because the earth spins, or rotates, there is day and night.

We can see the sunrise and the sunset.

We can see the sun, moon, and stars.

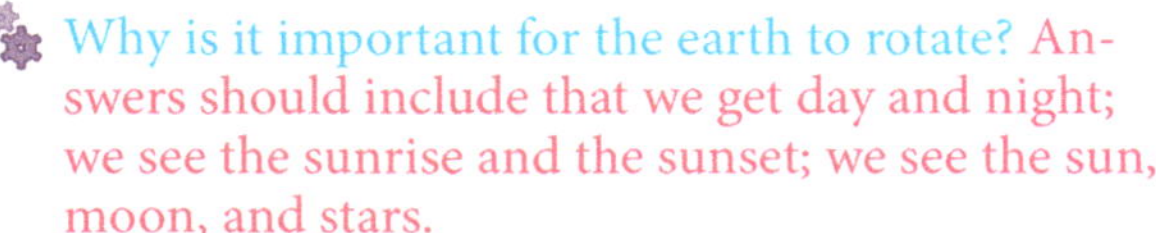

Meet the Scientist

Nicolaus Copernicus was a scientist who lived long ago. He studied the heavens. He observed that the earth moves. He proved that the earth goes around the sun. It took many years for other scientists to believe this discovery.

148

Scientist: Nicolaus Copernicus

Nicolaus Copernicus was a famous scientist born to Polish parents. He was born on February 19, 1473, and died on May 21, 1543. During his time, people believed that all planets revolved around the earth. They also believed that the earth did not move. This belief is called the *geocentric theory*. Copernicus proved that the earth does move and that the sun is in the center of our solar system. He also proved that the planets revolve around the sun and not around the earth.

What is another way the earth moves? It revolves. What does the earth revolve around? the sun

- Direct attention to the vocabulary term *revolve* and its definition. Ask a volunteer to read the word and definition aloud.

  How long does it take the earth to revolve around the sun? one year

- How many months are in one year? twelve
- Direct attention to the picture.
- What is this picture showing? The earth revolving around the sun.
- Direct each student to stand next to his desk.
- Guide the students as they act out the earth rotating on its axis.
- Pair the students to show the earth revolving around the sun.
- Instruct one student to be the "sun" and the other student to be the "earth." Monitor the students as the "earth" walks around the "sun."
- Direct students to switch positions and repeat the activity.

PREPARATION FOR READING

- Preview and pronounce the vocabulary terms *tilt* and *season,* as well as the word *direction.*
- Direct the students to read pages 150–51 silently to find out what a season is.

TEACH FOR UNDERSTANDING

What did God design the earth to do? tilt

- Display a desk globe and point out the tilt of the earth.

  What does *tilt* mean? to lean

- Ask the students to show what tilt means while they sit at their desks.

  Is the earth always tilted in the same direction? yes

  Is the earth tilted as it rotates? yes

  Is the earth tilted as it revolves? yes

- Direct attention to the picture.

  What do the red arrows on the earth show? The earth is rotating.

  What do the blue arrows around the sun show? The earth is revolving around the sun.

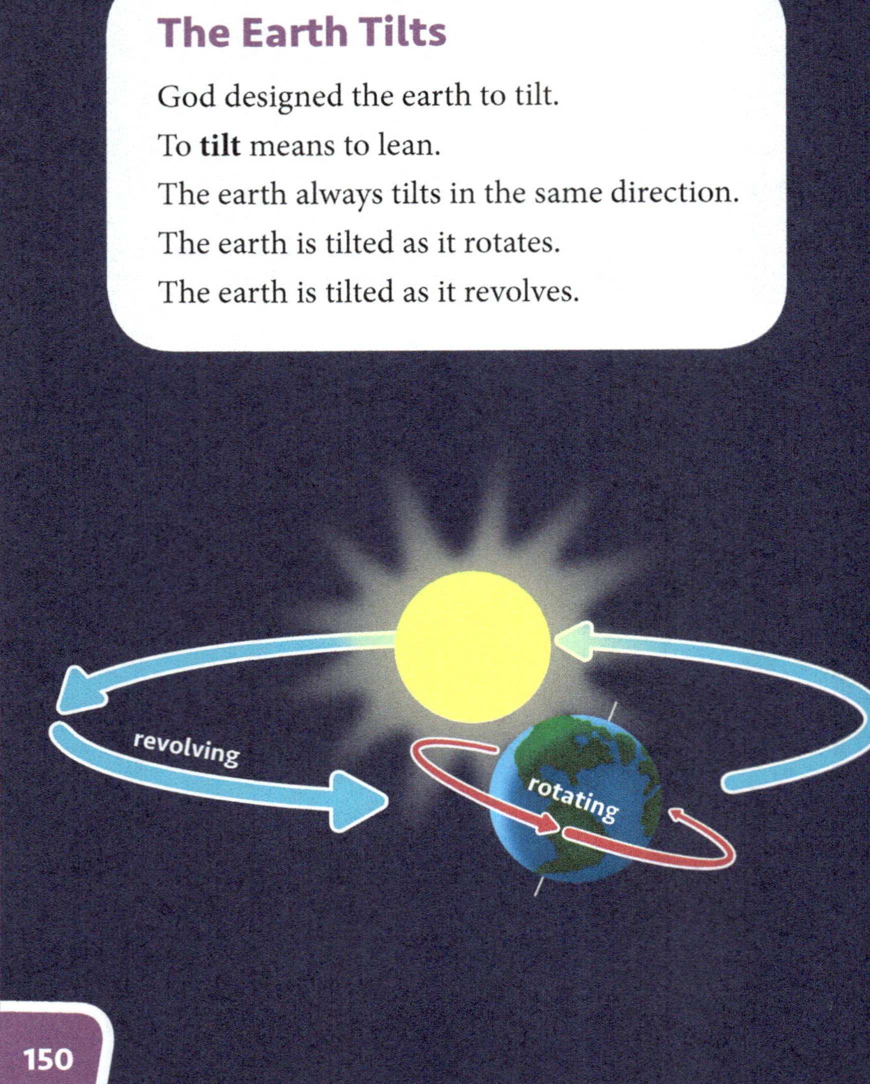

⚠ HELPS

Seasons Around the World
If you live in an area that does not have obvious seasonal changes, help the students see the changes that do occur in your area. It is also important for students to recognize what happens in the many places where four distinct seasons occur.

Cycle of Seasons
Although winter is the first season introduced in this chapter, students can also begin with *spring* as long as they get the correct order.

Seasons Illustration
The same view of earth is given in the illustration on page 151 to help the students get a better perspective about the amount of light on each hemisphere during each season.

The earth's tilt causes seasons.

The earth's revolving around the sun also causes seasons.

A **season** is a time of year.

The seasons are *winter*, *spring*, *summer*, and *fall*.

Seasons

spring

summer

winter

fall

What two things cause seasons?

151

What does the earth's tilt cause? seasons

What else causes the seasons? the earth revolving around the sun

What is a season? a time of year

What are the four seasons in order, starting with winter? winter, spring, summer, and fall

- Explain that the seasons happen in a cycle. You can begin with any season and still follow a cycle.

Name the four seasons in order, starting with summer. summer, fall, winter, spring

Name the four seasons in order, starting with spring. spring, summer, fall, winter

- Direct attention to Activities page 131. Point out that this illustration shows North America at the top and South America at the bottom of each earth.

Do you live in North America or South America? North America

- Explain that when North America gets more sun, it is summer there. When South America gets more sun, it is summer there.

- Direct the students to write *winter* by the earth to the right of the sun.

The sun isn't shining as much on North America as it is on South America. It is winter in North America.

- Direct the students to label the remaining earths.

The earth is revolving to the left or counterclockwise. Remember the seasons happen in a cycle. Which season is the earth above the sun showing? spring

Which season is the earth to the left of the sun showing? summer

Which season is the earth below the sun showing? fall

- Use the *Model the Cycle of the Seasons* activity to teach the motions associated with each season.

- Explain that the students will use these motions with a new song in the next lesson.

- Direct attention to the Big Question on page 146.

What causes the seasons? the earth's tilt and the earth's revolving around the sun

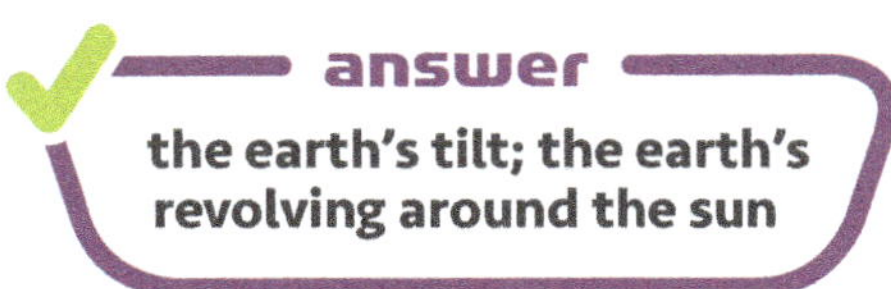

ACTIVITY

Model the Cycle of the Seasons

Assign the following motions to each season; spring (planting seeds), summer (swimming), fall (raking leaves), winter (shoveling snow).

Assign each of the seasons to a student or group of students. Invite the students to practice their motions.

Call out the seasons in order, starting with winter. Tell each student to do his motion when his season is named.

Explain that in the next lesson, they will sing a song about the seasons. They will use these motions while they sing the song.

Once the students have correctly associated the motions and seasons, repeat the activity, this time starting with a different season.

The word *cycle* is briefly mentioned in this lesson and will be reviewed in Lesson 59.

ACTIVITIES

Four Seasons, page 131

This page was completed during the lesson.

The Earth Moves, pages 133–34

These pages reinforce the concepts taught in this lesson.

Objectives
- Recall two things that cause the seasons
- Recall the thermometer as a scientific tool used to measure temperature
- Relate the movement of the red line on the thermometer to changes in temperature
- Measure temperature to record information
- Record temperature using a thermometer

Materials
- See Activities page 135.
- clear plastic cup with water
- alcohol thermometer *Note:* See Helps on the next page for additional information.

Vocabulary
- thermometer

INTRODUCTION

- Display an alcohol thermometer.
  What is the name of this tool? thermometer
  What do we use a thermometer for? to measure the temperature of something
- Remind the students that temperature is important. If people know that the temperature outside is too hot or too cold, they can choose the proper clothes to wear.

PREPARATION FOR READING

- Preview and pronounce the vocabulary term *thermometer.*
- Direct the students to remove pages 135–36 from their Activities books. Direct them to read page 152 and Activities pages 135–36 silently before beginning.

TEACH FOR UNDERSTANDING

- Direct attention to the **Process Skill**.
  What science process skill will you be using in this Exploration? Measure
- Review the process skill *measure* using information from Chapter 2.
- Choose a student to read the **Purpose** aloud.
- Direct attention to the **Procedure**. Use the clear plastic cup with water to demonstrate how to keep the red bulb in the water while reading the thermometer.
- Explain how you want the students to move to each station to read and record the temperature of the water in each beaker.
- Direct attention to the **Conclusions**.

EXPLORATION

Lesson 57

Using a Thermometer

How hot is it outside?

Temperature is how hot or cold something is.

A **thermometer** is a tool that measures temperature.

A thermometer tells how hot or cold something is.

In this Exploration, you will read and record temperature using a thermometer.

- Discuss the importance of knowing the temperature of some things.
-  What tool is used to measure temperature? a thermometer
- What happens to the thermometer's red line when the water gets hotter? The red line goes up.

ACTIVITIES

Exploration: Using a Thermometer, page 135
The students will explore measuring the temperature of water by using a thermometer.

Seasonal Activities, page 136
The students will match seasonal activities to the correct temperature in this Enrichment activity.

ASSESSMENT

Rubric
Use the prepared rubric or design a rubric to include your chosen criteria.

Daylight and Temperature

Name ___________________

Follow your teacher's directions to complete the chart.

Season	Months	Temperature	Hours of Daylight
(word bank)		cooler warmer hottest	shorter longer longest
winter	**December** **January** **February**	coldest — thermometer: 0 °C / 32 °F	shortest — About 9 hours
spring	**March** **April** **May**	*warmer* — thermometer: 18 °C / 65 °F	*longer* — About 13 hours

SCIENCE ① *Activities*

Lesson 58   **137**

⚠ **HELPS FOR LESSON 57**

Using a Thermometer

Set up three different stations. Each station will have a container with water at a different temperature. Group the students and direct them to rotate to each station. Students who finish early may work individually on the Enrichment activity on the back of the Activities page.

> You may bring hot water in an insulated cup or use a microwave to heat the water. Keep the water at or below 100°F to prevent possible burning. Add ice water to another container, and room-temperature water to the third container.

⚠ **HELPS FOR LESSON 58**

Daylight and Temperature

The average temperatures and daylight hours in this activity are based on averages in North America, excluding Hawaii and Alaska. Explain that not all places experience all four seasons in the same way or at the same time. Visit TeacherToolsOnline.com for links to find average temperatures and daylight hours for specific areas.

> This is an Activities lesson only. There are no corresponding Student Edition pages for this lesson.

Objectives
- Recall the cycle of the seasons by singing a song
- Compare and contrast temperature and amount of daylight among the seasons
- Infer the temperature and length of daylight hours for each season

Teacher Resources
- Visuals 8.2–8.4: *God's Four Seasons; Daylight and Temperature A; Daylight and Temperature B*

INTRODUCTION

- Review the motions for each season listed in Lesson 56.
- Display *God's Four Seasons.* Read through the words one time and then direct students to repeat after you. Sing both verses together.
- Explain that in this lesson the students will compare and contrast the most common temperatures and number of daylight hours among the four seasons.

PREPARATION FOR READING

- Direct the students to read Activities pages 137–38 silently to prepare for this activity.

TEACH FOR UNDERSTANDING

- Display *Daylight and Temperature A* and *B* to guide students as you complete the activity together.
- Direct attention to the *Season* column on Activities pages 137–38.
  The four seasons are listed in order, beginning with winter.
- Ask volunteers to read aloud the months that go with each season.
- Direct attention to the *Temperature* column. Explain to the students that they will write *cooler, warmer,* and *hottest* in this column.
- Direct attention to the thermometers. Guide the students as they write the temperature they see on each thermometer.
  Look at the thermometer for winter. What is the temperature on this thermometer? 0°C and 32°F
  Look at the thermometer for spring. What is the temperature on this thermometer? 18°C and 65°F

- Direct the students to turn to page 138.

  Look at the thermometer for summer. What is the temperature on this thermometer? 32°C and 90°F

  Look at the thermometer for fall. What is the temperature on this thermometer? 15°C and 60°F

- Direct the students to turn back to page 137 and look at all the thermometers.

  Why is winter the coldest season? Elicit that the thermometer shows it has the lowest temperature. The red line is the lowest of all thermometers.

  Look at the thermometer for spring. Compare it to the thermometer for winter. Is this temperature warmer or cooler than winter? warmer

- Direct the students to turn back to page 138.

  Look at the thermometer for summer. Compare summer's thermometer to all the other thermometers. Is this temperature cooler or the hottest? hottest

  Look at the thermometer for fall. Compare fall's thermometer to summer's thermometer. Is the red line higher or lower than in summer? It is lower.

  How would you describe the temperature? It is cooler.

- Direct attention to the *Hours of Daylight* column on page 137. The students will write *shorter, longer,* and *longest,* next to the given number of hours as you guide them through the illustration. (*Note:* The most common number of daylight hours has been given. The students only need to infer the length of daylight hours.)

- Display this illustration.

winter	9 hours
spring	13 hours
summer	14 hours
fall	11 hours

- Explain that each line shows how many hours of daylight the season has. Ask a volunteer to read the number of hours for each season.

  Which season has the shortest number of daylight hours? Winter has the shortest line, showing the shortest number of daylight hours.

  What is the number of daylight hours for spring? 13

  Is the line for spring longer or shorter than the line for winter? The line for spring is longer than the line for winter, showing spring daylight hours are longer than winter daylight hours.

- Direct the students to turn to page 138.

  How many daylight hours are there for summer? 14

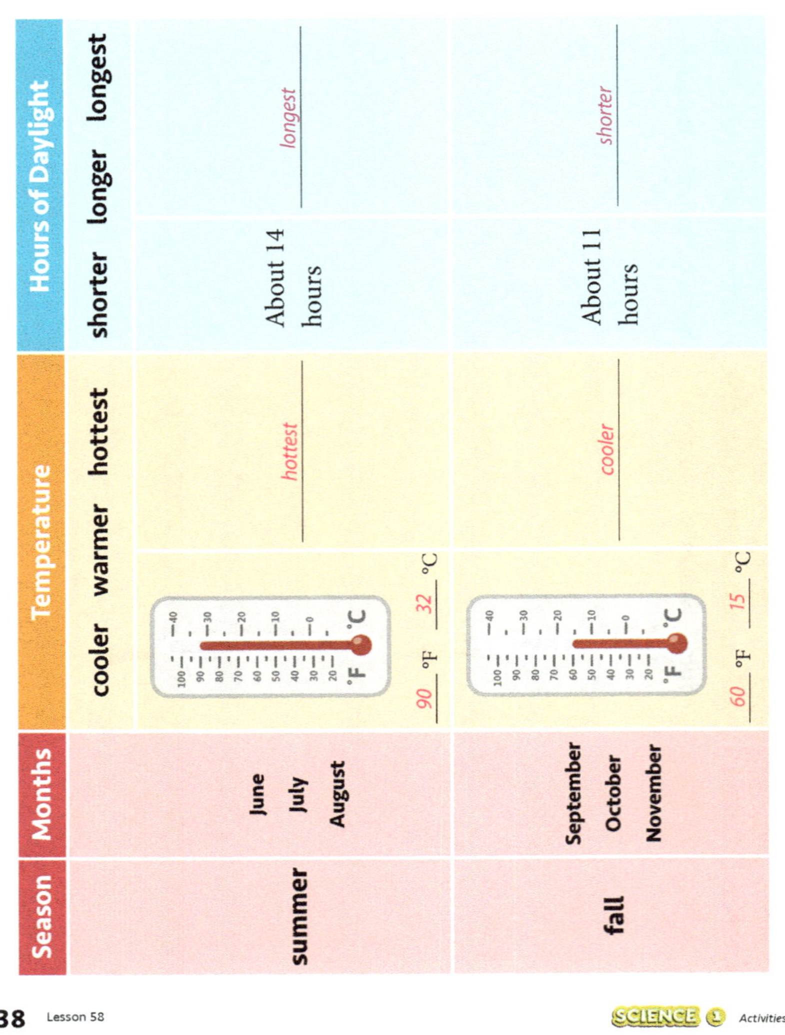

Is the line for summer the shortest or the longest of all seasons? The line for summer is the longest line, showing that summer has the longest number of daylight hours.

What is the number of daylight hours for fall? 11

Is the line for fall shorter or longer than the line for summer? The line for fall is shorter than the line for summer, showing fall daylight hours are shorter than summer daylight hours.

ACTIVITIES

Daylight and Temperature, pages 137–38
These pages were completed during the lesson.

Study Guide, pages 139–40
These pages review the concepts taught in Lessons 56–57. After completion, direct the students to keep them for review for Test 8.

Lessons 58 & 59

The Seasons

God gave us seasons.

In Genesis 8:22 God promises that there will always be seasons.

The seasons change.

God designed seasons to be a cycle.

Winter changes to spring.

Spring changes to summer.

Summer changes to fall.

And fall changes to winter again.

153

**LESSON
59**

Objectives

- Recall the cycle of the seasons by singing a song
- Explain, using Scripture, that seasonal patterns exist by God's design **BWS**
- Identify characteristics of winter and spring

Teacher Resources

- Visuals 8.1–8.2; 8.5: *Bible Verse; God's Four Seasons; A Time for Everything*

INTRODUCTION

- Review the cycle of seasons by singing "God's Four Seasons" together.
  You have been learning about the cause for the seasons; now you will learn about the different kinds of changes each season may bring.

PREPARATION FOR READING

- Direct the students to read page 153 silently to find out who designed the seasons to be a cycle.

TEACH FOR UNDERSTANDING

- Display *Bible Verse* and read Genesis 8:22 aloud.
  What does God promise that we will always have? seedtime and harvest; cold and heat; summer and winter; day and night

- Explain that seedtime is the time when the seeds are planted and that harvest is the time when fruits and vegetables are ready to be picked.
  Who gave us the seasons? God
  The seasons happen in a cycle. Who designed the seasons to be a *cycle*? God
  What does *cycle* mean? circle; when one step follows the next step in a circle

- Remind the students that lives of plants and animals follow a life cycle.

- Ask a volunteer to read the second section on page 153 aloud.
  What will always come after winter? spring
  What will always come after spring? summer
  What will always come after summer? fall
  What will always come after fall? winter
  What happens next? Winter comes again. The cycle starts over.
  What are the four seasons in order, beginning with winter? winter, spring, summer, fall

- Direct attention to the pictures. Ask a volunteer to name the seasons for each tree.
  Describe the tree that you might see during winter. The branches have no leaves; there is snow on the branches.
  Do you think it snows in all parts of the United States? no

- Why does it not snow in all parts of the United States? Elicit that not all parts of the United States have the seasons in the same way because of the earth's tilt.

- Describe the tree that you might see during spring. New leaves are growing.

- Describe the tree that you might see during fall. The leaves on the tree are yellow and some are on the ground.

PREPARATION FOR READING

- Preview and pronounce the word *winter*.
- Direct the students to read pages 154–55 silently to find out what happens to some plants in winter.

TEACH FOR UNDERSTANDING

What is the first month in winter? December Winter usually begins around December 21. The day winter begins has less daylight hours than any other day of the year.

- Direct attention to the winter scene picture.

What are some signs of winter that you see in this picture? The branches have no leaves; there is snow on the ground and on the branches; children have coats and look cold.

What happens to some plants in winter? Some plants die and other plants rest.

What makes winter a time of resting? It is too cold for some plants to grow; most plants and trees do not have leaves or fruit; some die and some rest; there is not as much food for animals, so they rest more in winter.

What season comes after winter? spring

When will plants start growing again? in the spring

- Explain that some trees and plants stop producing leaves and flowers to survive through winter. Some trees and plants have thin, needle-like leaves and do not lose their leaves during fall.

Winter

Winter comes after fall.

Winter begins in December.

The earth is tilted away from the sun.

The amount of daylight is shortest during winter.

This makes winter the coldest season.

Sometimes snow falls.

Winter is a time of resting.

Many trees have lost their leaves.

Some plants die.

Other plants rest.

They will start to grow again in spring.

154

🧪 ACTIVITIES

Different Leaves

Display a variety of leaves such as maple, oak, pine, and cedar. Point out the difference between leaves that fall and leaves that stay on the tree throughout fall and winter. You can also collect pictures of these leaves instead of bringing in real leaves.

Winter in Other Places

Materials: desk globe

Use a desk globe to point out the equator as the imaginary line that goes around the middle of the earth. Countries that are near the equator stay warm all year long because that part of the earth never tilts away from the sun. These places get the same number of daylight hours all year round.

God designed animals to survive during winter.

Some animals grow thicker fur to stay warm.

Some animals have fur that changes color in winter.

Some animals sleep through winter.

What is the difference between summer and winter fur?

Why is the temperature coldest in winter?

155

What happens to some animals during winter? Some animals grow thicker fur to stay warm; some animals' fur changes color; some animals sleep more and need less food.

- Direct attention to the picture at the top of the page.
- What are this bear and its cub doing? sleeping Another word for this kind of sleeping is *hibernating*.
- You may refer to the Background information if you choose to discuss hibernation.
- Direct attention to the picture of the rabbits and explain that they are snowshoe hares.
- Ask a volunteer to read the caption.
- What is the difference between the summer and winter fur? Summer fur is brown; winter fur is white.
- Why do you think the different color fur is helpful? Elicit that the change in fur color helps him blend in with his surroundings so that he can hide from danger.
- Direct attention to page 141 in the Activities book.
- Complete the winter and spring rows during this lesson and the summer and fall rows during Lesson 60. You may want the students to remove page 141 from their books and place it in their Science Notebooks for use in Lesson 60.
- Guide the students as they use the word bank to fill in the details that describe the winter season.
  What season does winter come after? fall
  How would you describe the temperature of this season? It is the coldest.
  How would you describe the length of daylight hours in winter? Daylight is shortest.
  What is winter a time of? resting

answer

The earth is tilted away from the sun and the amount of daylight is shortest.

BACKGROUND

Hibernation

Most animals that hibernate do not sleep uninterrupted for the entire winter. During hibernation, the animal's body temperature and metabolism are very low. As a result, the animals need little food to survive. In fact, many animals use only the fat they stored in their bodies before winter. Others wake up just enough to eat a little food and then go back to sleep. The important fact is that hibernation allows animals to survive when food is scarce during winter.

Black Bear and Cub Hibernating

Black bears usually give birth to their cubs in January or February. The mother bear will keep her cub warm and change her position to help the cub nurse. A cub will usually stay with its mother for the first two years of its life.

PREPARATION FOR READING

- Preview and pronounce the word *spring*.
- Direct the students to read pages 156–57 silently to find out what birds do in the spring.

TEACH FOR UNDERSTANDING

What season comes after winter? spring
What is the first month of spring? March

- Explain that spring usually begins about March 21. Spring begins when the hours of daytime and nighttime are the same.

What happens to the temperature in spring? It becomes warmer.

- Explain that although the days start to get warmer in the spring, not every day will be warmer than in winter.

What begins to grow in the spring? flowers
What happens to animals in the spring? Baby animals are born; birds build nests; bees buzz around flowers; some animal fur changes color.
What is spring a time of? It is a time of planting seeds.
What helps seeds grow in spring? rain
What happens to trees in spring? They grow new leaves.

🔍 **BACKGROUND**

The Spring Equinox
The spring equinox is when the daytime and nighttime hours are the same. Daytime hours begin getting longer in spring. The word *equinox* comes from two Latin words that mean "equal night."

What do birds do in the spring? They build nests and lay eggs.

What happens to the fur of some animals in the spring? It changes color.

- Direct attention to the barn scene on pages 156–57.

What are some signs of spring that you can see in this picture? bees buzzing around flowers; baby animals: lambs, chicks, kittens, foal; green grass, buds on the tree

What makes spring a time of planting? Answers should include that the temperature is getting warmer, and spring rain helps seeds grow.

- Direct attention to Activities page 141.
- Guide the students as they use the word bank to fill in the details for the spring season.

What season does spring come after? winter

How would you describe the temperature of this season? It is warmer.

How would you describe the length of daylight in the spring? Daylight is longer.

What is spring a time of? planting

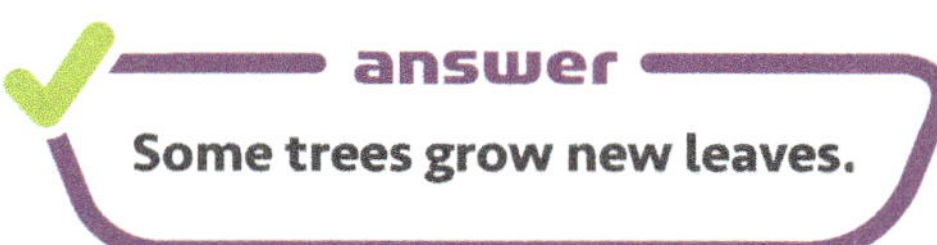

ACTIVITIES

Bible Verse, page 129
This Bible verse will be used in Chapter 8.

A Time for Everything, page 141
- Winter and spring rows were completed during this lesson.
- Summer and fall rows will be completed during Lesson 60.

READ-ALOUD BOOK

Making It Through Winter
You may read aloud *Chickadee Winter* by Dawn L. Watkins (available from JourneyForth Books, a division of BJU Press, at journeyforth.com).

Objectives
- Recall the cycle of seasons by singing a song
- Explain what a landscape architect does
- Identify characteristics of summer and fall
- Defend, using Scripture, that seasonal patterns exist by God's design **BWS**

Materials
- variety of fall leaves

Teacher Resources
- Visuals 8.1–8.2; 8.5: *Bible Verse; God's Four Seasons; A Time for Everything*

Vocabulary
- landscape architect

INTRODUCTION
- Display *God's Four Seasons.* Review the cycle of seasons by singing the song together.

PREPARATION FOR READING
- Preview and pronounce the vocabulary term *landscape architect* as well as the words *environment* and *fountains.*
- Direct the students to read page 158 to find out what a landscape architect does.

TEACH FOR UNDERSTANDING
- Explain that as the weather gets warmer, people will take on new projects in the areas outside their houses.

  What is a landscape architect? a type of engineer who helps people design their yards

- What do engineers like to do? They help others by finding answers to problems, and they come up with creative ideas.

  Besides yards, what are some other places a landscape architect can design? parks, gardens, outdoor spaces at houses or businesses

  What are some materials he can use in his designs? rocks, fences, benches, and fountains

  What does a landscape architect do? He helps people design their yards, he measures the space, he draws a plan of his idea, and he shares his plan with others.

STEM Careers

Lesson 60

Landscape Architect

A landscape architect is a type of engineer.

A **landscape architect** helps people design their yards.

He can design parks and gardens.

He measures the space.

He chooses the right plants for that environment.

He can use rocks and fountains.

He draws a plan of his idea.

He shares his plan with others.

158

- Direct attention to the picture at the top of the page.
- What do you think the two landscape architects are doing? Elicit that they may be measuring distance and taking notes of the type of ground and space they are working with.
- Direct attention to the picture at the bottom of the page.
- What is the landscape architect doing? Possible answer: he is drawing a plan of his design.

Summer

Summer comes after spring.

Summer begins in June.

The earth tilts toward the sun.

The amount of daylight is longest in summer.

This makes summer the hottest season.

✔ **Is daylight long or short in summer?**

159

PREPARATION FOR READING

- Preview and pronounce the word *summer*.
  In the last lesson you learned about things that happen in winter and spring. Today you will learn about things that happen in summer and fall.
- Direct the students to read pages 159–60 to find out what happens to plants in summer.

TEACH FOR UNDERSTANDING

What season comes after spring? summer

What month does summer begin in? June

Why is summer the hottest season? The earth's tilt toward the sun makes it the hottest season.

- Direct attention to the picture of the children standing in line.

What are these children doing? They are standing in line to buy a cup of lemonade.

Why do you think they want to drink lemonade? Answers may include that they are thirsty and sweaty; it is very hot outside.

Look at the picture of the boy on the grass. What is he doing? playing on a water slide

Why is summer the best time for water slides? It is the hottest season.

- Direct attention to the picture of the flower.

What kind of flower is this? a sunflower

Why do you think there is a picture of a sunflower on this page? Flowers grow during the summer.

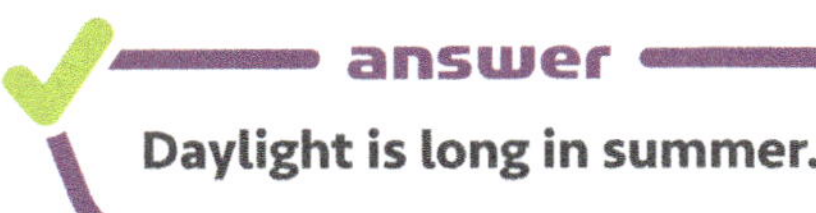

✔ **answer**

Daylight is long in summer.

What happens to plants in the summer? They grow bigger.

What do some trees and bushes do during the summer? They grow fruit.

Can you think of some trees or bushes that grow fruit during the summer? Possible answers: peach trees, apple trees, blueberry bushes

What makes summer a time for growing? The weather is warmer; there is more daylight.

What happens to vegetables in the summer? They grow and ripen.

- Invite a student to explain what it means to "become ripe." Explain that fruits and vegetables that are ripe are ready to be picked and eaten.

What do birds do during the summer? They learn to fly and leave their nests.

- Direct attention to the picture of the adult and children in a garden.

Where do you think this garden is? Possible answers: a backyard, a school, a park, a community garden

Where do you think that a garden can be grown? Elicit that a garden can be grown anywhere as long as there is enough dirt, water, and sunlight.

- Invite students to name the vegetables the children have picked. carrots, tomatoes, bell peppers

Look at the picture of the animals.

What kind of animals are they? lambs, sheep

- Direct attention to Activities page 141.
- Guide the students as they use the word bank to fill in the details that describe the summer season.

What season does summer come after? spring

How would you describe the temperature of this season? It is the hottest.

How would you describe the length of daylight in the summer? Daylight is the longest.

What is summer a time for? growing

Summer is a time for growing.

Plants grow bigger.

Some trees and bushes grow fruit.

Vegetables grow and become ripe.

Baby animals eat and grow.

Baby birds begin to fly and leave their nests.

160

Garden Photo
Point out that a garden can be planted in many different kinds of spaces. People who live in the city may plant raised gardens in their backyard, on the rooftop of an apartment building, or in a community garden.

Fall

Fall comes after summer.

Fall begins in September.

The amount of daylight gets shorter.

It gets dark earlier.

The temperature gets cooler.

Plants change in the fall.

Some leaves change colors.

Some leaves fall to the ground.

Some plants stop growing.

Fall is a time of harvest.

It is a time to pick fruits and vegetables.

161

PREPARATION FOR READING

- Display a variety of fall leaves.

  During which season do we begin to see these kinds of leaves? *in the fall*

- Display the word *fall*.

  Now, we are going to learn about the fall season.

- Direct the students to read pages 161–62 silently to find out what some animals do during the fall.

TEACH FOR UNDERSTANDING

What season does fall come after? *summer*

What month does fall begin in? *September*

What happens to the temperature during the fall? *It gets cooler.*

Describe the amount of daylight in fall. *Daylight is shorter.*

- Direct attention to the picture of the boy raking leaves.

Why does the boy need to rake the leaves? *Answers should include that the leaves have died; they have fallen off the trees.*

What happens to some plants in the fall? *They stop growing.*

- Direct attention to the leaves on the page.

These are a few examples of leaves that change color and fall off trees during the fall.

What does "harvest" mean? *It is the time to pick and gather fruits and vegetables that are ripe and ready to eat.*

What are some fruits or vegetables that can be picked during harvest? *Possible answers: apples, corn, pumpkins*

ACTIVITY

Write a Cinquain

Remind the students that we find out about God's world by using our senses.

Use the following pattern to write a cinquain poem about summer.

First line: one word (something you see)

Second line: two words (how something feels)

Third line: three words (something you hear)

Fourth line: four words (something you taste or smell)

Fifth line: one word (name of season)

Example

Sunlight

Hot, sticky

Children playing happily

Watermelon, freshly squeezed lemonade

Summer

Note: The activity may be completed together, in small groups, or individually.

- Direct attention to the picture of the chipmunk.
- ✿ What is this chipmunk doing? He is collecting food for winter.
- ✿ What is he collecting? peanuts
- Explain that God created animals with special ways to survive and grow in different seasons and places. Some animals gather and store food for the winter season, and other animals leave for warmer places and then go back in the spring.
- Point to the picture of the flying geese.
- ✿ What do you think these geese are doing? Answers should include that they are flying to a warmer place where they can find more food.
- Direct attention to Activities page 141.
- Guide the students as they use the word bank to fill in the details that describe the fall season.
   What season does fall come after? summer
   How would you describe the temperature of this season? It is cooler.
   How would you describe the length of daylight in the fall? Daylight is shorter.
   What is fall a time of? harvest
- ✿ What makes fall a time of harvest? Answers should include that fruits and vegetables have had enough time to grow and are ready to be picked.
- Display *Bible Verse*. Ask a volunteer to read aloud the verse.
- Direct attention to Activities page 142. Guide the students as they discuss the verse and complete this page.

✓ **answer**

Some animals gather and store food; other animals go to a warmer place where they can find more food.

ACTIVITIES

A Time for Everything, pages 141–42
These pages were completed during this lesson. You may want students to keep these completed pages in their Science Notebooks to help them with Study Guide pages 143–44.

Study Guide, pages 143–44
These pages review the concepts taught in Lessons 59–60. After completion, direct the students to keep them for review for Test 8.

Some animals gather food for winter.

Other animals move to a new place.

They go to a place that has more food.

They go to a place that is warmer.

The animals will return in spring.

✓ **How do animals prepare for winter?**

162

Migration
Many birds, such as geese and robins, are well known for their seasonal migrations. Other migrating animals include monarch butterflies, hummingbirds, elephants, zebras, salmon, ladybugs, and whales.

EXPLORATION

Lesson 61

Seasons Where I Live

The seasons are a cycle.

There may be four seasons where you live.

There may not be four seasons where you live.

What are seasons like where you live?

In this Exploration you will make a book.
Your book will show the seasons that you
observe where you live.

163

Process Skills
• Infer
• Communicate

LESSON
61

Objectives
• Compare and contrast the characteristics of seasons with the seasons in your local area
• Communicate by constructing a booklet that represents the seasons in your local area

Materials
• See Activities page 145.
• resealable sandwich bag, one per student *Note:* Students can use the bag to save any loose pieces that they cut.

INTRODUCTION

• Direct attention to the pictures.
• Explain that the boy and girl are showing how seasons can look different in other places.

How do you learn about God's world? by using your senses

Scientists infer and communicate what they observe. In this Exploration, you will infer and communicate what the seasons look like in your area.

PREPARATION FOR READING

• Direct the students to remove pages 145–51 from their Activities books. Direct them to read page 163 and Activities pages 145–51 silently before beginning.

TEACH FOR UNDERSTANDING

• Direct attention to the **Process Skills**.
• Review the process skills *infer* and *communicate*, using information from Chapter 2.
• Direct attention to the **Purpose**.

What is the Purpose of this Exploration? to make a booklet that shows what the seasons look like where I live.

• Direct attention to the **Procedure**.
• Explain that the students will need to look at page 151 and infer, based on what they learned about the seasons, which pictures represent the seasons where they live.
• Instruct the students to follow the steps listed on page 145. Direct them to cut and fold pages 147–49 first.
• Direct the students to find the kind of tree they see during winter in their area and then choose any pictures that go with this season. The students may choose to draw their own trees and any accompanying pictures.
• Instruct the students to follow the same procedure for the remaining seasons. Each season should have a tree that represents that season in their area.
• Explain that after they finish their pictures, they will write the name of their favorite season and one sentence explaining why they like it.
• Direct attention to the **Conclusions**. Choose several students to share one season from their booklets. They will describe how that season is the same or different from what they have learned in class.

ACTIVITIES

Exploration: Seasons Where I Live, pages 145–51
The students will infer and communicate what the seasons look like where they live.

ASSESSMENT

Rubric
Use the prepared rubric or design a rubric to include your chosen criteria.

Objective

• Recall terms and concepts from Chapter 8

REVIEW

• Material for Test 8 will come from the Study Guides on Activities pages 139–40 and 143–44. You may review any or all of the material during the lesson.
• You may choose to review Chapter 8 by playing *Dress the Trees* review game or another game from the Game Bank.

🏆 REVIEW GAME

Dress the Trees

Materials: cut-outs of leaves; cut-outs of flowers (optional); two drawings of a bare tree for display

Display two bare trees. The students will attach leaves as their team answers a question correctly from the Study Guides. You can add a second level to the game where the students attach flowers as their team answers more-challenging questions.

Divide the students into two teams. Alternate asking review questions to both teams. After each correct answer, allow the team member to place a leaf on his team's tree.

NOTES

Objective
- Recall and apply terms and concepts from Chapter 8

ASSESSMENT

- Administer Test 8.

NOTES

Chapter Objectives
- Identify the components (elements) of weather
- Sequence the water cycle
- Explain the role of a meteorologist in forecasting weather
- Use the science process skills of observe, measure and use numbers, and predict to collect and interpret weather data
- Evaluate a claim about the trustworthiness of science **BWS**
- Relate a weather prediction to Proverbs 22:3 **BWS**

Lesson Objectives
- Define *weather*
- Recall what temperature is
- Recall the scientific tool that measures temperature
- Define *wind*
- Identify the appearance of a flag when the wind is calm, light, and strong

Materials
- umbrella
- mittens or gloves
- thermometer, outdoor
- pinwheel
- picture of a sailboat

Teacher Resources
- Visual 9.2: *Bible Verses B*

Vocabulary
- weather
- wind

CHAPTER INTRODUCTION

- Direct attention to pages 164–65.
  What is the title of this chapter? "Weather"
- Direct attention to the picture.
- What is the girl holding? an umbrella
- Display an umbrella.
- When do you need an umbrella? when it is raining, snowing, or stormy
- Invite the students to share their experiences in the rain.
  Does it rain every day? no
  Is it dry every day? no
- Point out that weather is always changing.

Big Question
- Direct attention to the Big Question. Ask a volunteer to read the question aloud.
- Explain that the answer to the Big Question will be found as they read Chapter 9.

LOOKING AHEAD

Chapter 9 Science Materials
An outdoor thermometer, a pinwheel, and a small flag will be used in several lessons in this chapter to determine air temperature and the type of wind.

Displaying a thermometer outside where it is visible from inside will allow the students to easily participate in recording the temperature.

Chapter 10, Lesson 74 Science Materials
Pinhole boxes are required for the Investigation in Lesson 74. You may prepare these ahead of time. See the illustration on Student Edition page 196. Prepare the short end of a shoebox with a pinhole for observation. Along the long side of the shoebox make a hole the same size as the flashlight lens. Make one pinhole box for each science group or each student. Begin gathering enough objects and flashlights for each group or student. Without telling the students, make one object for each group or student be a glow-in-the-dark object, such as a sticker or glow rock.

Visit TeacherToolsOnline.com for resources to enhance the lessons.

Weather

Every day is different.

Will it be warm tomorrow, or will it be cold?

Will it rain today, or will it be dry?

If it is windy, you might fly a kite.

If it rains, you might splash in a puddle.

If it snows, you might build a snowman.

Living in God's world is exciting.

165

PREPARATION FOR READING

- Direct the students to read page 165 silently to find out what you might do if it is windy.

TEACH FOR UNDERSTANDING

Is every day the same? no

How are days different? It might be warm or cold; it might rain or be dry; it might be windy; it might snow.

What might you do if it is windy? fly a kite

- Invite the students to share their experiences of flying kites.

What might you do if it rains? splash in a puddle

- Direct attention to the picture.

How do we know that the girl might splash in a puddle? She is wearing yellow rubber boots or shoes.

What might you do if it snows? build a snowman

- Invite the students to share their experiences of playing in the rain or building a snowman.

Why is it exciting living in God's world? Every day is different; the weather is always changing.

- Preview and pronounce the vocabulary term *weather*.
- Direct the students to read pages 166–67 silently to find out how weather changes from day to day.

What do you call what the air outside is like? weather

What are some ways you think about weather? hot, windy, snowing

- Direct attention to the picture of the boy drinking water from the fountain.

Which sentence in the text describes the picture? It is hot.

 How do you know the boy is hot? He is sweating; he is thirsty.

- Explain that how hot it is outside is part of weather.
- Direct attention to the picture of the clothesline.

Which sentence in the text describes the picture? It is windy.

How do you know it is windy? The sheets and towels on the clothesline are blowing in the wind.

- Explain that wind is part of weather.
- Direct attention to the picture of the girl on page 167.

Which sentence in the text describes the picture? It is snowing.

What is the girl holding? a snowball

- Display a pair of mittens or gloves.

When do you wear mittens or gloves? when it is cold

Why is the girl wearing gloves? It is cold; the snowball is cold; to keep her hands warm

- Explain that the cold and snow are also part of weather.

The Weather

It is hot.

It is windy.

It is snowing.

These are ways you think about the weather.

The **weather** is what the air outside is like.

166

⚠ **HELPS**

Changing Weather

If you live in an area where the weather does not change much from day to day, help the students see the big picture of the yearly weather changes.

Is weather a part of God's world? yes

- Explain that the different kinds of weather are part of the world God made.

Does the weather stay the same each day? No; it changes from day to day.

- Choose a student to read the sentences aloud that describe how the weather may change.
- Invite a student to describe yesterday's weather.
- Invite another student to describe today's weather.

How would knowing what today's or tomorrow's weather will be help people? Possible answers: people can plan activities; people can plan what to wear.

- Point out that knowing what the weather will be today, tomorrow, or within the next hour is important to some people.

A farmer may want to know if it is going to rain so he will know when to plant seeds. You may want to know if it is going to snow so you can go sledding with your friends. Your parents may want to know if it is going to be sunny and dry so that they can plan a picnic in the park or a trip to the zoo.

- Explain that throughout history people have observed the weather and even made predictions.
- Display *Bible Verses B* and direct attention to Activities page 154. Read Matthew 16:2–3a aloud.

What does Jesus tell us about the color of the sky and the weather? You can look at the color of the sky in the evening or the morning to predict tomorrow's or today's weather; if the sky is red in the evening, it will be fair weather tomorrow; if the sky is red in the morning, it will be foul, or bad, weather today.

You can find out more about God's world as you study the weather.

PREPARATION FOR READING

- Preview and pronounce the words *Fahrenheit* and *Celsius*.
- Direct the students to read pages 168–69 silently to find out what the temperature of the air is measured in.

TEACH FOR UNDERSTANDING

What is temperature? how hot or cold it is

What science tool is used to measure the temperature of the air? a thermometer

What does the thermometer have on it to show the temperature? numbers

What is the temperature of the air measured in? degrees Fahrenheit or degrees Celsius

What do most scientists measure temperature in? degrees Celsius

- Display a thermometer that shows the temperature in °F and °C. Point out the Fahrenheit and Celsius markings and remind the students how to read an alcohol bulb thermometer.
- Direct attention to the picture. Explain that the thermometer on the left is a digital thermometer.

Where might you see a thermometer showing numbers like this? Possible answers: a bank, church, store, in the car, at home

- Remind the students that the thermometer on the right is an alcohol bulb thermometer.

What is the temperature that both thermometers are showing in degrees Fahrenheit? 73°F (23°C)

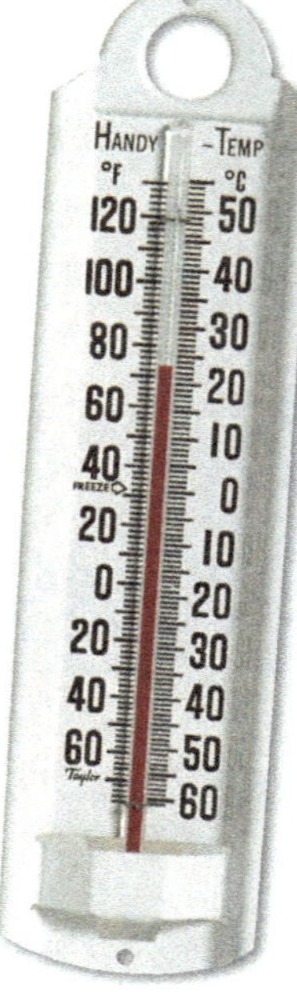

Temperature

Temperature is how hot or cold it is.

A thermometer is used to measure the temperature of the air.

The thermometer has numbers on it.

The numbers show the temperature of the air in degrees Fahrenheit or Celsius.

⚠ **HELPS**

Reading Thermometers

If your math program has already covered reading bulb thermometers, you may include additional questions related to reading temperature. The students should recognize that temperature is part of weather and that degrees are the standard unit for measuring temperature.

Fahrenheit or Celsius

You may explain more about the Celsius scale. In later grades the Celsius scale is used in BJU Press science books. However, on this grade level, temperature may be recorded using either the Fahrenheit scale or Celsius scale or both. Since the Fahrenheit scale will be more familiar to most first graders, they may record the daily temperature in Fahrenheit only.

The temperature changes during the day.

In the daytime the sun warms the air.

The temperature goes up.

At nighttime the sun does not warm the air.

The temperature goes down.

How does temperature change during a day?

169

What happens to the temperature during the day? It changes; it goes up.

What does the sun do to the air during the daytime? warms it

What happens to the temperature when it gets warmer? It goes up.

- Direct attention to the thermometer near the picture where the sun is shining.

What temperature does the thermometer show in degrees Fahrenheit? 70°F (21°C)

- Explain that 70°F is a mild temperature. It is warm, but it is not hot.

Does the sun warm the air during the nighttime? no
What happens to the temperature? It goes down.

- Direct attention to the thermometer near the nighttime picture.

What temperature does the thermometer show in degrees Fahrenheit? 40°F (4°C)

Is 40°F a warmer temperature or a cooler temperature than 70°F? cooler

How would knowing the temperature outside help you? Possible answers: it would help me plan activities; I can plan what to wear.

- Direct the students to complete the top of Activities page 156 to record today's outside temperature.

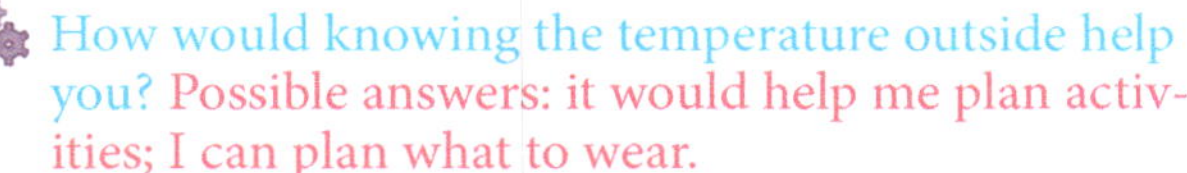

answer
The temperature goes up during the daytime, and it goes down during the nighttime.

BACKGROUND

Daily Temperature Changes

Many factors influence the temperature changes during the day and night. The student page (169) gives a simplified explanation to help the students understand the importance of the sun in warming the earth.

PREPARATION FOR READING

- Preview and pronounce the vocabulary term *wind*.
- Direct the students to read pages 170–71 silently to find out how much the wind blows when the wind is *strong*.

TEACH FOR UNDERSTANDING

- Display a pinwheel.
  - What does a pinwheel do? It turns when the wind blows.
  - What are some ways you could make it turn? Possible answers: blow on it; put it in front of a fan; take it outside in the wind.
  - What is wind? moving air
  - What happens to warm air? It rises.
  - What happens to cool air? It sinks.
  - What do the differences in the air cause? wind
  - What are some things that wind moves? leaves on trees, sailboats, windsurfers, a person's hair, pinwheels
- Direct attention to the picture on page 170.
  - What is the man in the picture doing? windsurfing
- Display a picture of a sailboat.
  - How is windsurfing like sailing? They both have a sail; the wind catches in the sail and moves the boat or the board through the water.

Windsurfing

Windsurfing is a sport that combines surfing and sailing. A sail that is attached to a board catches the wind and moves the board and the surfer through the water.

What does wind also move? flags
What does looking at a flag tell you about the wind? how hard the wind is blowing

- Direct attention to the circles with the flags. Explain that these are weather symbols showing how hard the wind is blowing.
- Mention that other examples of weather symbols will be used throughout this chapter.

What does it mean when you say the wind is calm? The wind does not blow much.

- Point out that the flag in the first circle is mostly hanging limp when the wind is calm.

What does it mean when you say the wind is light? The wind blows a little.

- Direct attention to the flag in the middle circle.

How is this flag different from the flag when the wind is calm? Possible answers: you can see more of the flag; the flag is stretched out more.

How much wind blows when the wind is strong? The wind blows a lot.

- Direct attention to the flag in the last circle.

How is this flag different from the flag in light wind? Possible answers: the flag in strong wind is stretched out all the way; you can see all of the flag; the flag in light wind is not stretched out all the way; you cannot see all of the flag.

- Direct the students to complete Activities pages 155–56.

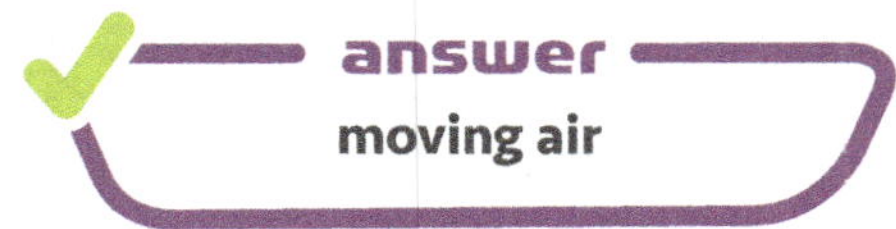

ACTIVITIES

Windy Days, pages 155–56
These pages were completed during the lesson.

Study Guide, page 157–58
These pages review the concepts taught in Lesson 64. After the pages are corrected, direct the students to keep them for review for Test 9.

Objectives
- Define *water cycle*
- Sequence the movement of water in the water cycle
- Identify the appearance of the sky on clear, partly cloudy, and cloudy days
- Identify types of precipitation
- Explain how the weather changes from day to day

Materials
- pinwheel
- self-stick dot
- video of a hailstorm (online Teacher Resources)

Teacher Resources
- Visual 9.1: *Bible Verses A*

Vocabulary
- water cycle
- cloud

INTRODUCTION

Clouds are interesting to watch.

- Poll the students to find out how many have sat and looked at clouds.
- Discuss the fact that clouds sometimes make shapes. Invite the students to share their experiences of looking at clouds.

What do you think clouds are made of?
Today we will learn what clouds are made of.

PREPARATION FOR READING

- Preview and pronounce the vocabulary terms *water cycle* and *cloud*.
- Direct the students to read pages 172–73 silently to find out what the water cycle is. Instruct the students to read the captions and to look at the diagrams as they read.

TEACH FOR UNDERSTANDING

How does the water from the earth move? Water moves from the earth to the sky and back to the earth again.

What is it called when water moves this way over and over? the water cycle

What is another word for *cycle*? circle

What do the steps in a cycle do? One step follows the next step in a circle.

- Display a pinwheel and place a self-stick dot on it. Slowly turn the pinwheel a full cycle, or circle.

What happens to the dot on the pinwheel? It moves; it goes back to the place where it started.

What has the dot done when it gets back to its

Lesson 65

The Water Cycle

The water moves from the earth to the sky and back to the earth again.

It moves this way over and over.

This is called the **water cycle**.

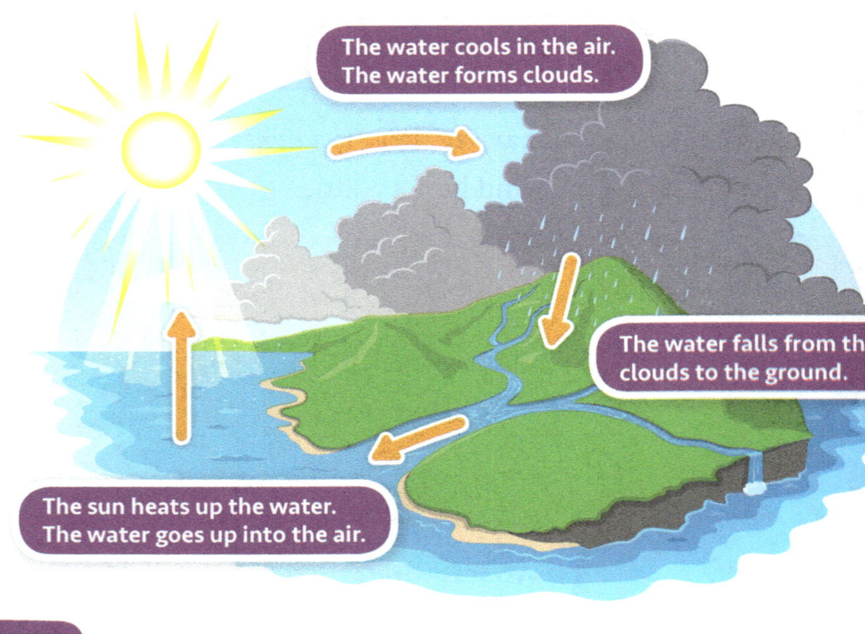

172

starting place? It has made a circle, or one "cycle."

- Point out that if a student blows on the pinwheel, the dot will do the same thing over and over.

There are many kinds of cycles.

What cycles have we already learned about this year? a plant's life cycle, a robin's life cycle, the cycle of the seasons

- Direct attention to the diagram of the water cycle.

What does the sun do to the water? heats it up
Where does the water go when it heats up? up into the air

- Point out that we cannot see water going into the air.

What happens to the water in the air? It cools; the water forms clouds.
What happens to the water in the clouds? It falls to the ground.

After the water falls to the ground, what happens to it? The sun heats up the water and it goes up into the air.

- Direct the students to put their fingers on the arrow pointing up at the sun. Instruct them to trace all of the arrows in the direction they are pointing, until they get back to the arrow that is pointing straight up.

What motion did you make when you traced the arrows? a circle, or cycle

Clouds

The water in the air causes clouds.
A **cloud** is made of tiny drops of water.
The drops of water cool in the air.
A cloud forms when the cooled drops of water gather together.

You observe clouds in the sky.
You may say it is *clear*.
There are no clouds.

You may say it is *partly cloudy*.
There are some clouds.

You may say it is *cloudy*.
There are many clouds.

What causes clouds to form?

173

What causes clouds? water in the air
What is a cloud made of? tiny drops of water
What happened to the tiny drops of water that caused the cloud to form? They have cooled in the air; the cooled drops of water have gathered together.
Where do you observe clouds? in the sky
What does it mean when you say that the sky is clear? There are no clouds.

- Direct attention to the cloud weather symbols.
- Direct attention to the sky in the first circle. Point out that there are no clouds in the sky and that the sun can be clearly seen.

What does it mean when you say that it is partly cloudy? There are some clouds.

- Direct attention to the sky in the middle circle.

How is the partly cloudy sky different from the clear sky? Possible answers: There are some clouds in the sky; you can see only part of the blue sky; you cannot see all of the sun.

What does it mean when you say that the sky is cloudy? There are many clouds.

- Direct attention to the sky in the last circle.

How is the cloudy sky different from the partly cloudy sky? Possible answers: there are many clouds; you can see very little of the blue sky; you cannot see the sun.

- Direct attention to the picture at the top of the page.

What word would you use to describe the sky? cloudy

How does the weather change from day to day? Possible answers: the sky may be clear, partly cloudy, or cloudy; the temperature of the air may be warm or cold; the winds may be calm, light, or strong; it may be dry, or water may fall from the clouds.

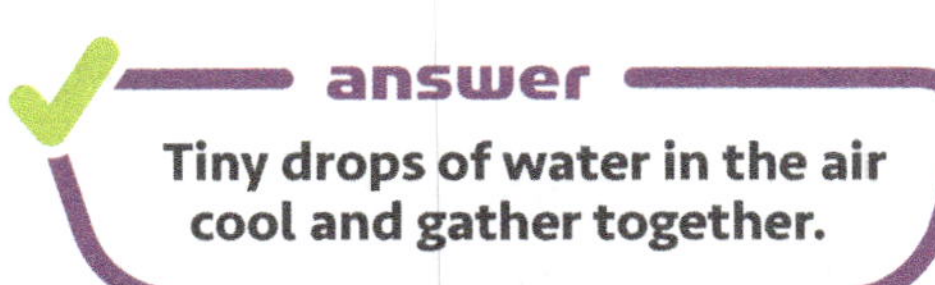
answer
Tiny drops of water in the air cool and gather together.

⚠ **HELPS**

Water Cycle Terminology
To ease the reading level, the terms *precipitation*, *evaporation*, and *condensation* will not be used on this grade level.

ACTIVITY

Partly Cloudy or Cloudy Day Picture
Materials: piece of paper for each student, crayons, cotton balls, glue
Instruct the students to draw trees, flowers, and grass on the bottom of their papers. Direct them to use cotton balls and glue to make a picture of a partly cloudy sky or a cloudy sky.
Invite several students to share their pictures and to explain the type of sky their pictures are depicting.

PREPARATION FOR READING

- Preview and pronounce the words *sleet* and *hail*.
- Direct the students to read pages 174–75 silently to find out what happens to drops of water in clouds.

TEACH FOR UNDERSTANDING

What are clouds made of? many drops of water

What happens when tiny drops in clouds join together? The drops get heavy and fall to the ground.

- Direct attention to the picture with the dog at the top of student page 174.

What kind of water is falling from the sky? rain

What are the raindrops doing when they hit the doghouse and the ground? They are splashing.

What are some words you can use to describe rain? Possible answers: It is misting; it is sprinkling; it is pouring.

- Direct attention to the weather symbol for rain. Point out the drops of water falling from the clouds.
- Display *Bible Verses A* and direct attention to Activities page 153. Read Psalm 135:7 aloud. Point out that this verse tells us that God is in control of the weather. Explain that "vapours to ascend" refers to the water cycle. Point out that *vapours* refers to water and *ascend* means to "go up."
- Direct attention to the Water Cycle on page 172.

Which arrow shows the "vapours ascending," or water going up? the arrow that points straight up from the water to the sun

- Ask a volunteer to read the caption below the arrow aloud.

What other weather terms are in this verse? *rain* ("he maketh lightnings for the rain") and *wind* ("he bringeth the wind out of his treasuries")

- Point out that rain is the water that students most often see falling to the ground, but it is not the only kind of water that falls from the sky.
- Choose a student to read the sentence beside the picture at the bottom of page 174.
- Direct attention to the weather symbol for sleet.

How does the weather symbol for sleet look different from the weather symbol for rain? There are small round balls falling from the clouds, instead of raindrops.

- Direct attention to the picture with the dog showing sleet.

How does this picture look different from the picture with the rain? The tree has no leaves; the sleet is bouncing off the doghouse and the ground.

- Use Background information to share details about sleet.

Water from the Sky

Clouds have many drops of water.

The tiny drops may join together.

This makes the drops get heavy.

The drops fall to the ground.

You may say the water that falls from the sky is *rain*.

You may say it is *sleet*.

Sleet and Hail

Sleet occurs during the winter, while hail usually occurs during the spring and summer. Sleet is smaller than hail, falling as tiny pellets of ice. Hail is formed by layers of water freezing in thunderstorms with strong updrafts. Hailstones that are too heavy to be carried by the updrafts will fall to the earth. The longer hail stays in a cloud, the larger the hailstone when it falls to the earth. Some hailstones can be the size of a baseball or even larger.

You may say the water that falls from the sky is *hail*.

You may say it is *snow*.

✓ **What are four kinds of water that fall from the sky?**

175

Sounds of Water from the Sky

Invite the students to make the sound of rain by tapping the ends of their fingers one at a time on their desks.

Direct the students to make the sound of sleet by tapping their fingernails one at a time on their desks. Sleet is ice, so it makes a little louder sound than rain.

Direct the students to make the sound of hail by keeping all their fingers together and tapping the ends of their fingers on their desks. Hail is usually larger pieces of ice than sleet and makes a louder sound.

Encourage the students to make the sound of snow by tapping very lightly on their desks with the ends of their fingers. Snow is usually very quiet when it falls. It hardly makes any sound at all.

Practice several times. Then say "It is sleeting." Instruct the students to tap the appropriate sound on their desks. Continue the activity with other kinds of precipitation. To add interest, you may add additional weather information to the statement, such as "It is cloudy and snowing" or "There are thunderstorms with hail."

- Direct attention to the picture with the dog at the top of page 175.
  What kind of water is falling to the ground? hail
  What does the hail look like? balls of ice; golf balls; baseballs
  What is the hail doing as it hits the doghouse and the ground? bouncing
  How does the picture with the hail look different from the picture with the sleet? The balls of ice are larger; there are leaves on the tree.
- Use Background information to share details about hail.
  Why is the cartoon dog shaking and covering his eyes with his ears? He is scared.
- Explain that hail can be loud as it hits things on its way to the ground.
- Show a video of a hailstorm (online Teacher Resources).
- Direct attention to the weather symbol for hail.
  How is the weather symbol for hail different from the weather symbol for sleet? The balls of ice falling from the clouds are larger.
- Direct attention to the picture with the dog at the bottom of the page.
  What kind of water is falling from the sky? snow
  Is the air temperature hot or cold when it snows? cold
  How can you tell that it is cold by looking at the picture? The tree has no leaves; there are icicles on the doghouse; there is snow on the ground; the cartoon dog is wearing a hat and scarf.
- Direct attention to the weather symbol for snow.
  How does the weather symbol for snow look different from the other weather symbols? There are six-sided flakes falling from the clouds.
- Explain that snow is made of ice crystals that fall as flakes.
- Direct the students to put their fingers on the weather symbols, as you name each symbol.
- Use the *Sounds of Water from the Sky* activity as you name each type of water from the sky.

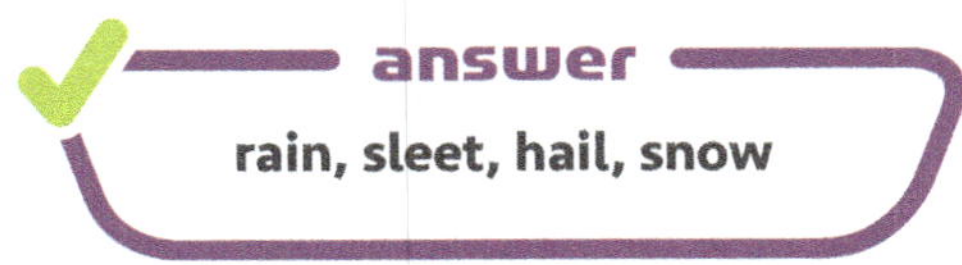

✓ **answer**

rain, sleet, hail, snow

The Water Cycle, pages 159–60
These pages reinforce the concepts in this lesson.

Objectives
- Define *meteorologist*
- Explain what a meteorologist does
- Contrast the trustworthiness of Bible promises with the trustworthiness of scientific predictions **BWS**
- Evaluate the statement that science gives us the most trustworthy information about our world **BWS**
- Practice using the tools of a meteorologist

Materials
- video of yesterday's local weather forecast
- thermometer, outdoor
- small flag

Teacher Resources
- Visual 9.2: *Bible Verses B*

Vocabulary
- meteorologist

INTRODUCTION

- Play a video of a local weather forecast.
  What things does a weather person tell about? Possible answers: today's weather; tomorrow's weather; air temperature; wind; clouds; water from the sky
  Today you will learn more about what a weather person does.

PREPARATION FOR READING

- Preview and pronounce the vocabulary term *meteorologist*.
- Preview and pronounce the word *trustworthy*.
- Direct the students to read pages 176–77 silently to find out what a meteorologist observes.

TEACH FOR UNDERSTANDING

What is a meteorologist? a scientist who studies the weather
What does a meteorologist observe? the weather; what the air outside is like
What does a meteorologist use to observe the weather? his senses; tools
- Direct attention to the pictures.
What science tools are the meteorologists using? a weather balloon; computers
What does the meteorologist do after he observes the weather? He predicts the weather.
What is a prediction? a careful guess at what may happen
Why does a meteorologist predict the weather? to help other people

---

STEM Careers

Lesson 66

Meteorologist

A **meteorologist** is a scientist who studies the weather.

He uses his senses to find out about the air outside.

A meteorologist *observes* the weather.

He uses tools to observe the weather.

A meteorologist *predicts* the weather after he observes it.

He makes a careful guess at what may happen.

A meteorologist predicts the weather to help other people.

176

Knowing about the weather helps people prepare for the future; it helps keep people safe.

⌕ BACKGROUND

Weather Person or Meteorologist?
A weather person is someone who reports the weather forecast. A meteorologist is someone who has training and experience in interpreting data to forecast the weather. Many weather people are also trained meteorologists.

Weather Balloon
Weather balloons carry instruments high into the air. The instruments send information about wind, temperature, humidity, and air pressure to the meteorologists.

Weather Reporting and Green Screens
Some television meteorologists report the weather using a blank green screen and monitors. With green-screen technology the meteorologist stands in front of a blank green screen and the weather graphics appear on the green screen behind him through a process called "chroma keying." He uses monitors in front of him to guide his movements as he reports the weather.

A meteorologist *records* what he observes about the weather.

He records, or writes down, what he finds out.

He makes maps and charts.

A meteorologist then *reports* the weather.

He reports, or tells, what he predicts.

A meteorologist tries to be trustworthy.

Many times what he predicts happens.

Sometimes what he predicts does not happen.

What four things does a meteorologist do?

177

What does a meteorologist record? what he observes about the weather
What does a meteorologist do when he records? He writes down what he finds out; he makes maps and charts.

- Direct attention to the picture of the female meteorologist on page 176. Point out that she is working on a map using a computer. Explain that this map is showing a hurricane, a large storm with very strong winds, over the Atlantic Ocean.

After a meteorologist observes, predicts, and records the weather, what does he do? He reports the weather.

How will hearing a weather report about a hurricane keep people safe? Possible answers: they will know to board up the windows on their homes; they will know to move away from the storm so they do not get hurt or drown.

What does a meteorologist try to be? trustworthy

What does it mean to be trustworthy? to be truthful; honest

Does the weather always happen the way a meteorologist predicts it will happen? No; many times it does, but sometimes it does not.

- Remind the students that a weather prediction is a careful guess at what the weather may be tomorrow.
- Display the statement, "Science gives us the most trustworthy information about our world."

- Explain that the students will test this statement using yesterday's weather forecast.
- Direct the students to Activities pages 161–62. Guide them as they cut out the weather symbols.
- Replay the video forecast used in the Introduction. Instruct the students to complete the weather prediction on Activities page 161 using information from the video.
- Instruct the students to complete Activities page 162. Guide the students as they practice reading the outdoor thermometer and as they use the small flag to tell the type of wind. They will also observe the sky for clouds and water. If there is no precipitation, the students will not put a "Water from the Sky" weather symbol on the chart.
- Direct the students to answer the question on Activities page 162.
- Direct attention to the displayed statement. Guide the students in determining whether science was right or wrong about today's weather prediction.
- Point out that meteorologists are often right but sometimes wrong. Explain that science is useful, but it is not always right.

Who is always right? God

Think of a Bible promise.

How many of you thought of John 3:16?

- Display *Bible Verses B* and direct attention to Activities page 154. Read John 3:16 aloud.

What promise does God make in John 3:16? If you believe that Jesus Christ died for you, God will give you eternal life; if you confess your sin (admit your sin and agree with God about your sin) and repent (turn from your sin), you will be saved.

- Point out that some people are more willing to trust modern science than to trust God and His Word. Explain that we can trust science up to a point because sometimes it is right. However, we can trust God and His Word all the time because God is always right.
- Display *Bible Verses B*. Read John 17:17 aloud. Emphasize the trustworthiness of God's Word.

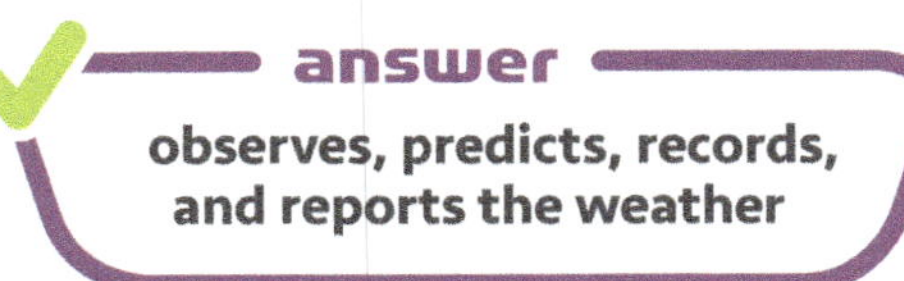

ACTIVITIES

Being a Meteorologist, pages 161–62
These pages were completed during the lesson.

Study Guide, pages 163–65
These pages review the concepts taught in Lessons 65–66. After completion, direct the students to keep them for review for Test 9.

Chapter 9: Weather

Objectives

- Recall what a weather prediction is
- Infer from Proverbs 22:3 that weather predictions help us to prepare for the future 
- Observe, collect, record, and report weather data using tools of a meteorologist
- Identify weather patterns in data collected to predict the weather
- Compare and contrast weather predictions with actual observations

Materials

- See Activities page 167.

Teacher Resources

- Visual 9.1: *Bible Verses A*

INTRODUCTION

- What is a weather prediction? a careful guess at what the weather may be tomorrow
- Display *Bible Verses A* and direct attention to Activities page 153. Read Proverbs 22:3 aloud.
- Explain that a wise person sees what will happen, and he prepares. Point out that when people rule over the earth well by making accurate weather predictions, life is better because people can be prepared.
- Pretend that you and your friend want to play outside. The meteorologist has predicted snow. According to Proverbs 22:3, why should you and your friend wear coats, hats, and mittens when you play outside? You want to be wise. A wise person listens to the weather report and takes steps to protect himself and others. A foolish person continues on and suffers for not being prepared. A prediction for snow means it will be cold. If it snows, it will be wet. A coat, hat, and mittens will keep you warm and dry.

TEACH FOR UNDERSTANDING

- Read aloud the *Science and the Bible* box on student page 179.

  What happened during the Genesis Flood? Water covered the whole earth.

  After the Flood, what promise did God make? God promised that He would never again cover all the earth with water.

- Display *Bible Verses A* and direct attention to Activities page 153. Read Genesis 9:13–15 aloud.

- Is God's promise trustworthy? Yes; God's Word is truth.

  What did God put in the sky to remind us that He always keeps His promises? a rainbow

Lessons 67–68

Weather Watching

Process Skills
- Observe
- Predict

The weather changes every day.

God tells us to rule over the earth.

A meteorologist studies the weather.

When he predicts the weather, he helps others to be safe.

You can be a meteorologist too.

In this Exploration you will observe and predict the weather.

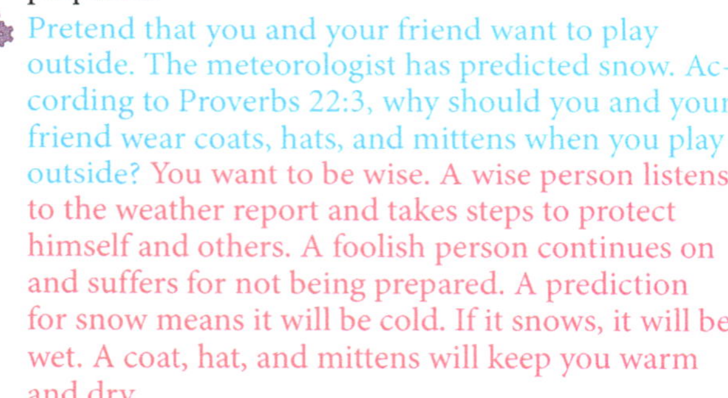

178

⚠ HELPS

Two-Day Lesson

This is a two-day lesson. The Exploration will be started in Lesson 67. Over a period of five days, the students will observe, predict, and record the weather. Lesson 68 will be conducted on Day 5, which will include the discussion of the Conclusions.

Weather Watching

The weather symbols needed for this activity are on Activities pages 171 and 173. The students may keep the pages and cut the columns apart themselves, or you may choose to collect their pages and give the students the weather symbols needed each day.

On the first day of the Exploration, take the students outside to directly observe the weather conditions. Instruct the students to record the weather. Record the information for display.

After the first day, direct the students to complete their charts by going outside to observe the weather, by staying inside and observing the weather from a window, or by getting the information from a weather website.

Each day, allow the students to report the weather.

Science and the Bible

During the Genesis Flood, it rained for many days. Water covered the whole earth. After the Flood, God made a promise. He said that He would never again cover all the earth with water. He put a rainbow in the sky. Noah and his family saw the rainbow. The rainbow reminds us that God keeps His promises.

179

PREPARATION FOR READING

- Direct the students to read student page 178 and Activities pages 167–73 silently before beginning the Exploration.

TEACH FOR UNDERSTANDING

What does the weather do every day? It changes.
What did God tell people to do? rule over the earth
Why does a meteorologist study and predict the weather? He helps other people to be safe.
Can you be a meteorologist? Yes; I can observe and predict the weather.

- Direct attention to the picture on student page 178.
  What type of sky is in the picture? partly cloudy
  What are the children doing in the picture? The girl is reading and recording the temperature of the air; the boy is checking the type of wind.
- Direct attention to the flag the boy is holding.
  What type of wind is there? light wind
- Direct the students to the **Process Skills**.
  What science process skills will you be using in this Exploration? observe and predict
- Review the process skills *observe* and *predict*.
- Direct attention to the **Purpose** on Activities page 167.
  What is the purpose of this Exploration? to observe and predict the weather
- Direct attention to the list of **Procedures**. Remind the students to follow each step. Encourage the students to check off each step as it is completed.
- Point out that this Exploration will be completed over the next five days. See *Two-Day Lesson* under Helps. Students may work individually, in pairs, or in science groups as they observe, predict, and record the weather.
- Discuss their **Conclusions**.

ACTIVITIES

Exploration: Weather Watching, pages 167–73

The students will explore by observing, predicting, and recording the weather.

ASSESSMENT

Rubric

Use the prepared rubric or design a rubric to include your chosen criteria.

Objective
- Recall terms and concepts from Chapter 9

REVIEW

- Material for Test 9 will come from the Study Guides on Activities pages 157–58 and 163–65. You may review any or all of the material during the lesson.
- You may choose to review Chapter 9 by playing *The Water Cycle* review game or another game from the Game Bank (Teacher Resources).

🏆 REVIEW GAME

The Water Cycle
Display a circle of four arrows, similar to Student Edition page 172. Display a sun in the upper left corner of the circle and a cloud in the upper right corner. Prepare five blue raindrops and five white raindrops.

Ask review questions. For each correct answer, place a blue raindrop under the cloud. When all five blue raindrops have been placed under the cloud, move the raindrops to the ground. Remind the students that rain falls to the ground.

Continue asking review questions. For each correct answer, change a blue raindrop to a white one and place the white raindrop under the sun. Explain that these are white since we cannot see the water that goes up into the air. Once all five white raindrops are in place, remove a white raindrop and place a blue raindrop under the cloud for each correct answer. Remind the students that water in the air cools and forms clouds. As time allows, continue asking questions and repeat the cycle.

Variation: Make two complete "water cycles" and two complete sets of "raindrops." Allow two teams to race to complete their water cycles.

NOTES

Objective
- Recall and apply terms and concepts from Chapter 9

ASSESSMENT

Administer Test 9.

NOTES

Materials
- video of different kinds of trains in motion

UNIT INTRODUCTION

- Direct attention to pages 180–81.

  What is the title of Unit 5? "Let's Learn About Energy"

  What do you see in the picture? a train and a bee

- Use Background information to share details about the image. Explain that this is a bullet train, or a train that runs at high speeds using electricity.

  Where have you seen trains? underground, in tunnels, on bridges, crossing streets and roads

  What do you think the bee is doing? flying, racing the bullet train

- Invite students to share about their train experiences.

- Play a video of trains in motion.

- Direct the students to look through Chapters 10, 11, and 12, pages 180–235. Guide the students through several pages, stopping to look at headings and pictures.

  What kinds of things will you learn about in Unit 5? electricity, light, sound, and communication

TEACH FOR UNDERSTANDING

- Direct attention to the word *energy* in the unit title.

  What is energy? It is what is needed to cause change.

  In Chapter 3 you learned that plants need energy.

  How do plants use energy? They use energy to grow and change.

  Where do plants get their energy? from food and the sun

  In Chapter 4 you learned that animals need energy.

  How do animals use energy? They use energy to grow and change.

  Where do animals get their energy? from food

- Explain that some nonliving things need energy to work.

- Direct attention to the picture.

  How does energy help this train? Energy helps it move along the tracks.

180

Unit Photo
The train pictured on pages 180–81 is a passenger bullet train in motion. This is a high speed train that uses electric energy. Bullet trains are popular in Europe and Asia.

LOOKING AHEAD

Science Materials
A pinhole box is required for the Investigation in Lesson 74. (See the illustration on page 196.) Prepare the short end of a shoebox with a pinhole for observation. Along the long side of the box make a hole the same size as the flashlight lens. Prepare one pinhole box for each group of students. Gather enough objects and flashlights for each group. Without telling the students, include one glow-in-the-dark object for each box.

Visit TeacherToolsOnline.com for resources to enhance the lessons.

NOTES

Chapter Objectives
- Describe energy and the properties of light
- Defend the statement that God created light `BWS`
- Apply science skills of observing, predicting, classifying, inferring, recording observations, graphing data, and drawing conclusions
- Draw conclusions as to when objects can be seen

Lesson Objectives
- Identify what energy is
- Identify light as energy
- Defend, using Scripture, the statement that God created light `BWS`
- Describe sources of light as natural or manmade
- Identify cause-and-effect energy relationships

Materials
- lamp with an electric chord

Teacher Resources
- Visuals 10.1; 2.2: *Bible Verse*; *Cause and Effect*

Vocabulary
- light
- electricity
- electrician

CHAPTER INTRODUCTION

- Direct attention to pages 182–83.
  What is the title of Chapter 10? Light
- Direct attention to the picture.
- What is this a picture of? a fair or carnival
- How does the picture give you a clue about what you will learn in this chapter? It shows a fair or carnival with all of its lights on at nighttime.

Big Question
- Direct attention to the Big Question on page 182. Read the question aloud.
- Explain that the answer to the Big Question will be found as they read Chapter 10.

BACKGROUND

Chapter Photo
The chapter photo shows a fair or carnival at nighttime. This picture highlights the use of lights at night.

Light

What is the first thing you do when you enter a dark room?

What helps you see to play a sport at night?

What do you see high in the sky on a bright day?

People, animals, and plants all need light to survive and grow.

God tells people to care for the earth.

Studying light helps you use light well.

Using light well helps you care for the earth.

183

Student Helpers
Create a rotating list of student helpers to turn off the lights each time all the students leave the room. This will encourage students to practice saving energy.

PREPARATION FOR READING

- Direct the students to read page 183 silently to find out how studying light helps people.

TEACH FOR UNDERSTANDING

What are some things you can do because there is light? see in a dark room; play a sports game at night; enjoy light from the sun; enjoy a picnic; read a book

What is one thing people, plants, and animals need to survive and grow? light

How does studying light help people? Light is part of God's creation, and we are to use it well to care for the earth.

PREPARATION FOR READING

- Preview and pronounce the vocabulary term *light*.
- Direct attention to the red heading on page 184.
  What is the heading? "What Light Is"
- Direct attention to the red heading on page 185.
  What is this heading? "Where Light Comes From"
- Direct attention to the black heading on page 185.
  What is this heading? "What Some Scientists Say"
  What will these pages be about? what light is, where it came from, and what some scientists say about light
- Direct the students to read pages 184–85 silently to name the lights that God created.

TEACH FOR UNDERSTANDING

What is energy? what is needed to cause change
Where do plants and animals get energy from? food
Where do you get energy from? food

- Ask a volunteer to read aloud the vocabulary term and definition for *light*.
  What do people need in order to see? light
  What lights did God create? sun, moon, and stars
- Direct attention to the picture.
- What do you see in this picture? the sun, moon, stars, and earth
  Which light did God create for the day? the sun
  Which lights did God create for the night? the moon and stars
- On what day did God create the lights? on the fourth day

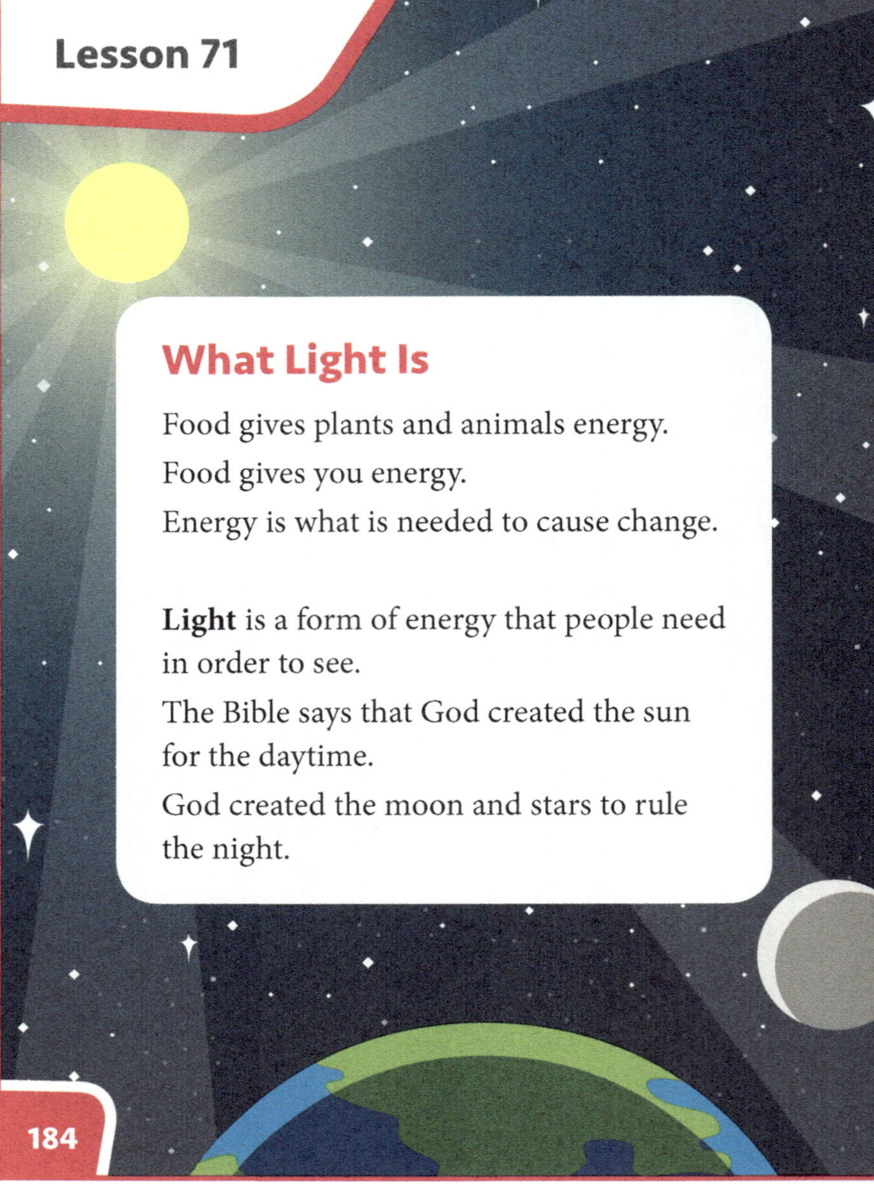

What Light Is

Food gives plants and animals energy.

Food gives you energy.

Energy is what is needed to cause change.

Light is a form of energy that people need in order to see.

The Bible says that God created the sun for the daytime.

God created the moon and stars to rule the night.

Where Light Comes From

Where does light come from?

Where will you look for that answer?

What Some Scientists Say

You look in a science book in the library.

You read that many, many, many years ago there was a tiny speck.

The speck went *bang!*

Then the speck got bigger and bigger.

There was light energy in the speck.

Some scientists explain that light came from this tiny speck.

What is light?

185

- Explain that some scientists have different world-views about where light came from.

Where can you find out about where light came from? in a science book from the library

What do some books say happened many, many, many years ago? There was a tiny speck.

What did the tiny speck do? It went bang!

Then what happened to the speck? It got bigger and bigger.

What was in the speck? light energy

Who believes this is where light came from? some scientists

- Direct attention to the picture.

What does this picture show? light energy coming from the tiny speck that went bang

answer

a form of energy that people need in order to see

The Big Bang

The Big Bang was also mentioned in Chapter 7. This lesson provides students with another opportunity to hear about the opposing view of the Big Bang and how to defend that God created light.

PREPARATION FOR READING

- Preview and pronounce the words *natural* and *manmade*.
- Direct the students to read pages 186–87 silently to find out the difference between natural and manmade light.

TEACH FOR UNDERSTANDING

You can look in the library to find out where light came from. Where else can you look? the Bible

- Display *Bible Verse.* Direct attention to Activities page 175. Ask a volunteer to read Genesis 1:3 aloud.
What does the Bible say about where light came from? God spoke and there was light.
Do some scientists and the Bible say the same thing about where light came from? no
When science and the Bible disagree, should you believe science or the Bible? the Bible
How do you know that you can believe the Bible? The Bible is God's Word and it is true.
Does God ever lie? No; God never lies.
- Direct attention to the picture.
What do you see in this picture? light and darkness
- Read Genesis 1:2 aloud to the students.
What was the earth like before God created light? The earth was empty and completely dark.
- Instruct the students to close their eyes until you ask them to open them again.
What can you see when your eyes are closed? nothing, black, darkness
This gives you an idea of what the earth may have looked like before Creation.
You may open your eyes now. What do you see with your eyes open? everything, light, color
- Tell the students that people are not sure exactly what kind of light God created in Genesis 1:3, but we know there was a difference.
You can praise God for creating light and making you able to see.

What the Bible Says

You look for the answer in the Bible.

You read Genesis 1.

It says, "And God said, Let there be light: and there was light."

Light came from God speaking.

When science does not agree with the Bible, the Bible is right.

The Bible is true because it is God's Word.

186

Make a Source of Light Display

Materials: yellow index cards, black crayons, paper clips (*Note:* Yellow cardstock may be used to make cards.)

Instruct the students to heavily color over the yellow with a black crayon. Direct the students to think of a natural or manmade source of light. Encourage them to be creative and think of things such as lightning bugs, candles, and cell phones.

Guide each student to use a paper clip to etch, or scratch off, the black crayon in the shape of the light source he chose. Wherever he scratches the black crayon off, the yellow "light" will shine through. Direct each student to etch the name of his source on his card. Provide a place to display the sources of light.

What are the two kinds of light? natural and manmade
What is a natural light? a light made by God
Name some examples of natural lights. sun, stars, some animals
How do the sun and stars give light? They make their own light.
Why do natural lights make their own light? God created them that way.
- Direct attention to the second kind of light.
What is a manmade light? a light made by man
Who gave man the mind to make these lights? God
Name some examples of manmade light. flashlights, lamps, candles, car headlights, bedroom lights
Can these objects make their own light? No; man has made the light in each of them.
Display the lamp, turned off and unplugged.
Why do we have lamps? to make light
Is it making light right now? no
What does this lamp need so that it can make light? electricity, to be plugged in, to be turned on

- Overemphasize plugging in the chord and turning on the lamp.
Is this lamp a manmade or natural light? manmade
Why does the lamp have light? It has electricity because it is plugged into the outlet.
- Explain that recognizing that the lamp is on because it is using electricity is called *inferring*.
What does it mean to infer? to use what you know to tell why things happen
- Display *Cause and Effect* as you complete Activities page 177 together.
- Ask for a volunteer to read the definition for *effect* aloud.
Put your finger on the arrow under *effect* and slide it down to the question.
What question is asked to find an effect? What happened?
- Ask a volunteer to read the definition for *cause* aloud.
Put your finger on the arrow under *cause* and slide it down to the question.
What question is asked to find a cause? Why did it happen?
- Remind the students that they can infer a *cause* from an *effect*.
What is the effect given on the page? The lamp is on.
- Ask the students to read silently the two choices for the cause.
What is the cause that you can infer from the lamp being on? The lamp is using electricity.
- Discuss the following examples of cause and effect as time allows.
Effect: The ice cream on the counter melted.
Cause: The air is hotter than the ice cream.
Effect: The sidewalk is hot to walk on.
Cause: The sun's energy has warmed the sidewalk.
- Direct attention to the picture on page 187.
What are some examples of manmade lights? table lamp, flashlight, cell phone screen, car headlights, street lamp
What are some examples of natural light? lightning, fireflies, stars, moon

STEM Careers

Electrician

Many manmade lights use electricity.
Electricity is a form of energy.
Because of electricity you can read a book at night.
A lamp uses electricity to give you light.
Because of electricity you can watch TV.

An **electrician** works with electricity.
He designs electrical systems for homes, stores, and offices.
He puts in the electrical wires.
An electrician keeps them working.
He knows how to work safely.
An electrician's job is very important.

188

PREPARATION FOR READING

- Preview and pronounce the vocabulary terms *electricity* and *electrician*.
- Preview and pronounce the word *systems*.
  What is a STEM Career? a job that uses science, technology, engineering, and math
- Direct the students to read page 188 silently to find out what an electrician does.

TEACH FOR UNDERSTANDING

- Discuss electricity before introducing an electrician's job.
  What do manmade lights use to work? electricity
  What is electricity? a form of energy
  What are some things you can do because there is electricity? read a book at night; watch TV; see to get dressed; cook a meal; use a computer
- Explain that it is important to use safety when working with electricity. Children should never play with electrical wires.
- Explain that this STEM Career job uses electricity.
  What is a person who works with electricity called? an electrician
  What does an electrician do? He designs electrical systems for homes, stores, and offices; he puts in the electrical wires.
- Direct attention to the pictures.
  Where are the two electricians working? They are in a lift, repairing the power lines on a light post.
  How are these electricians being safe while working? They are wearing hard hats and straps to keep them from falling out of the lift.
  How is the electrician on the floor being safe? He is wearing a hard hat, goggles, and gloves.
- Point out that the machine with the three zeros is an electrical tester that helps an electrician know whether the wires have electricity.

ACTIVITIES

Cause and Effect, page 177
This page was completed during the lesson.

Light Energy, page 178
This page reinforces the concepts taught in Lesson 71.

INVESTIGATION

Lesson 72

Observing Light

Light travels through some objects.

Light travels through a window.

Some objects block light.

Light does not travel through your desk.

In this Investigation, you will shine a flashlight on some objects.

You will observe how the light travels through the objects.

You will classify the objects by how much light travels through them.

Process Skills
- Predict
- Observe
- Classify

189

LESSON 72

Objectives
- Predict the amount of light that travels through different objects
- Record observations
- Graph data from observations
- Draw conclusions from the data

Materials
- See Activities page 179.
- object that allows most light through
- object that allows some light through
- object that allows no light through

INTRODUCTION

- Display the object that allows most light to travel through it. (*Note:* The words *transparent, translucent,* and *opaque* will be not be used in this lesson. They will be introduced in Lesson 73.)

  Do you predict that this object allows most light to travel through it, some light to travel through it, or no light to travel through it? Ask for a show of hands.

- Shine the flashlight on the object.

  Does this object allow most light to travel through it, some light to travel through it, or no light to travel through it? most light to travel through it

- Follow the same procedure with the object that allows some light to travel through it and the object that allows no light to travel through it.

PREPARATION FOR READING

- Direct the students to read page 189 and Activities pages 179–81 silently before beginning.

TEACH FOR UNDERSTANDING

- Direct the students to write the names of the objects they will use in the chart on Activities page 180. This activity may be completed in groups, individually, or together as a class.
- Direct attention to and discuss the **Process Skills**. If necessary, refer to Chapter 2.
- Choose a student to read the **Problem** aloud.
- Direct the students to predict how much light travels through each object by completing the **Hypothesis** column on the chart on page 180.
- Instruct the students to complete the **Procedure**. Be sure each student understands that these steps are recorded in the **Observations** column on the chart.
- In Step 4 the students will need to compare their hypothesis with their observations. Record the comparison in the Results column.
- Guide the students as they graph their results on the *Observing Lights* chart on Activities page 181. If each student is observing different objects, the graphs may look different.
- Direct attention to the **Conclusions**.

  How did you decide to classify each object as most, some, or none? by the amount of light that travels through

  Which kind of objects did you find the most of?

ACTIVITIES

Investigation: Observing Light, pages 179–81

ASSESSMENT

Rubric
Use the prepared rubric or design a rubric to include your chosen criteria.

Objectives

- Differentiate between objects that are transparent, translucent, and opaque
- Recognize that a shadow forms when light is blocked
- Explain that a shadow changes when a light source moves

Vocabulary

- transparent
- translucent
- opaque
- shadow

INTRODUCTION

- Discuss some of the objects that were used in the Lesson 72 Investigation.

- If you didn't have a flashlight, how could you tell if an object would allow most light, some light, or no light to travel through it? Accept any reasonable answer.

  Today you will learn how to tell how much light will travel through objects without a flashlight and how to classify these objects.

PREPARATION FOR READING

- Preview and pronounce the vocabulary terms *transparent* and *translucent*.
- Direct the students to read pages 190–91 silently to find the difference between transparent and translucent objects.

TEACH FOR UNDERSTANDING

- Ask a volunteer to read the vocabulary term *transparent* and its definition aloud.
- Invite students to identify transparent objects from the Lesson 72 Investigation.

  What are some other objects that are transparent? air, water, clear plastic, and clear glass windows

  Does light travel through these objects? yes, because they are transparent

- What are some transparent objects in the room? the windows, glasses, water bottle, magnifying glass
- Discuss the pictures.
- How do you know the piggy bank is transparent? You can see the coins clearly through the bank.
- How do you know the drinking glass is transparent? You can see the boy's hand clearly.
- How can you tell if an object is transparent without a flashlight? You can see other objects clearly through transparent objects.

Lesson 73

Light and Objects

Transparent

Light can travel through some objects. Anything that allows light to travel through it is **transparent**.

Air and water are transparent. Light travels through them. Clear plastic and clear glass windows are transparent. Light travels through them.

190

⚗ ACTIVITY

How Much Light Passes Through?

Materials: 4" × 6" cards, 4" × 6" pieces of plastic wrap, page protector plastic, wax paper, colored cellophane, dark construction paper, cardboard

Cut each index card to form a frame and glue one of the materials to the back of the frame, covering the opening. Number each frame. Place the frames in a learning center for the students to examine and determine whether each is transparent, translucent, or opaque. Provide a numbered key for self-checking. (*Note:* Additional frames may be made with other materials.)

Translucent

Anything that allows some light to travel through it is **translucent**.

Tissue paper is translucent.

Sunglasses and wax paper are translucent.

Some light travels through these objects.

Some light is blocked.

191

- Ask a volunteer to read the vocabulary term *translucent* and its definition aloud.

  What does it mean to say something is translucent? It allows only some light to travel through it.

  Name some translucent objects from the Lesson 72 Investigation. Answers will vary.

  What other objects are translucent? tissue paper, sunglasses, wax paper

- Direct attention to the sunglasses.

  How can you tell that the sunglasses are translucent? You can see the frames of the sunglasses through the lenses, but not clearly.

- Direct attention to the pencil holder.

  Is this pencil holder transparent or translucent? translucent; because you can see the pencils through the cup but not clearly

- Direct attention to the picture with a green leaf.

  Can you see what is on top of this leaf? a frog

  Can you tell what color or type of frog this is? No, because the leaf is translucent. All of the light does not travel through it, and you cannot see all the details of the frog.

  What is the difference between a transparent and translucent object? the amount of light that travels through them

  What happens to some of the light that travels through a translucent object? It is blocked.

PREPARATION FOR READING

- Preview and pronounce the vocabulary terms *opaque* and *shadow.*
- Direct the students to read pages 192–93 silently to find out how a shadow is formed.

TEACH FOR UNDERSTANDING

What does it mean to say that something is opaque? It does not allow any light to travel through it.

Name some opaque objects from the Lesson 72 Investigation. Answers will vary.

What other objects are opaque? a book, your body, desk, door, table, wall

- Direct attention to the picture of the tree trunk.

Is the sunlight traveling through the tree trunk? No, the tree is blocking the sunlight. It is opaque.

Can you see what is behind the tree trunk? No; the tree trunk is opaque.

- Direct attention to the blocks.

How do you know the blocks are opaque? No light is shining through; you cannot see anything behind them.

- Direct attention to the rocks.

Are the rocks opaque? Yes; there is no light shining through them.

Opaque

Some objects block all light.

Anything that does not allow light to travel through it is **opaque**.

Light does not travel through this book.

Light does not travel through you.

All light is blocked.

Your book and your body are both opaque.

192

Shadows

A **shadow** is a dark shape formed when an object blocks light.

The place where the shadow forms depends on where the light is.

Shine a light on the right side of an object.
The shadow forms on the left side.

Shine a light on the left side of an object.
The shadow forms on the right side.

What kind of object blocks all light?

193

Have you ever been frightened by a shadow? Maybe the shadow looked big and dark. But when you turned on the light, it was an object that you knew.

How is a shadow formed? when an object blocks light

Where does a shadow form? It depends on where the light is.

- Direct attention to the picture of the egg. Explain that shadows change when the light traveling to the object changes places.

  Put your finger on the top picture of the egg; where is the light? on the right side

  Where do you see the egg's shadow? on the left side of the egg

  Put your finger on the bottom picture of the egg; where is the light? on the left side

  Where do you see the egg's shadow? on the right side of the egg

  Why did the egg form a shadow? The egg blocked the light.

  Where does a shadow form? on the opposite side of the light

answer
opaque objects

Make Shadows

Materials: flashlight, stuffed toy, token

Darken the room. Shine the flashlight from the right of the toy.

Where is the shadow? on the left

Shine the flashlight from behind the toy.

Where is the shadow? in the front

Place the token near the toy. Instruct a volunteer to move the flashlight so the toy's shadow falls on the token. Move the token several times, allowing other volunteers to move the flashlight so that the shadow falls on the token. With each move, discuss how the student knew to hold the flashlight on the opposite side of the toy from where the token was.

PREPARATION FOR READING

- Direct the students to read pages 194–95 silently to find out how shadows change throughout the day.

TEACH FOR UNDERSTANDING

What shape does a shadow have? the shape of the object that blocks the light

- Direct attention to the picture of the boy playing tennis.

What shape is the shadow? a boy with a tennis racket in his hand, ready to hit the tennis ball

Where is the sun shining from? from behind the boy because you can see a shadow is in front of him

- Follow the same procedure with the other pictures on the page.

Where is the sun in the morning? low in the eastern sky

What kind of shadows does the morning sun make? long shadows

A shadow has a shape like the object that blocks the light.

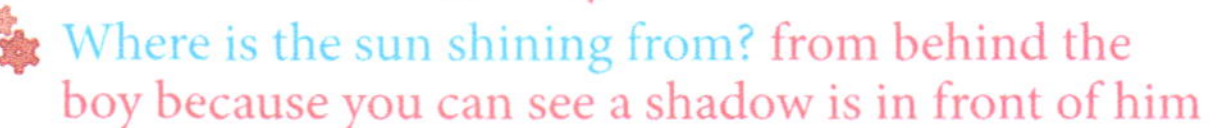

The sun is in different places during the day.

You can see shadows change.

In the morning the sun is low in the eastern sky.

This makes long shadows.

194

In the middle of the day, the sun is high above.
The shadows are short.

By evening, the sun is low in the western sky.
The shadows are long again.
Now the shadows are in a different place than they were in the morning.

During the day, when is your shadow the shortest?

195

- Direct attention to the diagram.
  Put your finger on the picture that shows the sun in the morning. Is the shadow long or short? long
  How does the boy's shadow change throughout the day? from long to short to long again
- What is giving the light for the boy's shadows? the sun is
- How is the shadow changing? The shadow changes as the sun moves across the sky or as the earth rotates.
- What is different about the long shadows in the pictures? They are facing different directions.
  Where is the sun by evening time? low in the western sky
  What kind of shadows does the evening sun make? long shadows
  Put your finger on the sun in the middle of the day. What kind of shadow does the boy see? a short shadow
  When does the boy see his long shadows? in the morning and evening
- Direct attention to the picture of the two people walking on the snow on student page 194.
  Can you tell what time of day it is by looking at their shadows? Yes; it is morning (or evening) because their shadows are long.
- Direct attention to the Big Question on page 182.
- What causes shadows to change? Shadows change as the light that hits the object changes places.

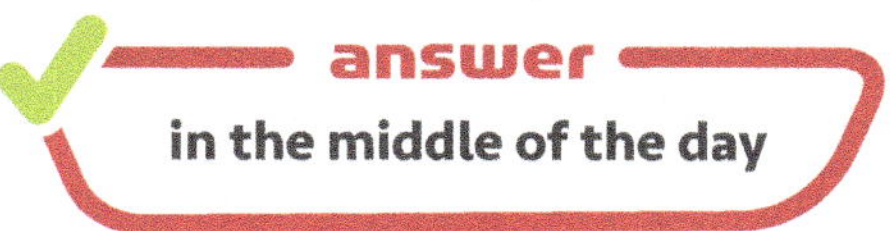
answer

in the middle of the day

ACTIVITIES

Study Guide, pages 183–85
These pages review the concepts in Lessons 71–73. After completion, direct the students to keep them for review for Test 10.

Objectives
- Predict if objects can be seen if light is available to illuminate them or if they give off their own light
- Observe objects in a pinhole box
- Infer that objects can be seen if light is available to illuminate them or if they give off their own light

Materials
- See Activities page 187. (*Note:* Use a glow-in-the-dark plastic star, a glow rock, or a glow-in-the-dark sticker so that one object will give off its own light.)

Vocabulary
- illuminate

INTRODUCTION

What happens when you turn on the light in a room? It lights up the room.

 Do all objects give off their own light? no
Today you will predict whether an object gives off its own light or needs another light so it can be seen.

PREPARATION FOR READING

- Preview and pronounce the vocabulary term *illuminate*.

> The word *illuminate* will be difficult for some students. Try to use it as often as possible in this lesson and the next lesson.

- Direct the students to read page 196 and Activities pages 187–89 before beginning.

TEACH FOR UNDERSTANDING

What does the word *illuminate* mean? light up
What illuminates our daytime? the sun
What illuminates a desk? a table lamp or overhead light

- Direct attention to and discuss the **Process Skills**. If necessary, refer to Chapter 2.
- Choose a student to read the **Problem** aloud.
- If one of the objects is being replaced, direct the students to draw a line through it on the chart on Activities page 188 and write the new object above it. For example, draw a line through the word *star* and write the word *sticker* above it.
- Instruct the students to record their predictions, or **Hypotheses**, in the first two columns on the chart.
- Instruct the students to follow the **Procedure** and record their **Observations** in the last two columns of the chart.

INVESTIGATION

Lesson 74

Illuminate Objects

When you turn a light on in a room, it **illuminates**, or lights up, the room.
The sun illuminates our daytime.
A table lamp illuminates a desk.

In this Investigation, you will predict whether an object is illuminated by its own light or by another light.

You will use a pinhole box to observe different objects.

196

- Direct the students to complete the **Conclusions**.
- Discuss the Conclusions together.
- Why was an object illuminated without the flashlight? The object gave off its own light.
- Why were some objects illuminated with the flashlight? The objects did not give off their own light.

ACTIVITIES

Investigation: Illuminate Objects, pages 187–89

ASSESSMENT

Rubric
Use the prepared rubric or design a rubric to include your chosen criteria.

Lesson 75

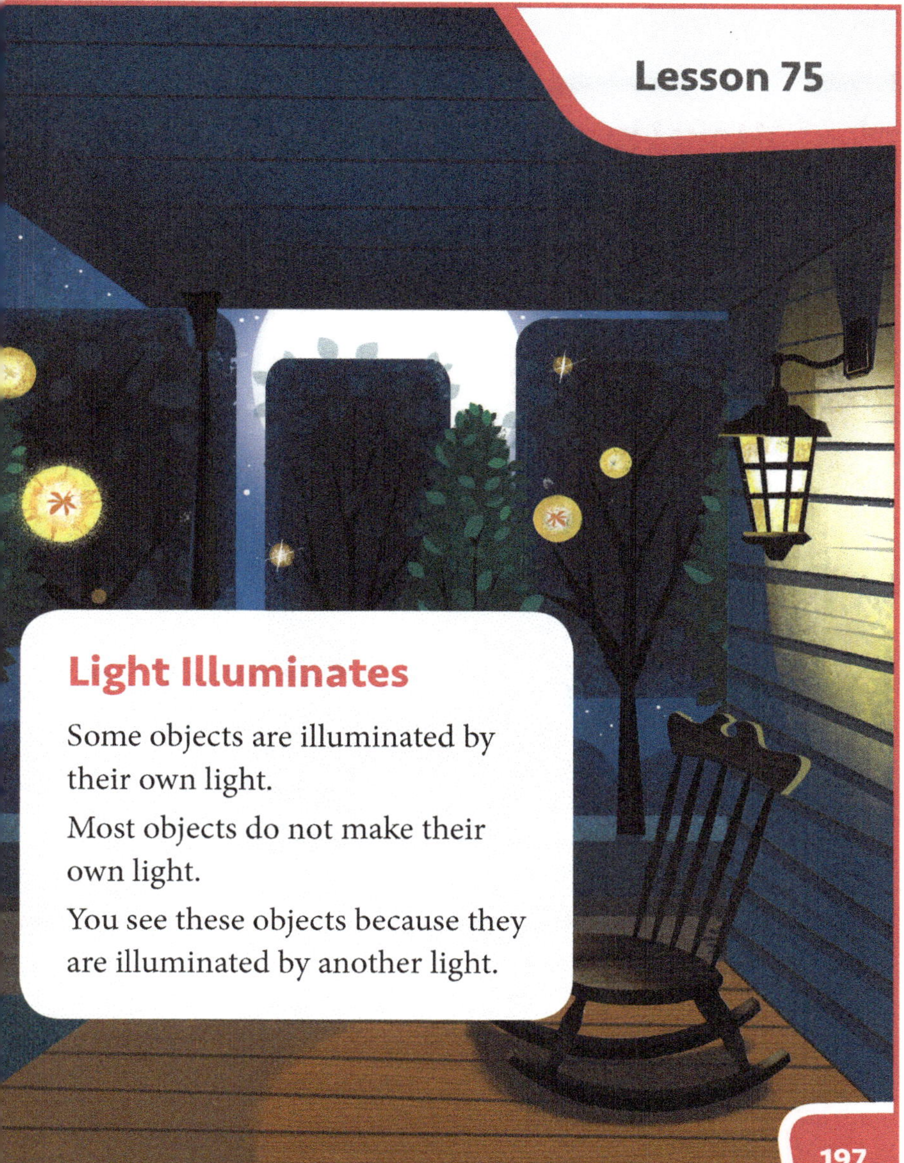

Objectives
- Recall that objects can be seen if light is available to illuminate them or if they give off their own light
- Identify that light travels in a straight line
- Infer that mirrors reflect light

Materials
- mirror

Vocabulary
- mirror

INTRODUCTION

What allows you to see most objects? light

What two ways can you see something? It is illuminated by its own light, or it is illuminated by another light.

Today you will learn how light allows you to see other objects.

PREPARATION FOR READING

- Ask a volunteer to read aloud the red heading.
- What is this lesson about? how light illuminates
- What does *illuminate* mean? to light up
- Direct the students to read page 197 silently to find out how light allows people to see most objects.

TEACH FOR UNDERSTANDING

Why are you able to see objects that do not make their own light? another light illuminates them

- Direct attention to the picture.

Put your finger on the rocking chair. Is it illuminated by its own light or by another light? another light

Put your finger on the moon. Can you see it because it is illuminated by its own light or another light? It is illuminated by the sun.

Are the trees illuminated by their own light? No; the trees are illuminated by the moon's light.

Why can you see the stars in the sky? They are illuminated by their own light.

Point to the circles of light by the trees. What is inside each circle? a firefly

How are the fireflies illuminated? by their own light

PREPARATION FOR READING

- Preview and pronounce the vocabulary term *mirror*.
- Preview and pronounce the word *image*.
- Ask a volunteer to read the red heading.
- ❋ How do you think light travels?
- Direct the students to read pages 198–99 silently to find out how light travels. (*Note:* The students do not need to read the *Health and Safety* box.)

TEACH FOR UNDERSTANDING

How does light travel? in a straight line

What happens when light hits an object? The object reflects the light.

❋ What does the word *reflect* mean? to bounce off

How are you able to see other objects? Your eyes see the light that reflects off other objects.

- Direct attention to the picture.

❋ What can the girl see? her doll

❋ Why is she able to see her doll? It is illuminated by the light from the flashlight.

Are there other objects around the girl? yes; a picture on the wall and a soccer ball

Does the flashlight illuminate these objects? No; the light is traveling in a straight line; they are not in the path of the light.

❋ What does the girl need to do in order to see the other objects? Point the flashlight at them.

- Read aloud the *Health and Safety* information as the students follow along.

Who made your eyes? God

❋ What happens to your eyes when you walk outside on a sunny day? They begin to close because the sunlight is so bright.

Should you look directly at the sun? No; the sunlight can damage your eyes.

How can you protect your eyes from sunlight? wear a hat and sunglasses

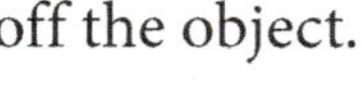

How Light Travels

Light travels in a straight line.
Light hits an object.
Then the object reflects the light.
Your eyes see the light that reflects off the object.

Health and Safety

God made your eyes. You should take care of them. Sunlight travels to your eyes. It can reflect off shiny things. Looking at the sun can damage your eyes. You should wear a hat and sunglasses. Both of these will keep most of the sunlight from your eyes.

198

Shiny and smooth objects reflect light well.

The shiny parts of cars reflect sunlight.

The rain on a road reflects light.

The moon reflects the light of the sun.

A **mirror** is shiny glass that reflects light.

Light travels from an object to the mirror.

The mirror reflects the image back to your eyes.

Light is a gift from God that allows us to observe His world.

Learning about light will help us care for the earth.

How does light travel?

199

What kind of objects reflect light well? shiny and smooth objects; shiny parts of a car; water or rain on the road

How does the moon give off light? It reflects the light of the sun.

- Direct attention to the top picture.

 What is making the road shiny? rain, or water

Why can you see light on the road? The wet road is reflecting the light from the cars' and trucks' headlights.

What other objects in the picture are reflecting light? the metal of the cars and trucks; the car mirrors

What is another name for a shiny glass that reflects light? a mirror

- Display the mirror.

How is a mirror useful? It helps you see yourself to make sure your hair is combed correctly and that your clothes are on properly.

How does a mirror reflect light? Light travels from an object to the mirror; the mirror reflects the image back to your eyes.

What is an image? a likeness

- Direct attention to the picture with the cat.

Why can the mirror reflect the cat's image? It is shiny and smooth.

What did God create on Day Four so that we can observe His world? light

How can we obey God by learning about light? learning about light helps us care for the earth

answer

in a straight line

Light Travels in a Straight Line

Materials: meter stick

Choose two students to perform the activity. Direct each student to hold one end of a meter stick. Explain that one student is the source of light and the other is where the light is shining.

Direct the students to move around the room. Ask them to try to move around corners, or ask one to try to hide behind something.

Can light shine around a corner? no

Emphasize that light travels in a straight line. It cannot bend around objects.

ACTIVITIES

Study Guide, pages 191–92

These pages review the concepts in Lesson 75. After completion, direct the students to keep them for review for Test 10.

Objective
- Recall terms and concepts from Chapter 10

REVIEW

- Material for Test 10 will come from the Study Guides on Activities pages 183–85 and 191–92. You may review any or all of the material during the lesson.
- You may choose to review Chapter 10 by playing *Shining Answers* review game or another game from the Game Bank (Teacher Resources).

🏆 REVIEW GAME

Shining Answers

Materials: 2 flashlights; words for display: sun, light, energy, transparent, translucent, opaque, shadow, illuminate, give off its own light, natural, manmade

Divide the students into two teams. Give each team a flashlight. Darken the room. Alternate asking questions to each team. Ask questions that can be answered using one of the words displayed. The student answering the question should shine the flashlight on the correct answer and say the word. If he is correct, the team gets a point. The team with the most points wins.

NOTES

Objective
- Recall and apply terms and concepts from Chapter 10

ASSESSMENT

- Administer Test 10.

NOTES

Chapter 10: Light

Chapter Objectives
- Identify the characteristics of sound
- Relate sound and the human ear to God's creational design **BWS**
- Apply the steps of the scientific method to determine how the thickness of a rubber band affects pitch
- Apply the engineering design process to make a stringed musical instrument with strings of varying pitch

Lesson Objectives
- Recall hearing as one of the five senses
- Identify sound as a form of energy
- Identify sound as a vibration that can be heard
- Infer different ways sound can be made

Materials
- rubber band

Vocabulary
- sound
- vibrate

CHAPTER INTRODUCTION

- Direct attention to pages 200–201.
  What is the title of Chapter 11? "Sound"
- Direct the students to look through the chapter.
  What are we going to be learning about in Chapter 11? how sound is made; how sound moves; how sound changes
- Instruct the students to close their eyes and listen very carefully to the sounds around them. Make a few sounds in addition to the normal classroom sounds.
- Direct the students to open their eyes.
- Invite the students to share what they heard.
- Explain that in this chapter the students will learn about sound and how it is made.

Big Question
- Direct attention to the Big Question. Choose a student to read the question aloud.
- Explain that the answer to the Big Question will be found as they read Chapter 11.

Materials for Lesson 81
Precut the rubber bands needed to conduct the Investigation in Lesson 81. Cut enough rubber bands into 10 cm lengths for the students to work in pairs for this Investigation.

Materials for Lesson 82
The following materials will be needed for each student as the students design and build a musical instrument that can be plucked: safety goggles, square or rectangular containers of various sizes (plastic containers, shoe boxes, tissue boxes, etc.), rubber bands of varying lengths and thicknesses, string, yarn.

Note: The instruments made in Lesson 82 will be used in Lesson 90.

Visit TeacherToolsOnline.com for resources to enhance the lessons.

Sound

Boom!

Drip.

Splash!

Ring.

You hear noises all around you.

What are you hearing?

How do you hear them?

We will find out!

201

PREPARATION FOR READING

- Direct the students to read page 201 silently to identify some words for sounds that they hear around them.

TEACH FOR UNDERSTANDING

- Direct attention to the picture.
- What is this a picture of? fireworks over the United States Capitol building
- When have you seen fireworks like these? Possible answers: on the Fourth of July; on New Year's Eve
- What sound do fireworks like these make? boom; bang
- Invite the students to share their experiences viewing fireworks displays, the sounds they heard, and their opinions of the sounds.
- Ask a volunteer to read the first four words aloud.
- Which word describes the sound a bell makes? ring
- What word describes the sound a frog makes jumping into a pond? splash
- What else makes a splashing sound? Possible answers: someone in a pool; someone jumping in a puddle
- What makes the "drip" sound? Possible answer: water

  What do you hear all around you? noises; sounds
- Explain that in this chapter the students will learn about many kinds of sounds, how they are made, and how the students hear them.

⚠ **HELPS**

Sound and Noise

Noise is sound. However, noise is defined as sound that is loud, harsh, unwanted, or surprising. There will be no distinction made on this grade level between the words *sound* and *noise*.

PREPARATION FOR READING

- Preview and pronounce the vocabulary terms *sound* and *vibrate*.
- Direct the students to the glossary to find the definition for each term. Ask volunteers to read the definitions aloud.
- Direct the students to read pages 202–3 silently to find out how they can make something vibrate.

TEACH FOR UNDERSTANDING

What are the five senses that God gave you? sight, touch, smell, taste, hearing

Why did God give you five senses? to observe His world

What do you use your ears to do? hear sounds

- Direct attention to the pictures.

What sounds are shown in these pictures? a barking dog, a ringing phone, a laughing friend, cheering people

What does hearing sounds help you to do? observe God's world

- Direct attention to the dog.

What can you observe about a barking dog? Possible answers: He is happy, hungry, lonely, or warning of danger.

What does a ringing phone mean? Someone is calling; someone wants to talk to you.

Is a laugh a happy or a sad sound? happy

- Direct attention to the picture of the girls on the playground.

Why are the girls laughing? They are having fun.

- Direct attention to the people cheering.

Why are these people cheering? Possible answers: they are excited; they want their team to win.

Lesson 78

How Sound Is Made

You hear sounds with your ears all the time.

A dog barks.

A phone rings.

A friend laughs.

People cheer.

Hearing sounds helps you observe God's world.

202

Sound is a form of energy that you can hear.
Sound is made when something vibrates.
To vibrate means to move back and forth
very quickly.
Sound is a vibration that you can hear.

How can you make something vibrate?
You can hit it or rub it.
You can shake it or pluck it.
You can blow air through it.

203

What is *sound*? a form of energy that you can hear; a vibration that you can hear

What is energy? what is needed to cause change

How is sound made? Sound is made when something vibrates.

What does *vibrate* mean? to move back and forth very quickly

- Direct the students to place their fingers gently on the front of their throats. Instruct them to hum or sing aloud a song together.

Did you feel any movement under your fingers when you hummed or sang the song? yes

- Explain that the movement the students felt was made when parts of their throats vibrated as they hummed or sang. Point out that they can talk and sing because those parts of their throats vibrate.

- Stretch a rubber band between your thumb and index finger. Instruct the students to watch the rubber band as you pluck it.

What happens to the rubber band when I pluck it? It vibrates; it makes a sound.

What are some ways that you can make something vibrate? hit it, rub it, shake it, pluck it, blow air through it or on it

- Direct attention to the pictures.

How is the boy in the red striped shirt making sound? He is blowing air through a flute.

- Instruct the students to infer the way each child in the remaining pictures is making sound. Clockwise: hitting a drum; rubbing violin strings; shaking maracas; plucking a guitar

answer

**a form of energy that you can hear;
a vibration that you can hear**

ACTIVITIES

Making Sounds, page 195
This page reinforces the concepts taught in the lesson.

Sounds I Hear, page 196
This Enrichment activity may be used to reinforce the concepts taught in Lesson 78.

Objectives

- Identify that sound travels in waves
- Observe that sound travels in all directions
- Observe that sound travels through matter
- Relate sound and the human ear to God's creational design **BWS**
- Relate sound to the vibration of materials

Materials

- objects or technology that can make sound
- rock
- balance scale
- tuning fork (C-256 Hz, or lower)
- rubber striking hammer
- plastic beaker filled with water
- bowl
- plastic wrap
- rubber band
- sugar (or salt)
- whistle

Teacher Resources

- Visual 11.1: *Bible Verses*

Vocabulary

- matter

INTRODUCTION

- Instruct the students to close their eyes. Stand in different places in the room. Make a few sounds or use technology to play a few sounds, such as the ticking of a clock, the dropping of a coin, and the chirping of a bird. Each time ask the students to point in the direction that they think the sound came from.
- Instruct the students to open their eyes. Ask volunteers to tell you what made each sound (clock, coin, bird).
- Show the students where you were located when each sound was made or played.
- If I was standing here, how did everyone in the room hear the sound? Sound moves in all directions.

  Today we will learn how sound moves.

PREPARATION FOR READING

- Preview and pronounce the words *sound waves*.
- Direct the students to read pages 204–5 silently to find out how sound moves or travels.

Lesson 79

How Sound Moves

Sound Waves

When you make a sound, you can hear it.

You cannot see sound moving.

But you can hear it moving.

Sound moves, or travels, in waves.
Sound waves travel up and down.

Sound waves travel to the right and the left.

TEACH FOR UNDERSTANDING

What happens when you make a sound? You can hear it.
Can you see sound moving or traveling? no
How do you know sound travels? You can hear it.
How does sound travel? in waves

- Direct attention to the first picture.
  What is the boy doing? hitting a drum
- Point out the curved lines.
- What do you think the curved lines are showing you? the direction that the sound is traveling; the direction of the vibrations caused by the drum being hit
  What direction is the sound traveling? up and down
- Direct attention to the picture of the second drummer.
  What direction is the sound traveling? right and left; sideways

Sound waves travel in front of you
and behind you.

Sound waves travel in all directions at
the same time.

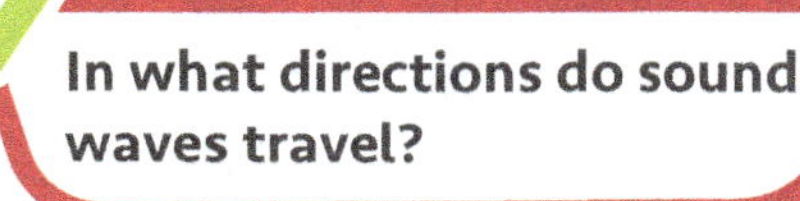

**In what directions do sound
waves travel?**

205

- Direct attention to the third drummer.
  What direction is the sound traveling? in front and behind
  What directions does sound travel in at the same time? all directions
- Explain that the pictures on these pages show the movement of sound at the same time. Sound does not travel just up and down, left and right, or backward and forward. Sound moves in all these directions at once.

answer

**in all directions; up and down, right and
left, in front of you and behind you**

BACKGROUND

Sound Waves

Sound is a vibration that can be heard. The vibration starts a sound wave, which moves outward in all directions. The vibration makes the surrounding molecules vibrate and bump into other molecules, keeping the sound moving on and spreading out. When there is no more energy for the molecules to vibrate, the sound stops.

PREPARATION FOR READING

- Preview and pronounce the vocabulary term *matter*.
- Direct the students to read pages 206–7 silently to find out what moving air makes the inside of your ear do.

TEACH FOR UNDERSTANDING

What does sound travel through? matter
What is *matter*? anything that has weight and takes up space
Where is matter? all around you
What are some examples of matter? my toys, my food, my body, the air I breathe

- Direct attention to the picture.

What in the picture is matter? Everything in the picture has weight and takes up space.

- Display the rock. Place the rock on a desk.

Does a rock take up space? yes

- Point out that the rock is taking up space on the desk. Nothing else can sit in exactly the same spot because the rock is there; the rock takes up space.

Can a rock be weighed? yes

- Use a balance scale to weigh the rock.

Do you take up space? yes

Can you be weighed? yes

What kind of matter do most of the sounds we hear move through? air

Why can you hear the sound of my voice? because the sound is traveling through the air

Can you see sound waves traveling through the air? No; sound waves cannot be seen because air cannot be seen; because air is invisible, sound waves are invisible.

> Teaching for page 207 begins here.

Who does the Bible say made your ears? God

- Display *Bible Verses*. Direct attention to Activities page 193. Choose a student to read Proverbs 20:12 aloud.

What did God design your ears to do? to hear sound
What makes sound? vibrating, or moving, air
What does the moving air make the inside of your ear do? vibrate
Where does your ear send the message that you heard? your brain
What does your brain tell you? what you heard

- Display *Bible Verses*. Direct attention to Activities page 193. Read Psalm 139:14 aloud.

How did God make you? fearfully and wonderfully

Sound and Matter

Sound travels through matter.

Matter is anything that has weight and takes up space.

Matter is all around you.

Your toys are matter.

Your food is matter.

Your body is matter.

Even the air you breathe is matter.

Most of what we hear is sound that moves through the air.

206

Hearing

A person hears sound when sound waves enter the outer ear and travel through the ear canal to the eardrum. The sound waves cause the eardrum and the three small bones of the middle ear to vibrate. Sound moves into the inner ear as the vibrations from the bones cause the fluid in the cochlea to vibrate. These vibrations are sent as messages to the auditory nerve, which sends them to the brain. The brain understands these messages as sound and identifies what the sound is.

Sound and Your Ears

Proverbs 20:12 tells us that God made your ears.

He designed your ears to hear sound.

Vibrating, or moving, air makes sound.

The moving air makes the inside of your ear vibrate.

Your brain knows what the vibrating inside your ear means.

Your brain tells you what you heard.

You are "fearfully and wonderfully made" (Psalm 139:14).

207

- Explain that to be "fearfully and wonderfully made" means that people have been designed by God in a wonderful way. Because of God's great design, we can hear. Because of God's great design, we praise, or thank, Him for who He is. People are the most important part of God's creation. We respect Him for making us just the way He planned.
- Direct the *Vibration and Sound* activity to demonstrate what vibration is and how vibration causes the sounds the students' ears hear.
- What did you observe about the tuning fork after it was hit? It was vibrating; it made a sound.
- When the tuning fork was hit the first time, why could you hear it hum or ring? Hitting the tuning fork caused it to vibrate, which caused the air around it to vibrate. The sound was caused by vibrating air. The vibrating air made the inside of my ear vibrate, and my brain told me what I heard.
- What did you observe when the tuning fork was placed in the water? The water moved or splashed.
- What caused the water to move? the vibrating tuning fork
- In what direction did the water move when the tuning fork was placed in the water? all directions
- Remind the students that sound waves travel in all directions, just like the water traveled in all directions.
- Explain that we cannot see sound, but we can see what it does. We can see how sound affects other materials.
- Direct the *Sound and Vibration* activity to demonstrate that sound can make materials vibrate.
  What happened to the sugar when the whistle was blown? It moved.
- Why did the sugar move? The sound energy from the whistle caused the plastic wrap to vibrate. The vibrating plastic wrap moved the sugar.
- Remind the students that something vibrating can make sound waves and sound waves can make something vibrate.

ACTIVITIES

Vibration and Sound

Materials: tuning fork (C-256 Hz, or lower), rubber striking hammer, plastic beaker filled with water *Note:* Use a plastic beaker for this activity. If a tuning fork touches the sides of a glass beaker, the glass may break.

Strike a tuning fork with a rubber striking hammer. Striking the tuning fork against a hard surface may damage the fork. If you do not have a rubber striking hammer, use the rubber sole of a shoe. You may repeat this several times, allowing the students to touch the tuning fork after it has been struck. Invite the students to share what they observe.

Place a plastic beaker of water where all students can observe it. Strike the tuning fork. As it is ringing, place the tuning fork in the water.

Sound and Vibration

Materials: bowl, plastic wrap, rubber band, sugar (or salt), whistle

Stretch a piece of plastic wrap tightly across the top of a bowl. Use a rubber band to secure the plastic wrap. Sprinkle a small amount of sugar or salt on the plastic wrap. Ask a volunteer to stand close to the bowl and blow the whistle without blowing air onto the plastic wrap. Instruct the students to observe the sugar or salt when the whistle is blown.

ACTIVITIES

Bible Verses, page 193
These Bible verses will be used in the teaching of this chapter.

Study Guide, pages 197–98
These pages review Lessons 78–79. After completion, direct the students to keep them for review for Test 11.

Objectives
- Identify the characteristics of volume
- List examples of loud and soft sound
- Identify the characteristics of pitch
- List examples of sound with high and low pitch
- Explain two ways that sound changes

Materials
- stringed instrument *Note:* A virtual violin or other stringed instrument may be used in place of an actual instrument (see Teacher Resources).

Vocabulary
- volume
- pitch

INTRODUCTION

- In a very soft voice, read the following question.
  Have you ever had a hard time hearing someone who was talking to you?
- In a very loud voice, read the following statement.
  Sometimes you do not have any trouble hearing someone who is talking to you.
- In a normal voice, read the following statement.
  Today you will learn about different ways that sound changes.

PREPARATION FOR READING

- Preview and pronounce the vocabulary term *volume*.
- Direct the students to read pages 208–9 silently to find out what sounds will not hurt their ears. The *Health and Safety* box will not need to be read at this time.

TEACH FOR UNDERSTANDING

Can sound change? yes
What is *volume*? how loud or soft a sound is
Are loud sounds easy or hard to hear? easy
What is an example of a loud sound? a shout, fireworks, a marching band, a siren
What can loud sounds do to your ears? hurt them
- Direct attention to the picture.
- Is the boy hearing a loud or soft sound? loud
- Explain that the boy in the background is using a bullhorn.

Lesson 80

How Sound Changes

Volume

Sound can change.

Volume is how loud or soft a sound is.

Loud sounds are easy to hear.
A shout is a loud sound.
Some loud sounds may hurt your ears.

208

How do you know the boy is hearing a loud sound? He looks scared; he is startled; he is covering his ears.

Why is the boy covering his ears? The loud sound hurts his ears so he is protecting them.

Soft sounds may be hard to hear.

A whisper is a soft sound.

Soft sounds will not hurt your ears.

What is volume?

Health and Safety

God made your ears. Your ears are important to God. You should take care of your ears. You should keep them clean. You should protect them from loud sounds. You should protect them from the cold and wind. Caring for your ears pleases God.

209

Are soft sounds easy or hard to hear? hard

- Direct attention to the pictures.

What is the first girl doing in the ear of the second girl? whispering

Why is she whispering so close to the second girl's ear? because a whisper is a soft sound and it is hard to hear

What is the boy doing? He is tiptoeing past a sleeping baby.

Why is he tiptoeing? He is trying to walk softly so that he will not wake the baby.

Will soft sounds hurt your ears? no

What are some soft sounds that will not hurt your ears? a whisper, a tiptoe, the tick of a clock

- Read aloud the *Health and Safety* box or ask a volunteer to read it aloud.

Who made your ears? God

Why should you take care of your ears? My ears are important to God; it pleases God when I care for my ears.

What should you do to take care of your ears? keep them clean, protect them from loud sounds, protect them from cold and wind

- Direct attention to the picture of the father and daughter.

What are the girl and her father doing? drilling a hole in a piece of wood

How are the girl and her father protecting their ears from the loud sound of the drill? They are wearing earmuffs.

- Explain that these earmuffs are used to muffle, or soften, loud sounds, such as the sound of a drill. Point out that there are also earmuffs that are worn to protect ears from cold and wind.

What other forms of protection are the girl and her father wearing? safety goggles, gloves

- Invite the students to share their experiences with loud sounds.

Bullhorn

A bullhorn is an electronic device used to raise the volume of the human voice so it can be heard at a distance. The closer a person holds a bullhorn to his mouth while speaking, the louder the sound of the person's voice.

- Preview and pronounce the vocabulary term *pitch*.
- Direct the students to the glossary to find the definition for *pitch*. Ask a volunteer to read the definition aloud.
- Direct the students to read pages 210–11 silently to find out what makes sound with a high pitch.

What is pitch? how high or low a sound is
What sounds have pitch? every sound

- Direct attention to the pictures.

Which animal makes a sound with a high pitch? the kitten

What is another example of a sound with a high pitch? Possible answers: tweeting bird; squeaking mouse; smoke detector alarm

What is an example of a sound with a low pitch? Possible answers: roaring lion; rumbling thunder; booming fireworks

Pitch

Pitch is how high or low a sound is. Every sound has pitch.

The meow of a kitten is a high pitch.

The roar of a lion is a low pitch.

210

A musical instrument can make sounds with high pitches and low pitches.

You can sing in a high or a low pitch.

211

Can musical instruments make sounds with high and low pitches? yes

- Direct attention to the picture of the girl and boy playing the musical instruments.

What instrument is the girl playing? a violin

How do you make sound with a violin? rub it; pluck it

What instrument is the boy playing? a recorder

How do you make sound with a recorder? blow air through it

- Direct attention to the picture of the girl with the violin. Explain that each string has a different pitch. Mention that the string closest to her hands has the highest pitch, and the string farthest from her hands has the lowest pitch.
- Display a stringed musical instrument, such as a violin, cello, or guitar. Direct the students to listen to the pitch as the strings are played or plucked.
- Play or pluck the strings again and ask the students to identify the pitch as either high or low.
- Direct attention to the picture of the boy with the recorder. Display a recorder or other wind instrument. Demonstrate pitch by playing several notes.
- Direct attention to the picture of the choir.
  What are the children doing? singing
  Can you sing in a high or low pitch? yes
- Demonstrate high and low pitch by singing a familiar song. Invite the students to join you as you sing. Ask the students to identify high pitch and low pitch by raising and lowering their hands as they sing.

How does sound change? The volume of a sound can be loud or soft. Sound can also have a high pitch or a low pitch.

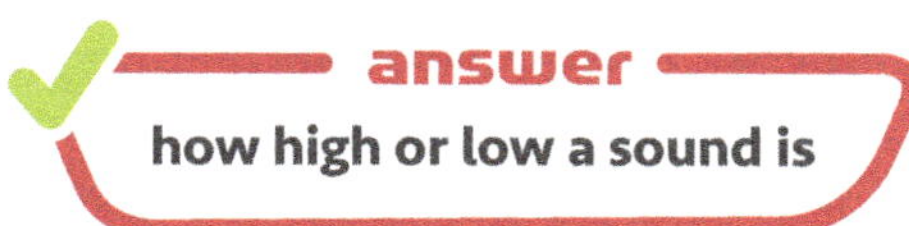

ACTIVITIES

Study Guide, pages 199–200
These pages review Lesson 80. After completion, direct the students to keep them for review for Test 11.

Objectives
- Formulate a hypothesis for how the thickness of a rubber band will affect pitch
- Measure with numbers the length of a stretched rubber band
- Observe that the pitch of a sound is affected by the thickness of a rubber band when the rubber band is plucked
- Infer that the thickness of a rubber band influences the pitch of the sound the rubber band produces
- Explain how the pitch of a stringed instrument can be changed

Materials
- See Activities page 201.

INTRODUCTION

What sounds are you hearing around you?
Are all of the sounds easy to hear?
Do all of the sounds have the same pitch?
Today you are going to investigate pitch.

PREPARATION FOR READING

- Direct the students to remove pages 201–3 from their Activities books.
- Direct the students to read page 212 and Activities pages 201–3 silently before beginning.

TEACH FOR UNDERSTANDING

- Direct attention to and discuss the **Process Skills**. If necessary, refer to Chapter 2.
- Choose a student to read the **Problem** aloud.
- Pair the students. Direct them to complete the **Hypothesis** to answer the problem.
- Distribute the **Materials** to each pair of students.
- Instruct the students to follow the **Procedure**. Point out that the 2 mm wide rubber band is thinner than the 5 mm wide rubber band. Review measuring in cm and mm.
- Instruct the students to record their **Observations** on the chart.
- Direct the students to draw **Conclusions** based on their observations. Discuss the conclusions.
- Relate the rubber bands to the strings on a musical instrument. Stressing this connection will be helpful as the students participate in the Lesson 82 STEM activity.

INVESTIGATION

Lesson 81

Hearing Pitch

Shhhh.

Listen.

There are sounds all around you.

Some sounds are loud.

Other sounds are soft.

Sounds are also high or low.

In this Investigation, you will find out what causes a rubber band to have a high or a low pitch.

212

ACTIVITIES

Investigation: Hearing Pitch, pages 201–3
This Investigation will explain how a rubber band's thickness or weight affects the rubber band's pitch.

ASSESSMENT

Rubric
Use the prepared rubric or design a rubric to include your chosen criteria.

⚠ HELPS

Making and Using Models
Models are physical representations that help us explain or clarify ideas. The instruments designed by the students in Lesson 82 are models of ways to make sound with different pitches. By using their instruments, the students can observe and compare how different pitches are made with a stringed instrument.

STEM

Lesson 82
Making Music

Some sounds make you happy.
Music can be a happy sound.

You may hum to make music.
Sometimes you may sing.
You may also play an instrument.

Music has high and low pitches.
In this STEM activity, you will design a stringed musical instrument.
You will design each string to have a different pitch.

213

LESSON 82

Objectives
- Design a musical instrument with four strings of varying pitch
- Draw and label the design of the stringed musical instrument
- Make a model of the stringed musical instrument
- Test and improve the stringed instrument model
- Explain how the design of the musical instrument solved the problem of having four strings of varying pitch

Materials
- square or rectangular containers of various sizes (plastic containers, tissue/cereal/shoe boxes), one per student *Note:* If the container does not already have a hole in it, cut a hole in one side.
- rubber bands of varying lengths and thicknesses
- string
- yarn
- safety goggles

INTRODUCTION

Today you will design an instrument that can be used to make music.

PREPARATION FOR READING

- Direct the students to remove pages 205–6 from their Activities books.
- Direct them to read page 213 and Activities pages 205–6 silently before beginning.

TEACH FOR UNDERSTANDING

- Direct attention to the picture on page 213.
- What instrument is the boy plucking? a cello
  How many strings does the cello have? four
- Choose a student to read aloud the problem on Activities page 205.
- Distribute the materials and explain that the students will use the engineering design process and the materials provided to find an answer to the problem. *Materials Note:* Allow the students to select any combination of rubber bands, string, and yarn as they design the strings of their instrument. The string and yarn will not produce the same acoustic results as the rubber bands. This should be noticeable to the students when they test and compare their models.
- Explain that for this STEM activity they will *Ask, Imagine, Plan, Make, Test and Make Better*, and *Share* a design to find an answer to the problem.

> For this activity, each student will design and make his own solution to the problem.

- Direct the students to compare their models of a four-stringed musical instrument to other student-made models.
- Instruct the students to test and make better their models.
- Invite the students to share their final designs. Direct the students to explain how their designs solved the problem.

> The instruments made in this lesson will be needed for Lesson 90. You may collect the instruments for later use.

ACTIVITIES

STEM Activity: Making Music, pages 205–6
In this STEM activity, the students will design a four-stringed musical instrument that can be plucked. The strings will have high and low pitches.

ASSESSMENT

Rubric
Use the prepared rubric or design a rubric to include your chosen criteria.

Objective
- Recall terms and concepts from Chapter 11

REVIEW

- Material for Test 11 will come from the Study Guides on Activities pages 197–200. You may review any or all of the material during the lesson.
- You may choose to review Chapter 11 by playing the *Tuning Up* review game or another game from the Game Bank (Teacher Resources).

🏆 REVIEW GAME

Tuning Up

Materials: tuning fork, plastic beaker filled with water

Divide the class into two teams. Ask review questions. For each correct answer, strike the tuning fork. For every five correct answers by the same team, strike the tuning fork and, as it is ringing, place the tuning fork in a plastic beaker of water.

Note: Use a plastic beaker for this review game. If a tuning fork comes in contact with the sides of a glass beaker, the glass beaker may break.

NOTES

Objective

Recall and apply terms and concepts from Chapter 11

ASSESSMENT

Administer Test 11.

NOTES

Chapter Objectives
- Explain how light and sound are used to communicate at home, at school, and in the community
- Apply the engineering design process by using light or sound to solve a communications problem
- Formulate a response to each biblical worldview question `BWS`
- Compose a song of praise for God's creation `BWS`

Lesson Objectives
- Identify ways light and sound are used to communicate at home and school
- Explain how various sources of light and sound communication at home and school can be used to help other people `BWS`
- Explain how to determine whether light and sound communication is good or bad `BWS`
- Evaluate uses of light and sound communication `BWS`

Materials
- fire truck siren sound clip

Teacher Resources
- Visuals 12.2–12.5: *Bible Verses B*; *Bible Verses C*; *Bible Verses D*; *Light and Sound at Home and School*

Vocabulary
- communicate

CHAPTER INTRODUCTION

- Direct attention to the Contents page.
  We are ready to begin Chapter 12. Find what page Chapter 12 begins on. 214 Turn to page 214.
- Direct the students to look through the chapter to decide what the chapter is about.
- What do you think the chapter is about? Possible answers: light, sound, where I can find light and sound, light and sound in different places
- As the students give their answers, ask them to tell you the page number and the item that helped them decide.
- Look at the picture on page 214. What do you see? fire trucks, fire station, fire chief truck
- Play the sound that a fire truck makes.
  You hear the siren from a fire truck. What do you see that lets you know to get out of the way? the flashing lights

Big Question
- Guide the students in reading the question.
  Where will you find the answer to this question? as the chapter is read

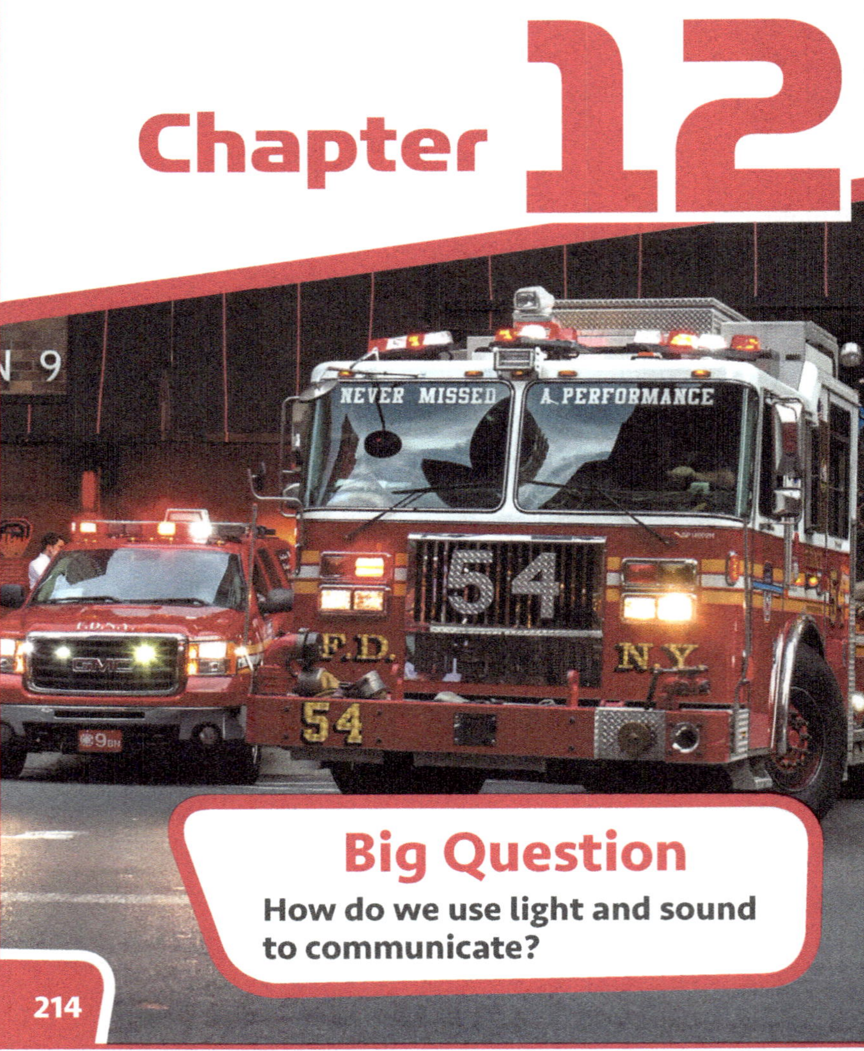

⚠ **HELPS**

Chapter Photo
The chapter photo is fire trucks at a fire station in New York City.

🔍 **BACKGROUND**

Assessments
There is not a test with this chapter. There are no Study Guides in the Activities book.

Communicating with Light and Sound

You hear sirens.

You see flashing lights.

A fire truck drives down your street.

You may think there is a fire nearby.

Different lights and sounds let you know something is happening.

Lights and sounds can help people make safe choices.

215

- Preview and pronounce the word *siren*.
- Direct the students to read page 215 silently to find out what helps people make safe choices.

 What is the sound from a fire truck called? a siren

Note: The siren is the device making the sound on a fire truck. Many students will describe the sound as a siren.

What do the lights and sounds let you know? Something is happening.

What can the lights and sound help people to make? safe choices

- Direct the students to listen to the following stories. Explain that you will read two endings to each story and they will choose the ending that shows how the sound's message is communicated correctly.

Story A: Mr. Ward had hunted and fished in the mountains since he was a young boy. As he made his way through the woods one day, he suddenly heard a rattling from the bushes near his feet.

Ending 1: Mr. Ward was so excited when he heard the noise that he ran right over to the bushes. He knew it was his lost dog, Scout, calling for help.

Ending 2: Mr. Ward froze in his tracks and slowly reached for his gun. He knew from the sound that a rattlesnake was only a few feet away. the correct ending

- Discuss the story, guiding the students to see that the sound from the rattlesnake helped Mr. Ward to make a safe choice.

Story B: Tim was on the front porch reading a book when he heard two different trucks with sirens pass by his house. He looked up and saw the flashing lights as the trucks parked at the end of his street.

Ending 1: Tim quickly took money from his piggy bank and ran to the curb to buy a treat from the ice-cream truck.

Ending 2: Tim stayed on the porch. He looked down the street to see where the fire was. the correct ending

- Discuss the story, guiding the students to see that the sound and lights from the fire trucks helped Tim to make a safe choice.

PREPARATION FOR READING

- Direct the students to the red heading.
  What is the heading? "Around My Home"
- Direct attention to the two black headings.
  What are these headings? "Light" and "Sound"
  What do you think you will be reading about on these two pages? light and sound around my home
- Preview and pronounce the vocabulary term *communicate*.
- Direct the students to read pages 216–17 silently to find out how light and sound communicate.

TEACH FOR UNDERSTANDING

What does it mean to communicate? to give information about something

> As you discuss this chapter, use the word *communicate* as often as possible. The more the word *communicate* is used when asking questions about light and sound the more the students develop a better understanding of its meaning.

What does a light on a stove communicate to you? that the stove may be on; that the stove may be hot

- Explain that the light on the stove does not use words to tell you it is hot, but the light is communicating that the stove is hot.
  What does the light on your front porch communicate to you? It lets you see if someone is at the door.
- What other lights in your home communicate something to you? Possible answers: Microwave light communicates that the microwave is on. A red light on the electric toothbrush communicates that it is charging. A light on the washing machine or the dryer communicates that the machines are working. The light on the laptop communicates that the computer is on. The light from the fireplace communicates that the fire is hot and to safely stay back.
- Discuss both pictures and how the light in each communicates something.

Lesson 85

Around My Home

Light and sound are all around you.
Many objects use light and sound to communicate.

To **communicate** means to give information about something.

Light

A light on the stove lets you know it is on.
The light lets you know the stove may be hot.

People turn on their porch light at night.
The light helps them to see who is at their door.

216

Sound

Ding dong!

A doorbell makes this sound.

It lets you know someone
is at your door.

An alarm clock beeps.

The sound lets you know it is time to wake up.

Some sounds are important for
you to listen to.

Your parents tell you how to
behave.

You should hear and obey what
they say.

The Bible tells children that it is
right to listen to their parents.

**What sounds at home are
important for you to listen to?**

217

Sounds in your home communicate as well.
What does the doorbell communicate to you? It
communicates that someone is at my door.

- Explain that the doorbell does not use words to
communicate to you; instead, it is the sound itself
that communicates that someone is at your door.
What does the beep of an alarm clock communicate? that it is time to wake up
What things do your parents communicate to you?
how to behave, to obey, truths from the Bible, things
you need to know for everyday life, how to dress,
how to eat properly
What does the Bible tell you about your parents? It
is right to listen to my parents.
What other sounds in your home communicate
something to you? Possible answers: The sound of
the hair dryer communicates that it is on and it is
blowing. The sound of the lawn mower communicates that it is on and that I should stay a safe
distance away.

- Discuss what the sound in each of the pictures is
communicating.
top picture: The doorbell communicates to the
people inside that someone is at the door.
bottom picture: A father is communicating to his
son what the Bible says.

answer

**the doorbell, the alarm clock,
and what a parent says**

PREPARATION FOR READING

- Preview and pronounce the word *favorite*.
- Direct the students to read page 218 silently to find out what objects communicate with light and sound.

TEACH FOR UNDERSTANDING

- Discuss how a TV communicates with light and sound.

  How does a TV communicate with light and sound? with voices, music, and pictures

  How does the light of the TV communicate? with pictures

  How does the sound of the TV communicate? with words or talking and music

  If you mute or turn off the sound on the TV, how is the TV communicating? with just pictures

- Discuss how a cell phone communicates with light and sound.

  How does a cell phone communicate with sound? It communicates with a ringing tone that someone is calling.

  Can a cell phone communicate with sound in other ways? through the games that are played, the videos you watch, the pinging sound that a text has come in, a wake up alarm

  How does a cell phone communicate with light? The light lets you see who is calling.

- Discuss how the pictures show communication with light and sound.

  Can a cell phone communicate with light in other ways? through the games that are played, through the videos you watch, through different apps

  Name some other things in your home that communicate with both light and sound. some alarm clocks, radios, video games, board games with light and sound, some doorbells, smoke detector, telephone, DVD players, computers or laptops, timer, digital clock

- Direct the students to listen to this story and choose the ending that shows how the sound or light's message is communicated correctly.

Light and Sound

A TV uses light and sound.
It communicates the words and pictures of your favorite show.

A cell phone uses light and sound.
The sound lets you know someone is calling.
The light lets you see who is calling.

218

Story: Emily was having a wonderful dream that she was a beautiful princess who lived in a crystal palace. In the middle of her dream she was awakened by a ringing sound coming from her night stand.

Ending 1: Emily reached over to turn off her alarm clock. It was time to get up. the correct ending

Ending 2: Emily got out of bed and went to the front door to see who was knocking.

- Discuss the story, guiding the students to see that the sound from the alarm clock communicated to or helped Emily know it was time to get up.

Around My School

There are many objects in your school that communicate with light and sound.

Light

An exit sign uses light.
This sign helps you find your way out of a building.

Light on a screen helps you see pictures.
These pictures help you learn more about God's world.

Which kind of sign helps you find your way out of a building?

PREPARATION FOR READING

- Preview and pronounce the words *exit* and *building*.
- Direct the students to read page 219 silently to find out what objects at school communicate with light.

TEACH FOR UNDERSTANDING

Let's look at how light and sound communicate at school.

- If there is an exit sign above the door in the classroom, direct attention to it. If not, direct attention to the picture of the exit sign.

  What sign is above every door to the outside in school and above some inside doors? an exit sign

  What is the exit sign communicating with light? It communicates that this is the way out of the building or room.

  Why is it important to know how to exit a building? in case there is a fire or other emergency

- Direct attention to any lighted screen in the classroom. If there is not one, direct attention to the bottom picture on the page.

  What does a light on a screen communicate to you? helps me see pictures, diagrams, and words

  What is the lighted screen in the bottom picture communicating? It is communicating through pictures how the water cycle works.

  Name some other lights in the school that communicate to you. spotlights in the auditorium point out something particular on stage; classroom computers or laptops; fire alarms

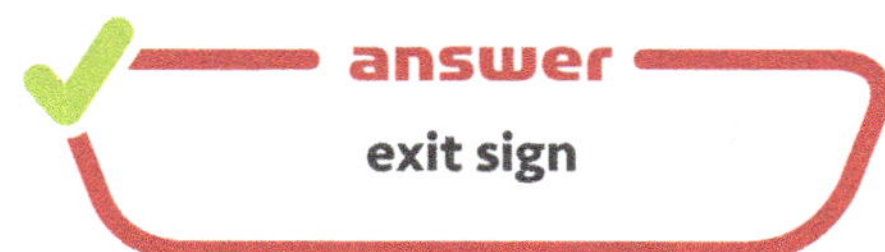

answer

exit sign

PREPARATION FOR READING

- Preview and pronounce the words *recess, video,* and *special.*
- Direct attention to the two black headings.
  What will you read about on these two pages? sound at school and light and sound at school
- Direct the students to read pages 220–21 silently to find out ways light and sound communicate at school.

TEACH FOR UNDERSTANDING

What does the sound of a bell at school communicate? time to begin the day, time to change classes, time to go home

What does a teacher's voice communicate at school? teaches me about God's world, reads a story to me, teaches me how to read, teaches me how to behave, sings to me about God

When you hear a whistle blow at recess, what does the sound communicate to you? Recess is finished. It is time to line up.

- Discuss how the pictures on the page show the communication of sound at school.
- 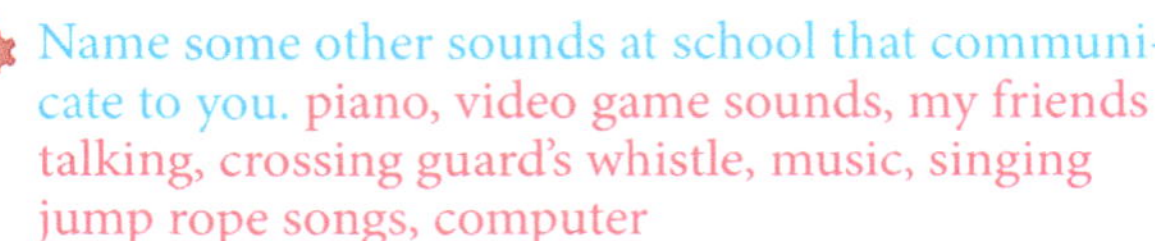 Name some other sounds at school that communicate to you. piano, video game sounds, my friends talking, crossing guard's whistle, music, singing jump rope songs, computer

---

Sound

You may hear a bell ring in your classroom.
This sound lets you know that it is time to begin the day.

Your teacher uses her voice to teach you about God's world.
Your teacher reads a book to you.
She blows her whistle at recess.
You know it is time to line up!

220

The discussion for Student Edition page 221 begins here.

Let's look at some things at school that communicate with both light and sound.

What does a computer use to communicate? Light communicates a video. Sound communicates the words.

What does a fire alarm use to communicate? Light and sound communicate there may be a fire. Light and sound communicate that you need to leave the building.

When there is a special program at school, what is used to communicate? Light and sound communicate a message to others.

- Direct attention to the bottom picture.

What do you see in the picture that is communicating to the audience? Spotlights are communicating where the audience should look. The microphone is communicating the sound of the actors' voices.

- Ask volunteers to share their experiences with communication of light and sound during special programs.
- Name some other things at school that communicate by using both light and sound. video games, DVDs, smart boards, TV screens

Light and Sound

A computer uses light and sound to communicate.

The light lets you see a video.

The sound lets you listen to the words.

Your classroom has a fire alarm.

It uses light and sound to let you know there may be a fire.

The light and sound mean that you need to leave the building.

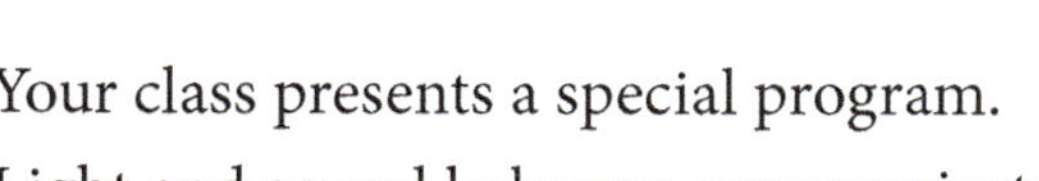

Your class presents a special program.

Light and sound help you communicate a message to others.

How do light and sound help you at school?

221

Display *Light and Sound at Home and School* chart. Direct attention to the chart on Activities page 211.

Ask the students to write the names of five objects at home or school that communicate with light, sound, or both light and sound. *Note:* You may direct all the students to write the same five objects, or you may want each student to write his own choices.

Direct the students to place an *X* under the *Light* column, the *Sound* column, or under both columns to tell what kind of communication the object has.

To complete the last two columns, guide students through questions such as the following with each object:

How are people helped by the communication of a doorbell? The doorbell communicates that someone is at the door.

How do you know if the doorbell is communicating something good? If the doorbell is used in a good way, someone will ring it once and wait for me to answer the door.

How do you know if the doorbell is communicating something bad? If the doorbell is used incorrectly, someone may ring the doorbell and run, or the person may repeatedly ring the doorbell.

- Direct attention to the fourth column on the chart. Can science tell me if the doorbell is communicating something good or bad? no
- Ask the students to circle the word *no* in the fourth column.
- Direct attention to the last column. Can the Bible tell me if the doorbell is communicating something good or bad? yes

> Some Bible verses explaining biblical principles that the students can apply to the last question are Psalm 101:3*a*, Matthew 7:12, Matthew 22:37–39, and Philippians 4:8. Choose the verse that applies to the communication being discussed. All of these verses can be found on *Bible Verses B, Bible Verses C,* or *Bible Verses D* and Activities pages 208–10.

What is a Bible verse that can tell us if the doorbell is communicating something good or bad? If the person ringing the doorbell uses it in a good way, he is showing love for his neighbor (Matthew 22:37–38); he is obeying God's commandment to love God and love others (Matthew 22:37–39); he is treating you like he wants to be treated (Matthew 7:12). If the person uses the doorbell in a bad way, he is not showing love for God or his neighbor, and he is not treating you the way he wants to be treated.

- Follow the same type of questions with each object of communication on the chart.
- Direct the students to remove Activities page 213 to complete with someone at home. Remind the students to bring the pages back to school tomorrow.

answer

communicate a message; communicate the need to leave the building because of a fire

ACTIVITIES

Light and Sound at Home and School, page 211
This page was completed as part of the lesson.

Light and Sound Communication, page 212
This page reinforces the concepts taught in the lesson.

Light and Sound in My Community, page 213
This page will be completed at home and used in Lesson 86.

Objectives
- Identify ways light and sound are used in the community to communicate
- Explain how various sources of light and sound communication in the community can be used to help other people **BWS**
- Explain how to determine whether light or sound communication is good or bad **BWS**
- Evaluate uses of light and sound communication **BWS**

Materials
- fire truck siren sound clip or video
- railroad crossing video
- tornado siren sound clip
- lighthouse fog horn sound clip

Teacher Resources
- Visuals 12.2–12.4, 12.6: *Bible Verses B; Bible Verses C; Bible Verses D; Light and Sound in My Community*

INTRODUCTION

- Play the fire truck siren sound clip or video.
- What is this sound? fire truck or ambulance siren
- What do you think about when you hear this sound? There is an emergency. Move out of the way. Someone needs help.

  Today we will talk about light and sound communication in our community.

PREPARATION FOR READING

- Preview and pronounce the word *community*.

  What is a community? The people living and working together in an area. My neighborhood is part of my community.
- Direct the students to read pages 222–23 silently to find out what the sound of a hissing snake communicates.

TEACH FOR UNDERSTANDING

What do light and sound in your community help you make? safe choices

What does God want you to use light and sound for? to help others

What did God give you to help you understand what light and sound communicate? a mind
- What color lights are on a traffic light? green, yellow, and red
- What does the green light communicate? It lets drivers know to go.
- What does the yellow light communicate? It lets drivers know to slow down or begin to stop.

Lesson 86

Around My Community

Light and sound help you make safe choices in your community.

God wants people to use light and sound to help others.

God gave people minds to understand what light and sound communicate.

Light

A traffic light lets drivers know when to stop and go.

A railroad crossing uses flashing lights to let people know a train is coming.

222

- What does the red light communicate? to stop
- Play the railroad crossing video.

  How does a railroad crossing communicate that a train is coming? by flashing lights, by making a warning sound, by lowering the gates
- How long does the railroad crossing use flashing lights? until the train has passed
- Direct attention to the top picture.
- What are the red lights communicating in this picture? A red traffic light lets drivers know to stop. Red car taillights let drivers know the car in front of them is stopped.
- Direct attention to the bottom picture.
- What do the red lights in this picture communicate? to stop for the train crossing the road
- Explain that the train cars are double-decker passenger train cars. People can ride on either level.

Sound

Some sounds do not use words but still tell you something.

A tornado siren warns people of danger.

It tells people that bad weather may be coming.

A hissing snake or a growling dog tells you to stay away.

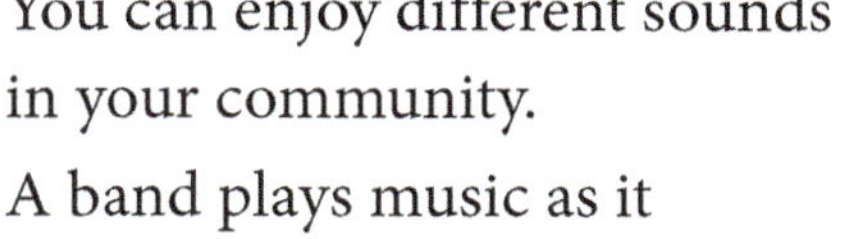

You can enjoy different sounds in your community.

A band plays music as it marches down your street.

People cheer as they watch your soccer game.

✓ **How do light and sound in your community keep you safe?**

223

Does a tornado siren warn people by yelling "A tornado is coming! A tornado is coming!"? No; it uses a siren to warn people.

- Direct attention to the tornado siren picture as you play the tornado siren sound clip. Explain that the siren is sounded when there is a possibility of danger coming. Sometimes the danger is a tornado, high winds, or large-size hail.

What does a hissing snake or a growling dog communicate? to stay away

- Explain that the snake pictured on the page is a rattlesnake.

❋ How does a rattlesnake warn you to stay away? It vibrates its tail.

- Point out that the hollow parts at the end of a rattlesnake's tail bang together making a rattle sound.

❋ Why is it important to stay away from a rattlesnake? It is poisonous.

- Direct attention to the last picture on the page.

What kind of communication is shown? A marching band is playing music.

Why do people cheer at soccer games or other types of sports? Their cheering lets the players and other people know that they are encouraging the players to win the game or make a certain play in the game.

✓ **answer**
Light and sound in the community warn people of dangers and help people make safe choices.

PREPARATION FOR READING

- Preview and pronounce the words *piano, ambulance,* and *celebrate.*
- Direct the students to read pages 224–25 silently to find out how an ambulance communicates. *Note:* The *Fantastic Facts* box does not need to be read at this time.

TEACH FOR UNDERSTANDING

How do you communicate with sound to praise God? Piano and other instruments communicate with music. My voice sings praises.

- Point out Psalm 147:1 on *Bible Verses C.* Ask a volunteer to read the verse aloud as the other students follow along on Activities page 209.

What does this verse say about praising God? It is good. It is pleasant. It is comely or fitting to praise God.

How is sound being used in the pictures to praise God? playing the piano, singing praise

You can also use sound to praise God.

Your friend plays the piano to make music.

You use your voice to sing at church.

The Bible says, "It is good to sing praises unto our God" (Psalm 147:1).

224

Light and Sound

A police car uses flashing lights and sounds.

It tells a driver to stop.

An ambulance uses flashing lights and sounds.

It tells drivers to move out of the way.

Light and sound can also help you celebrate.

You can see and hear fireworks on the Fourth of July.

Fireworks are bright and loud.

Why does a police car use lights and sounds?

Fantastic Facts

Lighthouses are built near the ocean. They use light and sound to communicate with ships at sea. Many ships travel at night. The lighthouse shines a bright light. The light helps guide captains. Heavy fog can make it hard to see the light. A lighthouse also uses a fog horn. It is a loud alarm. This alarm warns the captain of danger he cannot see. A lighthouse uses light and sound to help ships travel safely in the ocean.

225

How does a police car communicate? with flashing lights and a siren
What are the flashing lights and siren communicating? tells the driver to stop; tells the driver to get out of the way; tells the driver there is an emergency
How does an ambulance communicate? with flashing lights and a siren
What are the flashing lights and siren communicating? tells drivers to move out of the way; tells drivers there is an emergency
How can light and sound help you celebrate? see and hear fireworks and a parade on the Fourth of July; see and hear candles and singing on my birthday
Read the *Fantastic Facts* box. Play the sound of a lighthouse fog horn.
Where are lighthouses usually built? near the ocean, near water
What do lighthouses use to communicate? light and sound
Ask a volunteer to read the sentence that tells why light is not always a good way to communicate. Heavy fog can make it hard to see the light.
Who does the lighthouse communicate with? ships; the captain of a ship
Direct attention to *Light and Sound in My Community* and to Activities page 213. Invite students to share the objects of communication that they found and whether the object communicates with light, sound, or both. To complete the last two columns, guide students through questions such as the following.
How are people helped by the communication of a school bus?

It communicates with a stop sign and its blinking lights that the cars should stop until the children get on or off the bus.
How do you know if the school bus is communicating something good? The school bus is using its lights and stop sign to stop traffic and allow the children safety in getting on or off the bus.
How do you know if the school bus is communicating something bad? The school bus is using the lights and stop sign for something other than its purpose. It is stopping traffic, but there are no children.

- Direct attention to the fourth column on the chart.
  Can science tell me if the school bus is communicating something good or bad? no
- Direct attention to the last column.
  Can the Bible tell me if the school bus is communicating something good or bad? yes

Some Bible verses that explain biblical principles the students can apply to the last question are Psalm 101:3*a*, Matthew 7:12, Matthew 22:37–39, and Philippians 4:8. All of these verses can be found on *Bible Verses B, Bible Verses C,* or *Bible Verses D* and Activities pages 208–10.

What is a Bible verse that can tell us if the school bus is communicating something good or bad? The bus driver turns on the lights and uses the stop sign to show love to the children by trying to protect them (Matthew 22:37–38); he is obeying God's commandment to love God and love others (Matthew 22:37–39); he is treating you like he wants to be treated (Matthew 7:12). If the driver uses the lights and stop sign in a bad way, he is not showing love for God or his neighbor and he is not treating you the way he wants to be treated.

- Follow the same type of questions with a selection of objects from the students' charts.

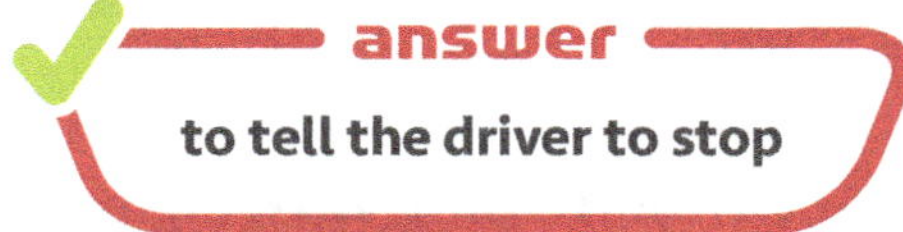

answer

to tell the driver to stop

ACTIVITIES

Light and Sound in My Community, page 213
This page was completed at home and used in the teaching of the lesson.

Communicating Something Good, page 215
This page reinforces concepts taught in this lesson.

Community Lights and Sounds, page 216
This Enrichment activity may be used to reinforce the concepts taught in Lesson 86.

Objectives
- Propose possible solutions to a real-life problem using light or sound
- Draw a design that uses light or sound to solve a real-life problem
- Communicate to others how the design solves the problem

Teacher Resources
- Visual 2.6: *STEM The Engineering Design Process*

INTRODUCTION

Today you will design a way to use sound or light to help someone in need.

PREPARATION FOR READING

- Direct the students to read page 226 and Activities pages 217–20 silently before beginning the STEM activity.

TEACH FOR UNDERSTANDING

- Display *STEM The Engineering Design Process*.
- Direct attention to the stories on Activities page 217. Instruct the students to choose one person in need for whom they will design a light or sound tool.
- Explain that for this STEM activity they will *Imagine*, *Plan*, and *Share* a design to find an answer to the problem. Point out that they will not use the steps *Make* or *Make Better* with their designs.
- For this activity, each student may design his own solution to the problem, or students may work in science groups to collaborate as they design a solution.
- Invite the students to share their designs with a partner or another science group.

ACTIVITIES

STEM Activity: Helping with Light and Sound, pages 217–20
In this STEM activity, the students will design one way to use light or sound to help someone in need.

ASSESSMENT

Rubric
Use the prepared rubric or design a rubric to include your chosen criteria.

STEM

Lesson 87

Helping with Light or Sound

Your friend has a need.

You see a light.

You hear a sound.

You have an idea!

In this STEM activity you will design a way to use light or sound to help someone in need.

226

Lessons 88–89

Biblical Worldview

Everyone has a worldview.

You have a worldview.

You have studied science to learn about God's world.

A worldview helps explain what you learn.

It is important to study science with a biblical worldview.

✓ **Why do you study science?**

227

Objectives

Recall what a worldview is

Summarize from the Bible where the world came from **BWS**

Construct a response explaining why things work the way they do in our world **BWS**

Determine who we are and why we are here **BWS**

Compare and contrast the importance of science with the importance of the Bible **BWS**

Materials

Activities page 13 from Chapter 1 Lessons 5–6

Teacher Resources

Visuals 1.6; 12.1–12.4; 12.7–12.8: *Worldview K-L Chart*; *Bible Verses A*; *Bible Verses B*; *Bible Verses C*; *Bible Verses D*; *Biblical Worldview A*; *Biblical Worldview B*

Vocabulary

Redemption

INTRODUCTION

- Direct attention to Student Edition page 17.
  What is pictured on the page? a scientist with dinosaur bones
  What helps the scientist explain what he has found? his worldview
  What is a worldview? the way a person thinks about the world
  Do all scientists believe in the Bible? no

PREPARATION FOR READING

What is the red heading on page 227? "Biblical Worldview"

- Direct the students to read page 227 silently to find out what it is important to study science with.

TEACH FOR UNDERSTANDING

Do you have a worldview? Yes; everyone has a worldview.

Why do you study science? to learn about God's world

What is it important to study science with? a biblical worldview

- Return the Worldview K-L Charts from Lessons 5–6 to the students. Give the students time to read what they wrote under the Know column.

- Display *Worldview K-L Chart*. Discuss each question orally, allowing the students to share what they have learned in *Science 1*.

- Write an answer for each question on the *Worldview K-L Chart*. A one- or two-sentence answer is acceptable. Give the students time to complete their charts.

✓ — **answer** —

to learn about God's world

- Direct the students to read pages 228–29 silently to find out why pain, sickness, and death happened.

Where did our world come from? God created a perfect world in six days.

How did God create people? in His image

What does the word *image* mean? likeness

What does the Bible say you should do because God created the world? I am to praise Him.

How can you praise God? Pray and thank God for making the world and us. Sing praises to God.

- Direct attention to Psalm 8:1 on *Bible Verses B*. Ask a volunteer to read the verse aloud.

  What is one reason you should praise God? for His excellent name

- Ask a volunteer to read Psalm 8:3–4 aloud.

  What do these verses tell you that God made and that you should praise Him for? the heavens, the moon and the stars

- Invite another student to read aloud Psalm 8:9.

  What is the psalmist praising God for? His excellent name

- Direct attention to the picture in the Student Edition.

  What is the boy doing? praying to God, thanking God, studying his Bible

  Can the dog pray to God? No; God did not make animals in His image.

- Direct attention to *Biblical Worldview A* and Activities page 221.

- Guide the students as they complete the sentences under the question "Where did our world come from?"

- Display the completed *Worldview K-L* chart or direct attention to the students' completed charts. Guide a discussion, comparing and contrasting their answers in the Learned column to what they have just read about the first worldview question.

Where did our world come from?

God created a perfect world in six days.

He created the plants and animals.

He created the sun, moon, and stars.

God created light and sound.

God created people in His image.

The Bible says we are to praise God.

We are to thank God for making the world.

We are to thank God for making us.

228

Why do things in our world work together the way they do?

God designed the parts of the earth to work together.

Some people see the weeds and thistles.

They see pain, sickness, and death.

These things happened because of Adam's sin, or the Fall.

✓ **What is the Fall?**

229

Why do things in our world work together the way they do? God designed them to work together. God is an Engineer. He is an Artist.

How do the gears of a machine show how the parts of the earth work together? Each part works with the other parts. Everything works together.

Why are there weeds and thistles in the world? because of Adam's sin

What was Adam's sin? He ate of the tree of the knowledge of good and evil. He disobeyed God.

- Direct attention to Genesis 3:17–18 on *Bible Verses A* and *Bible Verses B.* Ask a volunteer to read the verses aloud.

What did God do because Adam ate of the tree? He cursed the ground. The ground grew thorns and thistles.

What else happened because of Adam's sin? pain, sickness, and death

What is another name for Adam's sin? the Fall

- Direct attention to the picture at the bottom of the page.

What are Adam and Eve doing? leaving the Garden of Eden

Why must they leave? They sinned against God.

- Direct attention to Romans 3:23 on *Bible Verses D.* Ask the students to read the verse aloud together.

Is Adam the only person to ever sin? No; we all have sinned.

Does that mean you have sinned? yes

✓ **answer**

when Adam sinned and brought death into the world

PREPARATION FOR READING

- Preview and pronounce the vocabulary term *Redemption*.
- Direct the students to read pages 230–31 silently to find out what the greatest commandment is.

TEACH FOR UNDERSTANDING

What will God do to the world one day? make it perfect again

Who will rule the earth? Jesus

- Direct attention to John 3:16 on *Bible Verses D*. Ask a volunteer to read the verse aloud.

Who did God give you because He loved you? His only begotten Son, Jesus

We are all sinners. What must a person do to live with God forever? turn away from sin and trust Jesus

What is turning away from sin and trusting Jesus called? Redemption

- Guide the students in finding the definition for *Redemption* in the glossary.
- Direct attention to the picture.

What is the picture showing? the return of Jesus

- You may want to use this opportunity to explain the gospel to the students. *Note:* See *Explaining the Gospel* in the Teacher Resources in the back of this book for additional helps.
- Direct attention to *Biblical Worldview A* and Activities page 221.
- Guide the students as they complete the sentences under the question "Why do things work together the way they do?"
- Display the completed *Worldview K-L* chart or direct attention to the students' completed charts. Guide a discussion, comparing and contrasting their answers in the Learned column to what they have just read about the second worldview question.

What is the third question? Who are we? Why are we here?

Some books teach that God did not make people. What do these books teach about where people came from? Nature; Nature is the mother of all living things.

Where does the Bible say people came from? God created people.

Do these science books agree with the Bible? no

What do we do when science disagrees with the Bible? believe the Bible because the Bible is God's Word and it is right

The Bible teaches that God will make this world perfect again.

One day everything will work together as it should.

Jesus will rule the earth.

Everyone who turns away from sin and trusts in Jesus will live with God forever.

We call this **Redemption**.

Who are we? Why are we here?

Some books say that God did not make people.

These books teach that people came from nature.

They teach that nature is the mother of all living things.

230

God says that people are the most important part of His creation.

People are to love God more than they love anything or anyone else.

The Bible teaches that the greatest commandment is to love God with all our hearts.

We are to try to be like Jesus.

We are to love other people because they are made in God's image.

We are to take care of the earth that God has made.

> **What is the most important part of God's creation?**

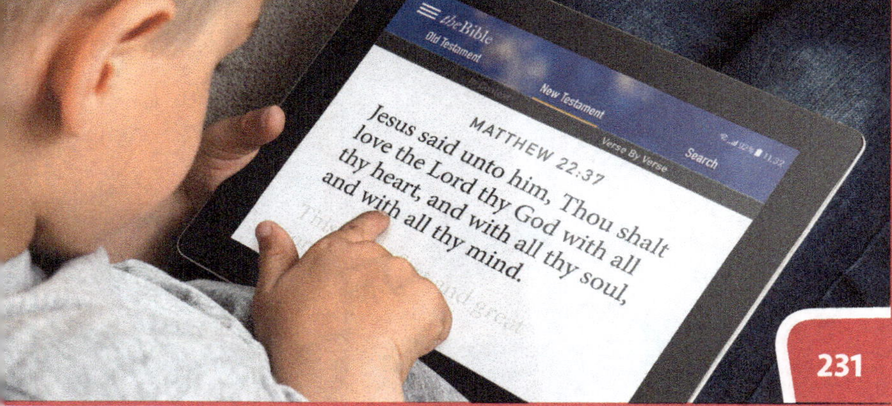

231

Who does God say are the most important part of His creation? people

- Direct attention to Genesis 1:26–28 on *Bible Verses A*. Ask a volunteer to read the verses aloud.

How did God make man? in His image

What job did God give man to do? have dominion over the earth

What does it mean to *have dominion*? Elicit that God wants people to rule over His world by caring for the earth and the animals. He wants people to be in charge of all the things He made. He wants people to manage the things He made. It is our job.

- Direct attention to Matthew 22:37–39 on *Bible Verses C*. Ask a volunteer to read the verses aloud.

What is the greatest commandment? to love God with all our hearts

Who are we to try to be like? Jesus

What is the second commandment? to love your neighbor the way you love yourself

Why are we to love other people? because they are made in God's image

- Direct attention to the picture.

What verse is the boy reading? Matthew 22:37, the greatest commandment

- Direct attention to *Biblical Worldview B* and Activities page 222.
- Guide the students as they complete the sentences under the question "Who are we? Why are we here?"
- Display the completed *Worldview K-L* chart or direct attention to the students' completed charts. Guide a discussion, comparing and contrasting their answers in the Learned column to what they have just read about the third worldview question.

> **answer**
>
> **people**

PREPARATION FOR READING

What is the question these pages are answering? Why is science important?

- Direct the students to read pages 232–33 silently to find out what some books teach about science.

TEACH FOR UNDERSTANDING

What is a powerful tool to help you answer questions? science

Name four things that some books teach about science. Science can answer all important questions. Science is the only way to know something for sure. Science proves the Bible is wrong. Science proves there is no God.

Which of those things are right? none of them; science can answer some questions but not all important questions.

Why is science important?

Science can be a powerful tool to help us answer questions.

Some books teach that science can answer all important questions.

They say that science is the only way to know something for sure.

Some books teach that science proves the Bible is wrong.

They even say that science proves there is no God.

232

Science is good at answering some questions.

What are the needs of plants?

Why does the sun seem to travel across the sky?

Why are there seasons?

How are light and sound used to communicate better?

Name some questions that science is good at answering. What are the needs of plants? Why does the sun seem to travel across the sky? Why are there seasons? How are light and sound used to communicate better?

answer

questions that you have while observing and studying God's world

What kinds of questions can science answer?

233

PREPARATION FOR READING

- Direct the students to read page 234 silently to find out what questions science cannot answer.

TEACH FOR UNDERSTANDING

Name some questions that science cannot answer. Who is God? Why am I here? Where did light come from? Did a flood cover all the earth with water?

- Direct attention to John 17:17 on *Bible Verses D.* Ask a volunteer to read the verse.

 Why is what the Bible says true? because it is God's Word

What do you do when science disagrees with the Bible? Believe that the Bible is right because it is God's Word.

Are there questions that both science and the Bible can answer? yes

- Direct attention to the pictures on pages 232–34.
- Ask the students to think of a question about each picture and whether science or the Bible can answer the question. Examples are given below.

page 232
What will happen when I pour the colored water into another beaker? Science can answer this question.

page 233, top
What do I feed chickens? Science can answer this question. Where did chickens come from? The Bible can answer this question.

page 233, middle
Why do we have seasons? Science and the Bible can answer this question. Why do the leaves change colors during the fall? Science can answer this question.

page 233, bottom
How do I take care of plants? Science can answer this question. Where did plants come from? The Bible can answer this question.

page 234
Who built the ark? The Bible can answer this question. Did water cover the whole earth? The Bible and science can answer this question.

- Direct attention to *Biblical Worldview B* and Activities page 222.
- Guide the students as they complete the sentences under the question "Why is science important?"
- Display the completed *Worldview K-L* chart or direct attention to the students' completed charts. Guide a discussion comparing and contrasting their answers in the Learned column to what they have just read about the last worldview question.

There are other important questions that science cannot answer.

Who is God?

Why am I here?

Where did light come from?

Did a flood cover all the earth with water?

The Bible can answer these questions.

What the Bible says is true.

When science disagrees with the Bible, the Bible is right.

What should you believe when science disagrees with the Bible?

234

answer
What the Bible says is always true.

ACTIVITIES

Worldview K-L Chart, page 13
This page is completed during the teaching of the lessons.

Biblical Worldview, pages 221–22
These pages were completed during the teaching of the lessons.

EXPLORATION

Process Skills
Observe
Communicate

Lesson 90

A Song of Praise

You can sing songs to praise God.

You can play music to praise Him.

You praise God to thank Him.

You thank God for who He is and what He does.

In this Exploration you will write a song.

Your song will praise God for His creation.

Science and the Bible

Music is one way people communicate with sound. There are many songs in the Bible. The book of Psalms is filled with songs of praise. King David wrote many of the psalms. Sometimes music was played when they were sung. It is good to praise God with our music.

235

LESSON 90

Objectives
- Create a song of praise for God's creation **BWS**
- Formulate a sentence explaining how the song of praise will be used **BWS**
- Explain how to determine whether the words of the song of praise are good or bad **BWS**

Materials
- See Activities page 223.
- recording equipment

Teacher Resources
- Visuals 12.2–12.3: *Bible Verses B; Bible Verses C*

INTRODUCTION

Music that praises God is important. Music in worship is important to God.
- Read aloud the *Science and the Bible* box.
- Explain that the lyre, or harp, that David played was a stringed instrument. The Bible tells us that even as a young man, David played the harp very well (1 Samuel 16:16–18, 23).

What is the book of Psalms filled with? songs of praise
Who wrote many of the psalms? King David
What was sometimes played with the songs when they were sung? music, musical instruments

- Display *Bible Verses C* and read Psalm 144:9 aloud. Explain that a psaltery was a stringed instrument. Today you will write and sing your own song of praise to God.

PREPARATION FOR READING

- Direct the students to read page 235 and Activities pages 223–24 silently before beginning.

TEACH FOR UNDERSTANDING

- Direct attention to the **Process Skills**. Review the process skills *observe* and *communicate* using information from Chapter 2.
- Direct attention to the **Purpose**.
  What is the purpose of this Exploration? to write a song of praise to God for His creation
- Direct the students to complete the purpose by telling how they will use their songs of praise.
- Direct attention to *Bible Verses C* and read Psalm 147:1 aloud. Explain that "comely" praise is pleasant or beautiful.
- Direct attention to the **Procedure**. Explain that the four-stringed musical instrument made in Lesson 82 will be used to complete Steps 3 and 4.
- You may record and share the students' original songs of praise.
- Discuss the students' **Conclusions**.
  How can you decide if the words of your song of praise are good or bad? Possible answers: if my song can be used to praise God; if my song shows love for God; if my song agrees with the Bible

ACTIVITIES

Exploration: My Song of Praise, pages 223–24

The students will explore by observing God's creation and by communicating a song of praise to God for His creation.

ASSESSMENT

Rubric

Use the prepared rubric or design a rubric to include your chosen criteria.

Teacher Resources

Materials Checklists are available, by chapter, at TeacherToolsOnline.com.

Explaining the Gospel

One of the greatest desires of Christian teachers is to lead children to faith in the Savior. God has called you to present the gospel to your students so that they may repent and trust Christ, thereby being acceptable to God through Christ.

Relying on the Holy Spirit, you should take advantage of the opportunities that arise for presenting the good news of Jesus Christ. Ask questions to personally apply the Ten Commandments to your students (e.g., What is sin? Have you ever told a lie or taken something that wasn't yours? Are you a sinner?). You may also ask questions to discern the child's sincerity or to reveal any misunderstanding he might have (e.g., What is the gospel? What does it mean to repent? Can you do anything to save yourself?). Read verses from your Bible. You may find the following outline helpful, especially when dealing individually with a child.

1. I have sinned (Romans 3:23).
- Sin is disobeying God's Word (1 John 3:4). I break the Ten Commandments (Exodus 20:2–17) by loving other people or things more than I love God, worshiping other things or people, using God's name lightly, disobeying and dishonoring my parents, lying, stealing, cheating, thinking harmful and sinful thoughts, or wanting something that belongs to somebody else.
- Therefore, I am a sinner (Psalm 51:5; 58:3; Jeremiah 17:9).
- God is holy and must punish me for my sin (Isaiah 6:3; Romans 6:23).
- God hates sin, and there is nothing that I can do to get rid of my sin by myself (Titus 3:5; Romans 3:20, 28). I cannot make myself become a good person.

2. Jesus died for me (Romans 5:8).
- God loves me even though I am a sinner.
- He sent His Son, Jesus Christ, to die on the cross for me. Christ is sinless and did not deserve death. Because of His love for me, Christ took my sin on Himself and was punished in my place (1 Peter 2:24a; 1 Corinthians 15:3; John 1:29).
- God accepted Christ's death as the perfect substitution for the punishment of my sin (2 Corinthians 5:21).
- Three days later, God raised Jesus from the dead. Jesus is alive today and offers salvation to all. This is the gospel of Jesus Christ: He died on the cross for our sins according to the Scriptures, and He rose again the third day according to the Scriptures (1 Corinthians 15:1–4; 2 Peter 3:9; 1 Timothy 2:4).

3. I need to put my trust in Jesus (Romans 10:9–10, 13–14a).
- I must repent (turn away from my sin) and trust only Jesus Christ for salvation (Mark 1:15).
- If I repent and believe in what Jesus has done, I am putting my trust in Jesus.
- Everyone who trusts in Jesus is forgiven of sin (Acts 2:21) and will live forever with God (John 3:16). I am given His righteousness and become a new creation, with Christ living in me (2 Corinthians 5:21; Colossians 1:27).

If a child shows genuine interest and readiness, ask, "Are you ready to put your trust in Jesus and depend on only Christ for salvation?" If he says yes, then ask him to talk to God about this. Perhaps he will pray something like the following:

> God, I know that I've sinned against You and that You hate sin, but that You also love me. I believe that Jesus died to pay for my sin and that He rose from the dead, so I put my trust in Jesus to forgive me and give me a home with You forever. In Jesus' name I pray. Amen.

Show the child how to know from God's Word whether he is forgiven and in God's family (1 John 5:12–13; John 3:18). Encourage the child to follow Jesus by obeying Him each day. Tell the child that whenever he sins, he will be forgiven as soon as he confesses those sins to God (1 John 1:9).

Biblical Worldview Shaping Scope and Sequence

Worldview Topic	CH 1	CH 2	CH 3	CH 4	CH 5	CH 6
Where did our world come from? (Origins)	R		E, E			
Why do things work the way they do? (Design in Nature)	R		E, F	R, E	A, E, E	
Who are we? Why are we here? (Human Element)	R	E			EV	A
Why is science important? (Purpose of Science)	E, E					

Worldview Topic	CH 7	CH 8	CH 9	CH 10	CH 11	CH 12
Where did our world come from? (Origins)	E, EV, A			A		E
Why do things work the way they do? (Design in Nature)		E, F			E	F
Who are we? Why are we here? (Human Element)			E			E, EV, E, A, F
Why is science important? (Purpose of Science)			EV			E, E, EV, E

Above are the biblical worldview themes that are important for first grade science students to know.

Early in the course students will recall and explain these themes. However, as these themes are repeated, the students will evaluate ideas within them, formulate a biblical understanding of them, and apply what they have learned about these themes to real-life situations. High levels of internalization are expected wherever students are required to apply their learning.

KEY

R: Recall biblical teaching/science details

E: Explain biblical teaching/science concepts

EV: Evaluate controversial concepts

F: Formulate a biblical understanding of a controversial concept

A: Apply this biblical understanding to life

Getting Started

Scheduling

Science 1 provides material for a one-semester program. Assessment and review days are included, and some lessons have been allotted two days for completion. The instruction of all textual information, Investigations, Explorations, and STEM activities will adequately fill a half-year science program. Consider the following while planning the schedule.

- The chapters should be taught in order. Vocabulary, concepts, skills, and reading difficulty build throughout the book.
- This one-semester course may be taught on alternate days or for two quarters.
- The lesson plans are designed for 20–30 minute lessons. The text pages may be read as part of the lesson or prior to the lesson at home or in class. Adjust the presentation of the Preparation for Reading sections to precede the students' assigned reading schedule.
- Some Exploration lessons will be introduced in class and then assigned to be completed outside of class.
- Many of the Explorations and additional activities can be adjusted for use in learning centers.

Science Notebook

It is recommended that each student keep a three-ring notebook for science. The Activities pages are three-hole punched to accommodate this suggestion. A notebook will allow the students to organize Instructional Aids, Study Guides, Investigations, Explorations, STEM activities, and other useful information.

A section of the notebook should also serve as a science journal for recording notes, drawings, observations, and thoughts about concepts and activities.

Teaching a Text Lesson

The teacher is the key to unlocking students' understanding. The Teacher Edition provides a variety of resources to customize instruction for each student's needs.

The introduction of each lesson provides a short activity or relevant questions that focus on the main topic. Most lessons cover four pages of the Student Edition. The Preparation for Reading section should be presented before the students begin to read the text. The students should then read the material silently on their own. Assign the students to read the text as part of the lesson or prior to the lesson at home or in class. You may find it beneficial to read the text to the students until they have the ability to do so on their own. Oral reading of the text should be reserved for short passages only as they relate to the discussion.

The main part of each lesson plan provides questions to help you and the students interact as you identify students' level of comprehension of the material they have read. This discussion provides an auditory reinforcement of key ideas. Higher-level thinking questions, marked with the gear icon ( ), help the students apply the information that they have read by relating it to previous knowledge and everyday situations.

Background information and activities located at the bottom of the lesson pages provide additional material for use with the lesson based on the class dynamics and available time. An activity may reinforce or extend a concept taught in the lesson. Enrichment information and activities may also be used to apply the science concepts to other subjects.

Materials

The materials needed to teach *Science 1* can be found at TeacherToolsOnline.com and in the Teacher Edition in the Materials section at the beginning of each lesson.

The Materials List at TeacherToolsOnline.com specifies by chapter the materials needed to teach the entire program.

School supplies typically found in a student's desk (Bible, pencil, paper, glue stick, scissors, etc.) are not usually listed in the materials list.

Teacher Resources

The Teacher Resources lists the Instructional Aids and Visuals used in teaching the lesson. These resources can be found in the back of the Teacher Edition or at TeacherToolsOnline.com. Instructional Aids are pages designed to be used by the students during a lesson. Visuals will be displayed by the teacher for use during the lesson.

Vocabulary

The Vocabulary section lists the key terms for the lesson. These terms are bold in the Student Edition. The terms and their definitions will be reviewed on the Study Guides in each Activities chapter.

Preparation for Reading

Reading for information is an essential skill for reading success. The Preparation for Reading section prepares the students for reading in the following ways.

- Read and pronounce together vocabulary and other words that may slow the students' reading fluency. Definitions are not discussed at this time to allow the students to gain an understanding of the meanings from sentence context.
- Emphasize skills used for reading in the content area by directing the students' attention to headings, captions, diagrams, feature boxes, and boldfaced and italicized words as aids to understanding the text and as sources of additional information.
- The reading for each lesson is usually divided into two-page segments. One focus statement is provided to direct the students as they read.

Teach for Understanding

In the Teach for Understanding section, you will find questions and comments to help you evaluate the students' understanding of concepts. The questions listed are not exhaustive but are a springboard to help you begin the discussion.

- A gear icon ( ) is used to help you identify the higher-level thinking questions. The answers to a higher-level thinking question are not taken directly from the pages being discussed, thus requiring some analysis by the students. Supply any prompts or background needed to guide the students to answer these questions.
- Develop in the students the expectation of supporting yes/no responses with an explanation.

Some lessons incorporate brief activities with questions to provide hands-on opportunities to aid students' understanding.

Charts, graphs, photos, illustrations, and diagrams are forms of communication that enhance students' reading and understanding. Guide the students in understanding, interpreting, and using these features.

Graphic organizers are tools to aid the students' understanding. The organizers can help the students see the organization of facts and details. Organizers may be completed during the discussion or as reinforcement following the teaching. They may also be used to help the students take a closer look at the information by making comparisons.

Teaching Investigation, Exploration, and STEM Lessons

It is important to emphasize the value of the Investigation, Exploration, and STEM lessons in SCIENCE 1. In order for the students to truly know and understand science, they must be able to use it. Knowing when and how to apply science is the basic premise behind the term *scientific literacy*.

The Investigation, Exploration, and STEM lessons allow the students to demonstrate their understanding of science concepts through hands-on activities. However, these activities also allow the students to apply knowledge from other subject areas, such as measurement skills from math, and writing and communication skills from Language Arts.

Many activities in SCIENCE 1 are open-ended. These require the students to incorporate basic science process skills. Because the goal of these activities is to teach a mental process as opposed to specific procedures, the activities may not follow an exact format. An activity may have more than one correct result.

The Explorations, STEM activities, and some of the other activities are project oriented. They sometimes require the students to work outside of class and usually require more creativity than the activities that involve basic experiments. While doing Explorations and STEM activities, the students will use and develop skills such as designing solutions to problems, writing, and sharing orally with a group or partner.

Most of the Investigation, Exploration, and STEM lessons can be completed by students working in groups. Using science groups helps students learn cooperation and management skills as they collaborate. Brainstorming ideas among group members is usually a good problem-solving technique. As they handle materials and interact with one another, the students also demonstrate what they are learning about biblical worldview.

Groups for Investigations and Explorations

Place students in groups of three or four to allow for maximum participation.

A science group should be a mixture of students from all achievement levels. You may find that students who are not good "book learners" are much better at doing hands-on activities.

At first, you may choose to assign tasks to each group member to ensure active participation by all. Later in the year, the students within each group can decide the tasks of each member. Tasks may include handling materials, measuring, recording information, and communicating results.

Groups for STEM Activities

Using groups for STEM Activities may require extra input from you. Because the STEM lessons are less structured, students may have a difficult time assigning themselves appropriate tasks. However, since these lessons often require a high level of creativity, students may be less intimidated in a group setting.

Management Tips

Being prepared is the key to a successful activity. A printable copy of the Materials Checklist is available, by chapter, at TeacherToolsOnline.com. Materials are also listed, by lesson, in the Teacher Edition and the Activities book. You will need to determine specific quantities based on the number of science groups or students you have.

Organize materials needed for group Investigations, Explorations, and STEM Activities ahead of time. Place the materials needed on a nonbreakable tray or in a nonbreakable container. Distributing and collecting materials in this way will save time and will help with cleanup.

Keep newspaper or plastic on hand to cover work surfaces and buckets or containers in which to dispose of waste material. It is helpful to keep paper towels or rags readily available.

Review and Test Lessons

Included in the Activities book are Study Guide pages identified with a colored Review tab. Each chapter has Study Guide pages to help the students review the information in smaller segments. One Study Guide in each chapter includes a Write About It application question that requires a longer written response.

A Review lesson is provided for each chapter. Use of the suggested game or an alternate game from the Game Bank from the Teacher Resources will review vocabulary and concepts in a fun way.

The Review lesson also gives the teacher an opportunity to check that each student has accurately completed the Study Guides for the chapter. The material for each chapter assessment is taken from the Study Guide pages. A student who understands the material (and does not just know the answers) covered on the review pages will be adequately prepared to take the assessment.

The assessment day has been given its own lesson number. There is no material assigned to be taught in this lesson. (*Note:* The assessments and keys must be purchased separately.)

Science Process Skills

Science is not just a collection of facts. It is also a demonstration of processes that shows understanding of how science works. Each Investigation and Exploration lesson in *Science 1* emphasizes at least one basic process skill in the discussion. The students will use some of the other skills during each activity as well.

Although process skills are emphasized in the Investigation and Exploration lessons, they are also used in lesson discussions, STEM activities, and other activity lessons. The basic science process skills are the skills emphasized in the *Science 1* activities.

Observe	use the senses to gather information regarding objects and events
Measure Using Numbers	use standard devices and techniques to quantify information
Infer	draw conclusions based on previous knowledge or observation
Classify	group or order objects based on similarities or differences
Predict	forecast an expected result based on prior experience or knowledge
Communicate	use written, oral, or graphic means to transmit information to others

Most of the integrated science process skills are introduced in *Science 1*. While they will not be mastered, you may choose to emphasize the teaching of some of these skills in preparation for their use in later grades.

Hypothesize	formulate a statement that can be tested by experimentation
Identify and Control Variables	recognize the changing and unchanging factors in an investigation and adjust one factor to obtain data
Define Operationally	explain an object or event properly but in terms of each student's observation and experience
Experiment	set up and follow a procedure to test a hypothesis
Collect, Record, and Interpret Data	gather information about objects and events in an organized and systematic manner
Make and Use Models	create a physical or mental representation to explain or clarify ideas, objects, or events

Game Bank

Baseball Challenge

Identify four areas in the classroom to use as bases. Divide the class into two teams and flip a token to determine which team bats first. Ask a question to the first student, or "batter." If the batter gets the question right, he may proceed to first base. Players advance to the next base each time a batter answers a question correctly. Correct answers to difficult questions could allow the batter to advance more bases. A "run" is scored when a batter gets to home base.

If the batter answers incorrectly, someone from the opposing team (the outfield) may answer the question. If the outfielder answers correctly, the batter is out. If the outfielder's answer is incorrect, the batter receives a second chance. Three incorrect answers from the batter equal an out. When a team has received three outs, the teams switch sides.

Basketball

Divide the class into two teams. Each team should choose a spokesperson. Give the teams time to make up several questions about the lesson or chapter. They must know the correct answers to the questions.

The teams will take turns asking each other questions. The team may discuss the answer to a question, but the final answer should come from the team's spokesperson.
If the team answers correctly, it receives two points. If desired, the team may also get a chance to make a "basket" by shooting a foam ball into a trash can or other container.

Concentration

Divide the class into two teams. Display a grid with various point values. There should be two of each value. Cover the point values with consecutively numbered squares. Ask a review question. If a student answers correctly, he may choose two squares. If he finds two of the same point value, he may add those points to his team's score. If he does not find a match, play switches to the other team.

Ducks and Decoys

This game is best used with a multiple choice review. Designate the four corners of the classroom as A, B, C, and D. Count out enough blank index cards so that you have one for each student. Label three-fourths of the index cards "duck" and the remaining cards "decoy." Give one card to each student. The students should not tell others whether they are ducks or decoys. As the review questions are read, ducks should go to the corner that corresponds with the answer they believe is correct. Decoys should intentionally pick a corner that does not have the correct answer. This will encourage students to think of the answer for themselves and not to "follow the flock."

Football

Display a football field with yard lines and end zones marked. Divide the class into two teams. Use a token or marker to represent each team. Place the markers at the 50-yard line. Decide which team goes first. This team receives the first question. If the team answers correctly, its marker advances 10 yards. The opposing team's marker also moves to the new mark (losing ground). If the answer is incorrect, the markers stay in the same place. When two consecutive questions are answered incorrectly, the ball switches to the other team. The second team now receives the questions.

When a team advances its marker to the end zone, a touchdown is scored (6 points). The team has the option of receiving an additional question for an extra point while in the end zone. After a touchdown, place the team markers back at the 50-yard line. The team that did not score the touchdown should now receive the questions.

Four in a Row

Display a grid of five horizontal lines and five vertical lines. Divide the class into two teams: Xs and Os. As team members answer the review questions correctly, they may place their team symbol in a section of the grid. The first team to get four symbols in a row wins.

Jump Start

Group the class into teams. Provide a "jump" chair at the front of the room for each team. The first jumper from each team should sit in his team's chair, keeping both feet flat on the floor and his back against the chair. After the question is read, the seated students who know the answer should jump to their feet and remain standing. The first student to jump up and give the correct answer receives points for his team. Rotate jumpers after each question.

Mix and Match

Group the class into teams. Write several questions on strips of paper. Write the answers to those questions on separate strips of paper. Place the questions in one container and the answers in another container. Mix up the papers. Draw a question from the container and read it aloud. Choose a student to draw an answer. The student should determine whether the answer matches the question he heard. If it does match, the student receives a point for his team. If it does not match, he has the option of stating the correct answer. If he can give the correct answer, he receives a point for his team. Questions and answers should be placed back into their respective containers before the next question is drawn.

Puzzling Questions

Group the class into teams. Write several questions on strips of paper. Write the answers on separate strips of paper. Mix up the papers and give each team a set of questions and corresponding answers. At a given signal, each team should start organizing its papers and matching up the questions with the correct answers. The first team to display all the questions and answers correctly wins.

Instructional Aids

Bookmarks

Copy on cardstock and cut out. Distribute one pair to each student.

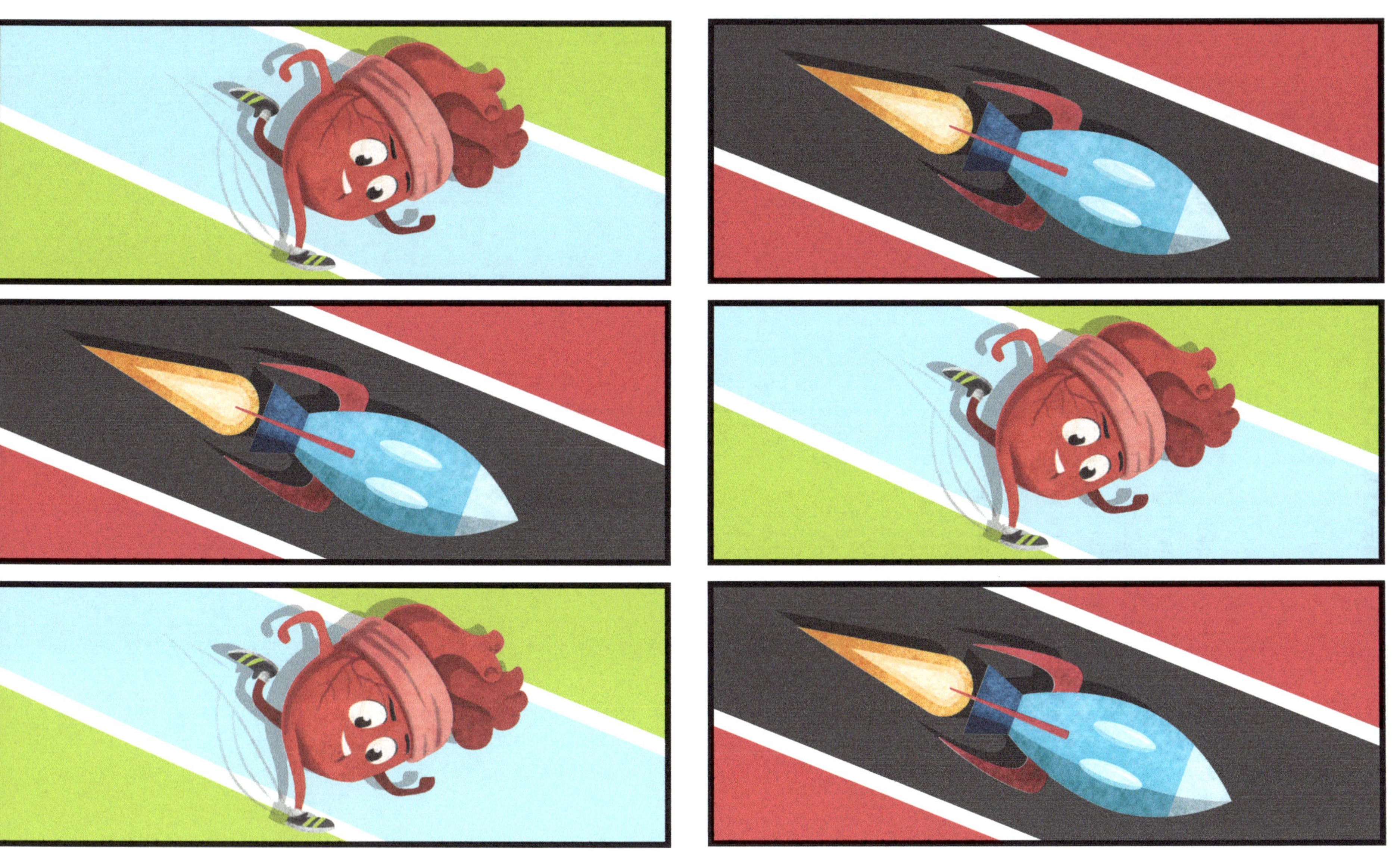

STEM
The Engineering Design Process

Instructional Aid 1.2
Introduced in Lesson 1

Dress the Scientist

Copy on cardstock and cut out, one set for each team.

Science Process Skills

Observe | to use your senses to find out something

Classify | to group things that are alike

Measure | to find the size, number, or amount of something

Infer | to use what you know to tell why things happen

Predict | to make a careful guess at what may happen after observing; using what you know to say what may happen

Communicate | to share what you know; to give information about something

Steps
of the
Scientific Method

Problem

1 An investigation begins with a problem.
The **problem** is a question that needs to be answered.

Hypothesis

2 A **hypothesis** is a possible answer to the problem.
The investigation tests the hypothesis to find whether it is true.

Materials

3 The **materials** are the supplies needed to do the investigation.

Procedure

4 The **procedure** is the steps of an investigation.
The procedure must be followed carefully.

Observations

5 **Observe** what happens using the five senses.
Scientists record what they observe in charts, lists, or pictures.

Conclusions

6 The **conclusions** are the answer to the test.
They tell what was learned from the investigation.

Instructional Aid 2.2
For use with Lesson 13

Parent Letter

Dear Parents,

Your child will need a paper grocery bag for ________________________________,
Day

________________________________. It will be helpful if you would cut the paper
Date

bag according to the directions below.

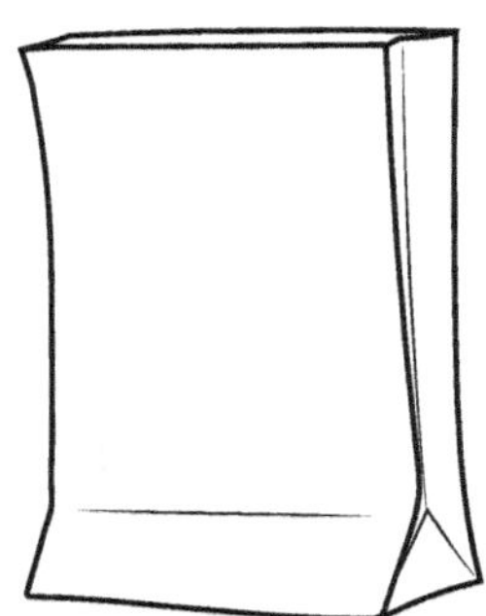

Thank you,

Directions

1. Cut out both sides of the paper bag.

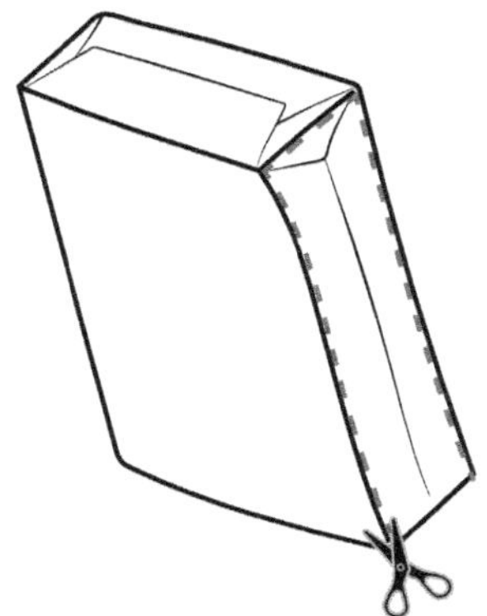

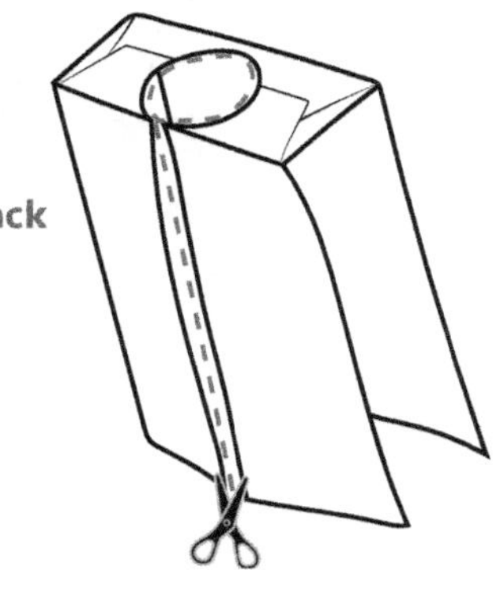

2. Cut a slit up the back of the paper bag for ease in putting the paper vest on.

3. Cut a hole for your child's head in the bottom of the paper bag.

4. You will have a paper bag vest for your child.

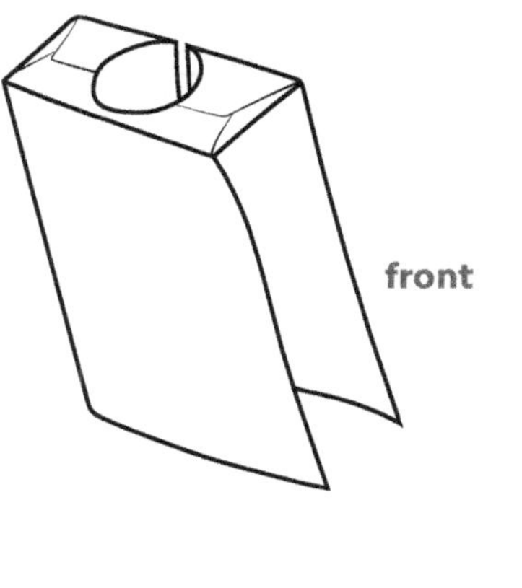

Science 1 • *Teacher Resources*

Instructional Aid 5.1
For use with Lesson 32

Hand Washing Poster

Name _______________________

1 Wet your hands.

2 Add soap.

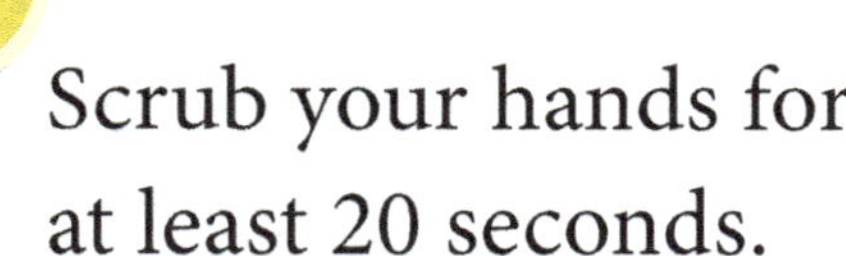

3 Scrub your hands for at least 20 seconds.

4 Rinse your hands.

5 Dry your hands.

Fire Safety Cards

Color each picture. Cut out each card.

Do not put clothes on top of lamps.

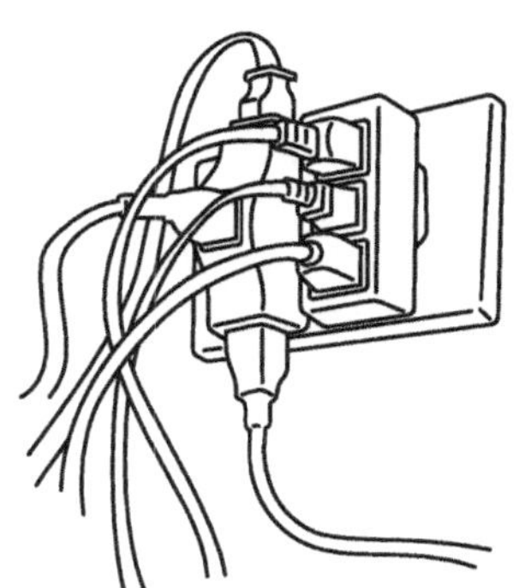

Do not plug too many things into an outlet.

Keep potholders and towels away from the stove.

Do not lay an iron face down on an ironing board.

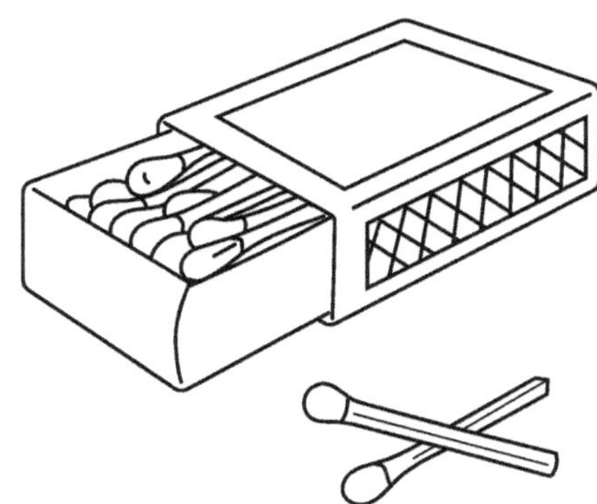

Do not touch matches.

Do not put stuffed animals or towels on top of a heater.

Visuals

Bible Verses A

Genesis 1:1

In the beginning God created the heaven and the earth.

Genesis 1:26–28

26 **And God said, Let us make man in our image, after our likeness: and let them have dominion over the fish of the sea, and over the fowl of the air, and over the cattle, and over all the earth, and over every creeping thing that creepeth upon the earth.**
27 **So God created man in his own image, in the image of God created he him; male and female created he them.**
28 **And God blessed them, and God said unto them, Be fruitful, and multiply, and replenish the earth, and subdue it: and have dominion over the fish of the sea, and over the fowl of the air, and over every living thing that moveth upon the earth.**

Bible Verses B

Psalm 139:14

I will praise thee;
for I am fearfully and wonderfully made:
marvellous are thy works;
and that my soul knoweth right well.

Proverbs 20:12

The hearing ear,
and the seeing eye,
the Lord hath made even both of them.

Mark 12:30–31

30 And thou shalt love the Lord thy God
with all thy heart,
and with all thy soul,
and with all thy mind,
and with all thy strength:
this is the first commandment.
31 And the second is like, namely this,
Thou shalt love thy neighbour as thyself.
There is none other commandment greater
than these.

Is It Science? A

Is It Science? B

Worldview

Dinosaurs and people lived together.
Dinosaurs lived a long time before people lived.
God created the world in six days.
The world began over a million years ago.
Only parts of the Bible are true.
There was a flood that covered all the earth.

Worldview K-L Chart

Where did our world come from?	
Know	**Learned**

Why do things work together the way they do?	
Know	**Learned**

Who are we? Why are we here?	
Know	**Learned**

Why is science important?	
Know	**Learned**

Visual 1.6
For use with Lessons 5–6

A Biblical Worldview

Using My Senses to Observe

Sight	The color of the pretzel is ____.	yellow brown
	Its shape is ____.	curvy round square straight
Touch	The pretzel feels ____.	hard soft
Smell	The pretzel smells like ____.	dough fruit
Hearing	I can find out how a pretzel sounds when I ____.	shake it break it
	The sound it makes is a ____.	snap boom
Taste	The pretzel tastes ____.	salty sour

Cause and Effect

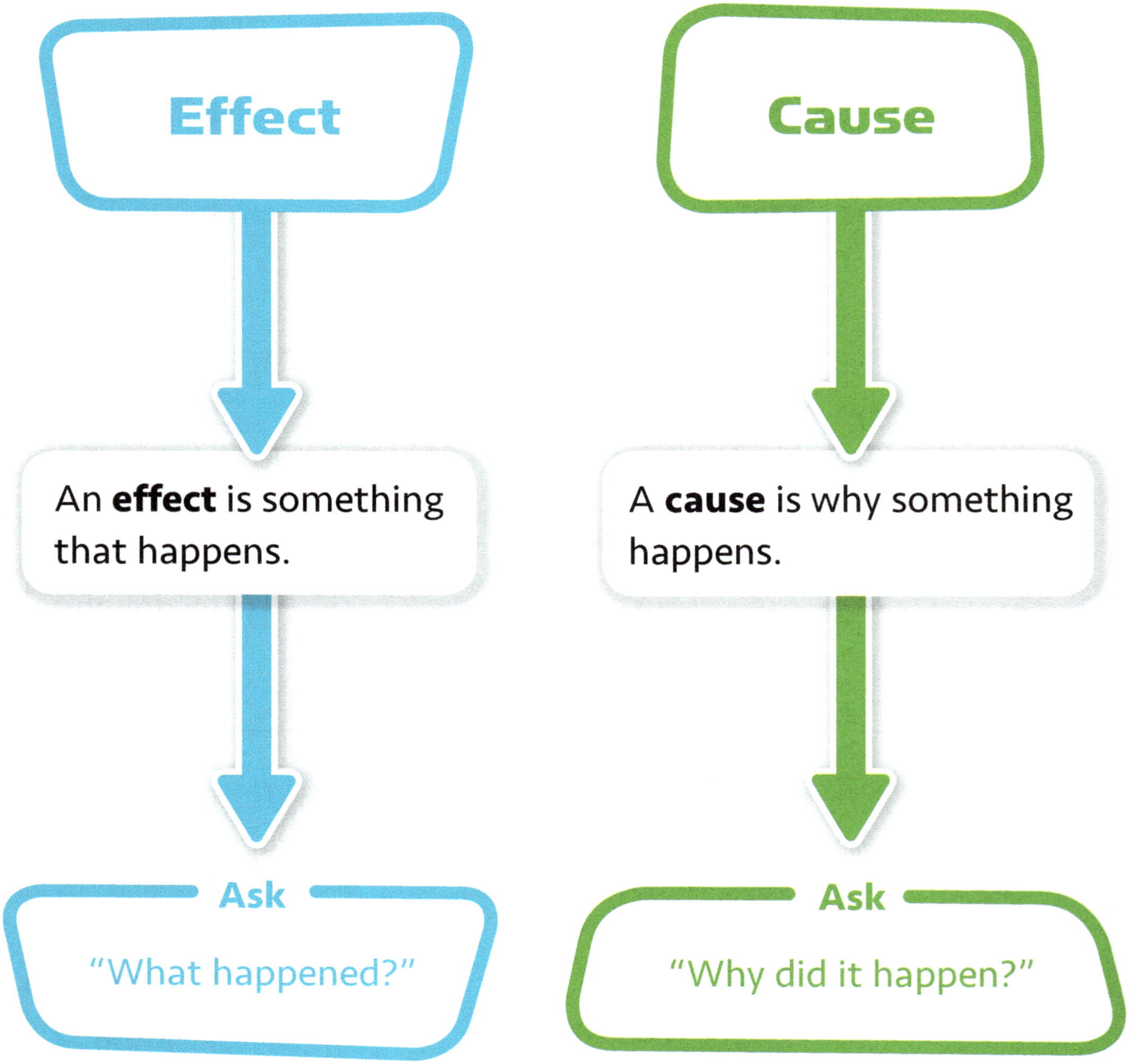

Effect

Cause

An **effect** is something that happens.

A **cause** is why something happens.

Ask

"What happened?"

Ask

"Why did it happen?"

Bible Verse

Genesis 1:28

And God blessed them,
and God said unto them,
Be fruitful, and multiply,
and replenish the earth,
and subdue it:
and have dominion over
the fish of the sea, and
over the fowl of the air, and
over every living thing
that moveth upon the earth.

Science Safety Tips

1. **Listen to your teacher's directions.**

2. **Use tools the way they were made to be used.**

3. **Handle supplies with care.**

4. **Wear safety goggles when needed.**

5. **Tell your teacher about accidents.**

6. **Wait your turn.**

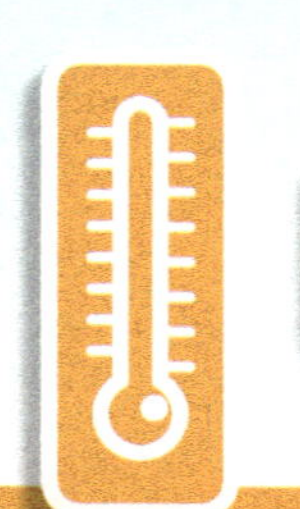

Visual 2.4
For use with Lesson 12
Photo: Sergey Novikov/Shutterstock.com

Using Science Tools Chart

Tool	My Measurements	Teacher's Measurements
ruler	_________ cm	_________ cm
measuring cup	_________ mL	_________ mL
balance scale	_________ g	_________ g
thermometer	_________ °C _________ °F	_________ °C _________ °F

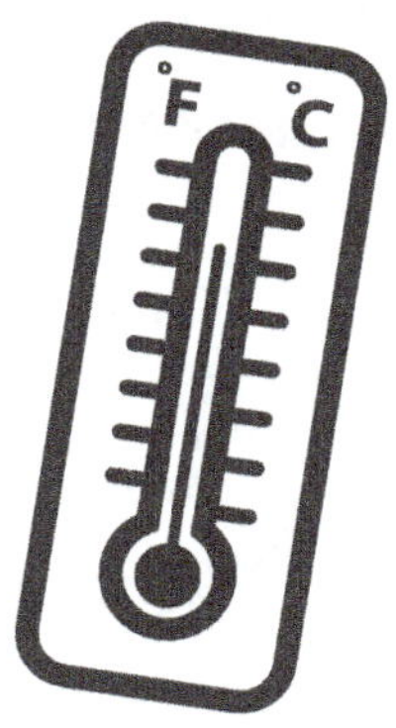

STEM
The Engineering Design Process

Bible Verses A

11 And God said, Let the earth bring forth grass, the herb yielding seed, and the fruit tree yielding fruit after his kind, whose seed is in itself, upon the earth: and it was so.
12 And the earth brought forth grass, and herb yielding seed after his kind, and the tree yielding fruit, whose seed was in itself, after his kind: and God saw that it was good.
13 And the evening and the morning were the third day.

Bible Verses B

17 And unto Adam he said, Because thou hast hearkened unto the voice of thy wife, and hast eaten of the tree, of which I commanded thee, saying, Thou shalt not eat of it: cursed is the ground for thy sake; in sorrow shalt thou eat of it all the days of thy life;

18 Thorns also and thistles shall it bring forth to thee; and thou shalt eat the herb of the field;

Plant Parts Song

Sing to the tune of "The Muffin Man."

We hold the plant in the ground,
In the ground, in the ground.
We take in water all around.
We are called the roots.

We move water through the plant,
Through the plant, through the plant.
We move water through the plant.
We are called the stems.

Our job is to make the food,
Make the food, make the food.
Our job is to make the food.
We are called the leaves.

We stay busy making seeds,
Making seeds, making seeds.
We stay busy making seeds.
We are called the flowers.

Life Cycle of a Pumpkin Plant

Adult Tomato Plant and Seedling Venn Diagram

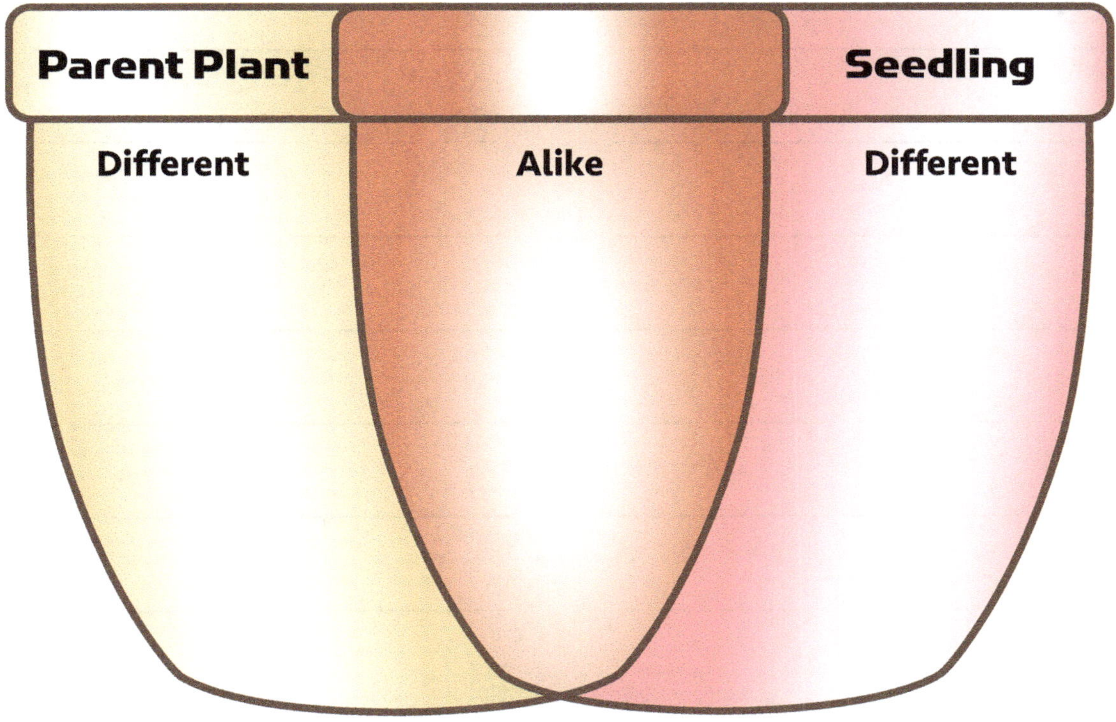

Adult Roses T-Chart

<table>
<tr><th>Alike</th><th>Different</th></tr>
<tr><td>________________________</td><td>________________________</td></tr>
<tr><td>________________________</td><td>________________________</td></tr>
<tr><td>________________________</td><td>________________________</td></tr>
<tr><td>________________________</td><td>________________________</td></tr>
<tr><td>________________________</td><td>________________________</td></tr>
<tr><td>________________________</td><td>________________________</td></tr>
</table>

Bible Verses

29 And God said, Behold, I have given you every herb bearing seed, which is upon the face of all the earth, and every tree, in the which is the fruit of a tree yielding seed; to you it shall be for meat.
30 And to every beast of the earth, and to every fowl of the air, and to every thing that creepeth upon the earth, wherein there is life, I have given every green herb for meat: and it was so.

A righteous man regardeth the life of his beast: but the tender mercies of the wicked are cruel.

Be thou diligent to know the state of thy flocks, and look well to thy herds.

Elephant

Classify Animals

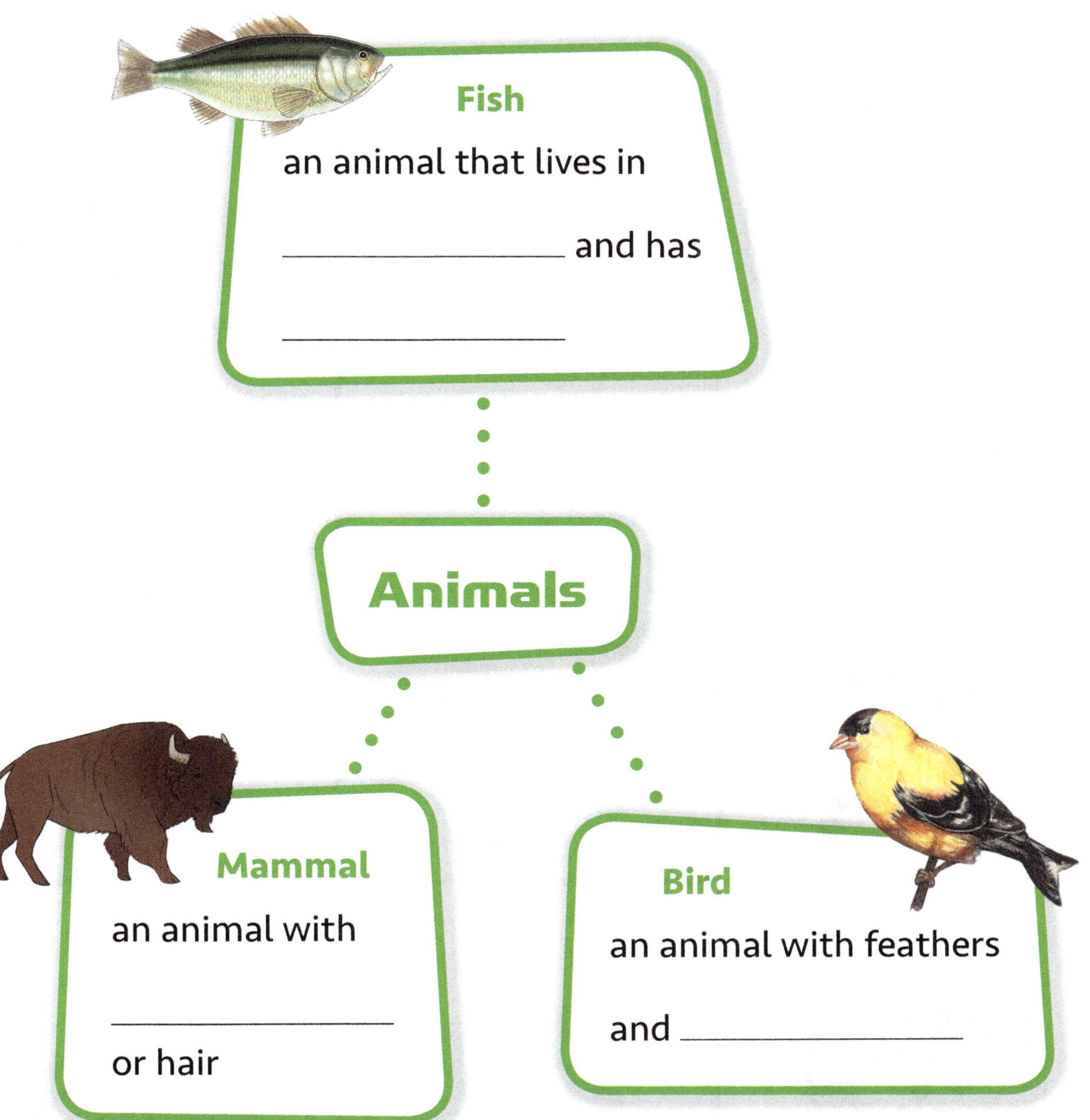

Animal Cards

Bible Verses A

Genesis 1:26–27

26 And God said, Let us make man in our image, after our likeness: and let them have dominion over the fish of the sea, and over the fowl of the air, and over the cattle, and over all the earth, and over every creeping thing that creepeth upon the earth.
27 So God created man in his own image, in the image of God created he him; male and female created he them.

Genesis 2:7

And the Lord God formed man of the dust of the ground, and breathed into his nostrils the breath of life; and man became a living soul.

Visual 5.1
For use with Chapter 5

Bible Verses B

Genesis 2:21–22

21 And the LORD God caused a deep sleep to fall upon Adam, and he slept: and he took one of his ribs, and closed up the flesh instead thereof;
22 And the rib, which the LORD God had taken from man, made he a woman, and brought her unto the man.

Psalm 139:14

I will praise thee; for I am fearfully and wonderfully made: marvellous are thy works; and that my soul knoweth right well.

Animals and People
Venn Diagram

People **Animals**

Different (People)

- are created in

  God's _____________

- _____________ with
  God and each other

- God _____________
  with us

- can _____________
  God and each other

- will live somewhere

  ____

Alike

- are _____________
  things

- have the same

  ____

Different (Animals)

- are not created in

  God's _____________

- do not _____________
  with God

- God does not

  _____________ with them

- do not _____________ God

- will not live somewhere

  ____

Path of Food

_______ The food travels down a tube.

_______ Your teeth chew food.

_______ The food reaches your stomach.

_______ The stomach breaks down the food.

_______ You swallow the broken-up food.

Path of Food Key

_3_ The food travels down a tube.

_1_ Your teeth chew food.

_4_ The food reaches your stomach.

_5_ The stomach breaks down the food.

_2_ You swallow the broken-up food.

Model of Butterfly

Copy the parts of the butterfly onto orange cardstock or construction paper.
Glue the wings onto a clothespin.
Attach two short pieces of chenille wire for the antennae.

Bible Verses A

Genesis 1:26–27

26 And God said, Let us make man in our image, after our likeness: and let them have dominion over the fish of the sea, and over the fowl of the air, and over the cattle, and over all the earth, and over every creeping thing that creepeth upon the earth.
27 So God created man in his own image, in the image of God created he him; male and female created he them.

Proverbs 16:24

Pleasant words are as an honeycomb, sweet to the soul, and health to the bones.

Bible Verses B

Matthew 7:12

Therefore all things whatsoever ye would
that men should do to you, do ye even so
to them: for this is the law and the prophets.

Mark 12:31

And the second is like, namely this,
Thou shalt love thy neighbour as thyself.
There is none other commandment greater
than these.

Ephesians 4:32

And be ye kind one to another,
tenderhearted, forgiving one another,
even as God for Christ's sake hath
forgiven you.

Hand Washing Song

Sing to the tune of "Row, Row, Row Your Boat"

Wash, wash, wash your hands,
Wash your hands all clean!
Carefully, carefully, carefully, carefully
Wash your hands all clean.

Wash, wash, wash your hands
Often every day!
Carefully, carefully, carefully, carefully
Wash those germs away.

Tooth Decay

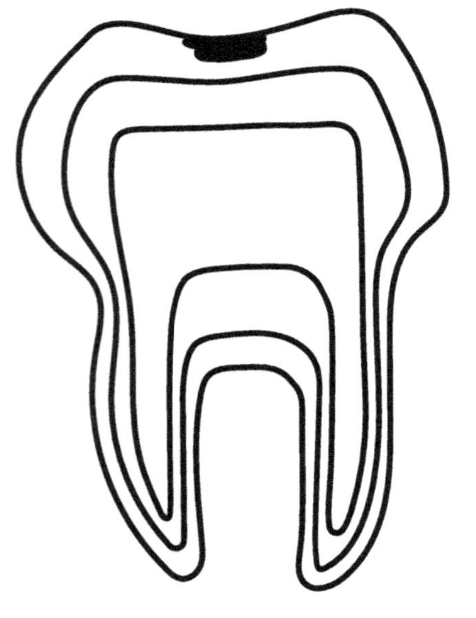

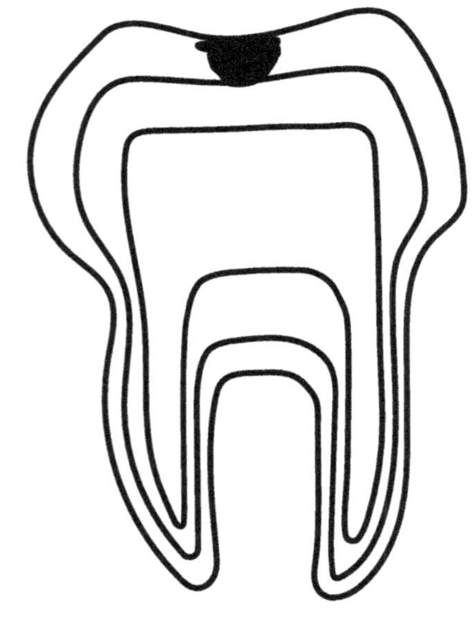

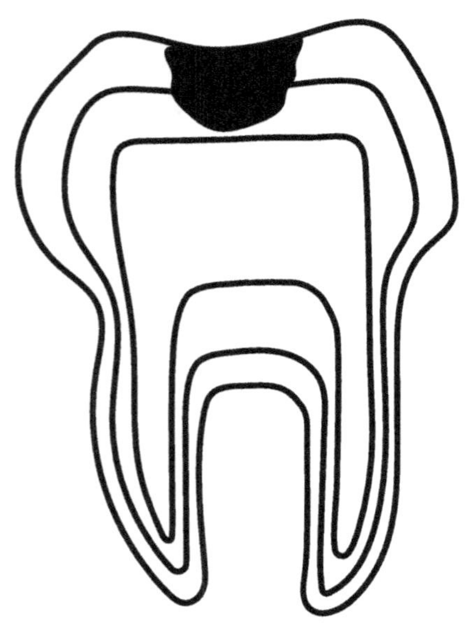

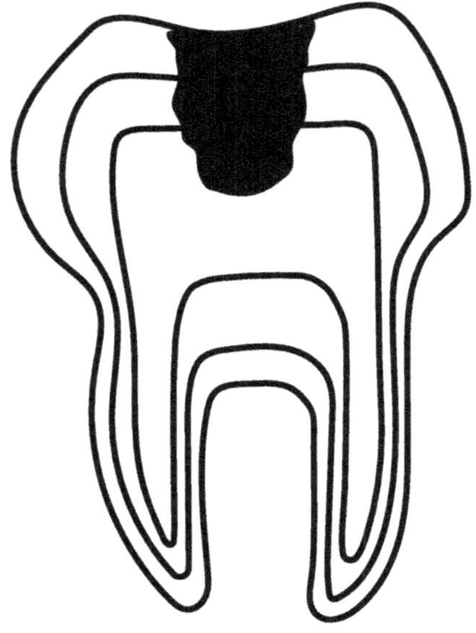

Safety Song

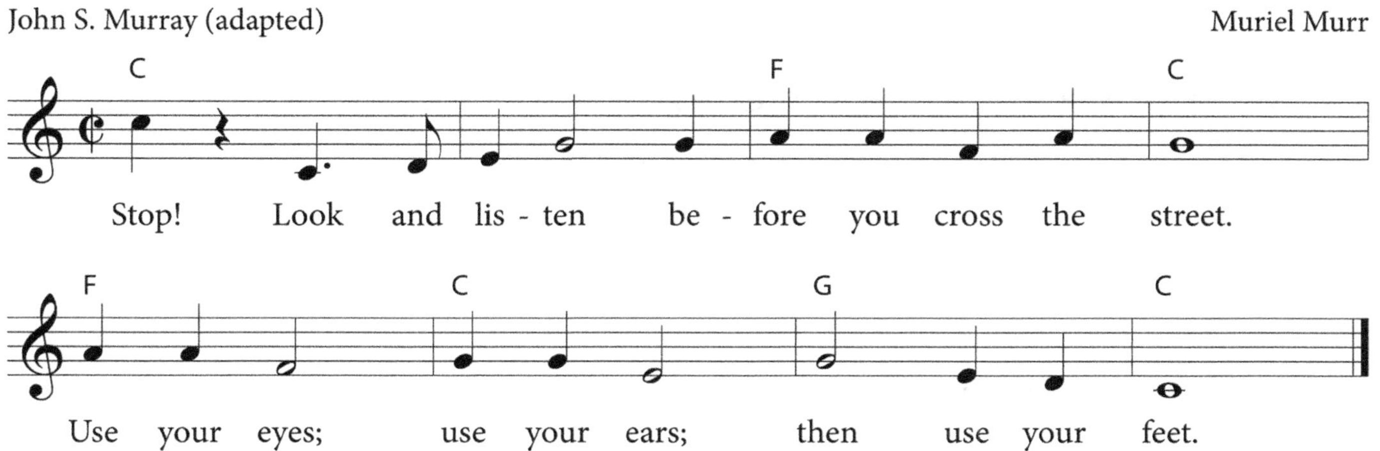

Music © 1980 BJU Press. All rights reserved.

Bible Verses

Genesis 1:1

In the beginning God created the heaven and the earth.

Genesis 1:16

And God made two great lights; the greater light to rule the day, and the lesser light to rule the night: he made the stars also.

Bible Verse

Genesis 8:22

While the earth remaineth, seedtime and harvest, and cold and heat, and summer and winter, and day and night shall not cease.

God's Four Seasons

Jan Joss and Ann Larson

Nicole Strickland

Words and music © 2004 BJU Press. All rights reserved.

Daylight and Temperature A

Season	Months	Temperature			Hours of Daylight		
		cooler	warmer	hottest	shorter	longer	longest
winter	December January February	_____ °F _____ °C		coldest	About 9 hours	shortest	
spring	March April May	_____ °F _____ °C	___________		About 13 hours	___________	

Daylight and Temperature B

Season	Months	Temperature			Hours of Daylight		
		cooler	warmer	hottest	shorter	longer	longest
summer	June July August	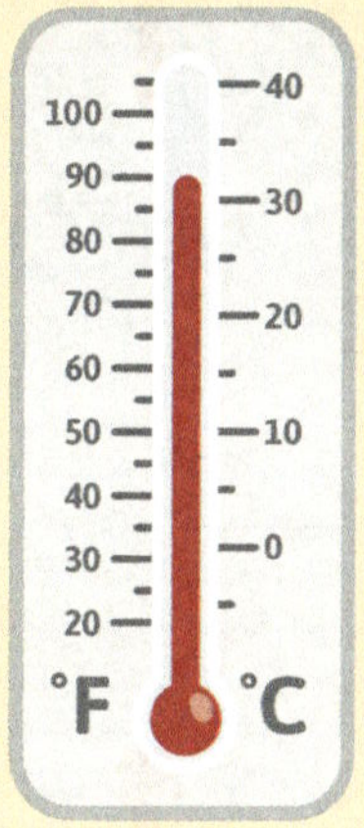 _______ °F _______ °C	____________		About 14 hours	____________	
fall	September October November	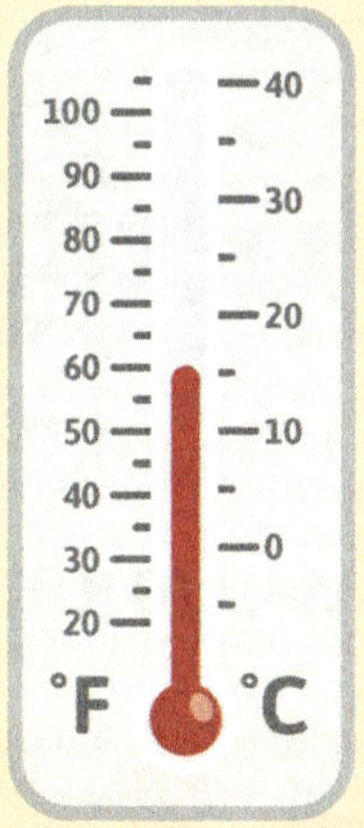 _______ °F _______ °C	____________		About 11 hours	____________	

A Time for Everything

Season	Temperature	Daylight Hours	Time
fall	coldest	longer	growing
spring	cooler	longest	harvest
summer	hottest	shorter	planting
winter	warmer	shortest	resting

	Season	Temperature	Daylight Hours	Time
winter	comes after ________	temperature is ________	daylight is ________	time of ________
spring	comes after ________	temperature is ________	daylight is ________	time of ________
summer	comes after ________	temperature is ________	daylight is ________	time of ________
fall	comes after ________	temperature is ________	daylight is ________	time of ________

Bible Verses A

Genesis 9:13–15

[13] I do set my bow in the cloud, and it shall be for a token of a covenant between me and the earth.
[14] And it shall come to pass, when I bring a cloud over the earth, that the bow shall be seen in the cloud:
[15] And I will remember my covenant, which is between me and you and every living creature of all flesh; and the waters shall no more become a flood to destroy all flesh.

Psalm 135:7

He causeth the vapours to ascend from the ends of the earth; he maketh lightnings for the rain; he bringeth the wind out of his treasuries.

Proverbs 22:3

A prudent man foreseeth the evil, and hideth himself: but the simple pass on, and are punished.

Bible Verses B

Matthew 16:2–3*a*

[2] He answered and said unto them, When it is evening, ye say, It will be fair weather: for the sky is red.
[3a] And in the morning, It will be foul weather today: for the sky is red and lowring.

John 3:16

For God so loved the world, that he gave his only begotten Son, that whosoever believeth in him should not perish, but have everlasting life.

John 17:17

Sanctify them through thy truth: thy word is truth.

Bible Verse

**And God said,
Let there be light:
and there was light.**

Bible Verses

Psalm 139:14

I will praise thee; for I am fearfully and wonderfully made: marvellous are thy works; and that my soul knoweth right well.

Proverbs 20:12

The hearing ear, and the seeing eye, the LORD hath made even both of them.

Bible Verses A

Genesis 1:26–28

26 And God said, Let us make man in our image, after our likeness: and let them have dominion over the fish of the sea, and over the fowl of the air, and over the cattle, and over all the earth, and over every creeping thing that creepeth upon the earth.

27 So God created man in his own image, in the image of God created he him; male and female created he them.

28 And God blessed them, and God said unto them, Be fruitful, and multiply, and replenish the earth, and subdue it: and have dominion over the fish of the sea, and over the fowl of the air, and over every living thing that moveth upon the earth.

Genesis 3:17–18

17 And unto Adam he said, Because thou hast hearkened unto the voice of thy wife, and hast eaten of the tree, of which I commanded thee, saying, Thou shalt not eat of it: cursed is the ground for thy sake;

Bible Verses B

in sorrow shalt thou eat of it all the days
of thy life;
18 Thorns also and thistles shall it bring
forth to thee; and thou shalt eat the herb
of the field;

Psalm 8:1

O LORD our Lord, how excellent is thy
name in all the earth! who hast set thy
glory above the heavens.

Psalm 8:3–4

3 When I consider thy heavens, the work
of thy fingers, the moon and the stars,
which thou hast ordained;
4 What is man, that thou art mindful of
him? and the son of man, that thou
visitest him?

Psalm 8:9

O LORD our Lord, how excellent is thy
name in all the earth!

Psalm 101:3a

I will set no wicked thing before mine eyes:

Bible Verses C

Psalm 144:9

I will sing a new song unto thee, O God: upon a psaltery and an instrument of ten strings will I sing praises unto thee.

Psalm 147:1

Praise ye the LORD: for it is good to sing praises unto our God; for it is pleasant; and praise is comely.

Matthew 7:12

Therefore all things whatsoever ye would that men should do to you, do ye even so to them: for this is the law and the prophets.

Matthew 22:37–39

37 Jesus said unto him, Thou shalt love the Lord thy God with all thy heart, and with all thy soul, and with all thy mind.
38 This is the first and great commandment.
39 And the second is like unto it, Thou shalt love thy neighbour as thyself.

Bible Verses D

John 3:16

For God so loved the world, that he gave his only begotten Son, that whosoever believeth in him should not perish, but have everlasting life.

John 17:17

Sanctify them through thy truth: thy word is truth.

Romans 3:23

For all have sinned, and come short of the glory of God;

Philippians 4:8

Finally, brethren, whatsoever things are true, whatsoever things are honest, whatsoever things are just, whatsoever things are pure, whatsoever things are lovely, whatsoever things are of good report; if there be any virtue, and if there be any praise, think on these things.

Light and Sound at Home and School

Object of Communication	Light	Sound	Can science tell me if it is good or bad?	Can the Bible tell me if it is good or bad?
			yes	yes
			no	no
			yes	yes
			no	no
			yes	yes
			no	no
			yes	yes
			no	no
			yes	yes
			no	no

Light and Sound in My Community

Object of Communication	Light	Sound	Can science tell me if it is good or bad?	Can the Bible tell me if it is good or bad?
			yes no	yes no
			yes no	yes no
			yes no	yes no
			yes no	yes no
			yes no	yes no

Visual 12.6
For use with Lesson 86

Biblical Worldview A

designed	Fall	perfect
praise	Redemption	

1 Where did our world come from?

God created a _______________________ world in six days.

We are to _______________________ God for making the world.

2 Why do things work together the way they do?

The parts of the earth work together because God

_______________________ them to work together.

Weeds, pain, sickness, and death happened because of

the _______________________.

God will make this world _______________________ again.

Everyone will live with God forever if they have

_______________________ through Jesus Christ.

Biblical Worldview B

3 Who are we? Why are we here?

The most important part of God's creation is _______________.

We are to love _______________.

We are to love people because they are made in God's _______________.

We are to take _______________ of the earth.

4 Why is science important?

Science can be a powerful _______________ to help us answer questions.

Science is good at answering some _______________.

Some questions can be answered by _______________ science and the Bible.

Other questions can be answered only by the _______________.

When science disagrees with the Bible, the _______________ is right.

Index

engineer, 15, 19, 43, 158
engineering design
    process, 16, 43
environment, 68–69,
    70–71, 75, 91, 92, 158
eyes, 8, 76, 86, 96

F

Fall, 85
fall, 151, 153, 161–162
fish, 72–73, 76–77, 79
Flood, 179
flower, 6, 156–157
    definition, 59
    plant part, 54–55, 62–
        63, 86
full moon, 143–144

G

Galilei, Galileo, 138
germs, 113–117

grow, 5–6
    animals, 66, 70–71, 75,
        76, 79, 86, 155
    living things, 48, 61, 91
    people, 90, 92, 96, 98,
        100, 106–107
    plants, 46–47, 50, 54,
        56–59, 60, 64

H

habit, 112–115, 118–123
hail, 175
hand lens, 32
hands, 99, 117
head, 95, 96, 100, 105
healthy, 112–115, 118–
    122
hearing, 8, 11, 201–213
heart, 104, 106
human body, 90–107,
    108–111
hypothesis, 39–40, 42, 60

safe, 71, 75, 82, 100, 108,
  118–122, 123, 188,
  215, 222
science, 3–6, 15, 17, 21,
  38–42, 43, 185–186,
  227, 232–234
science process skills,
  24–30
  classify, 26, 72, 189
  communicate, 30, 216
  infer, 28, 196, 212
  measure, 27, 31, 33–36,
    168, 212
  observe, 25, 28–29, 37,
    60, 145, 163, 178, 196,
    212, 235
  predict, 29, 60, 176–
    178, 189, 196
science tools, 31–36, 37
scientific method, 38

scientist, 13–14, 23–24,
  31, 33–34, 36, 38, 72
  Bacon, Francis, 21
  Carver, George
    Washington, 64
  Copernicus, Nicolaus,
    148
  Galilei, Galileo, 138
  kind of, 75, 176
  worldview, 17–21
season, 146–163
seed, 59, 61–63, 156
seedling, 61–63
senses, 2, 8–12, 13, 25,
  41, 95, 176
shadow, 193–195
shelter, 6, 71, 77, 79,
  82, 92
Shepard, Alan, 141
sight, 8
sleet, 174
smell, 8–9, 25, 76, 78, 98

T

Teacher Edition Index

Activities Features, x–xi
additional activity lessons. *See under* enrichment
assessment, xiv, 14
    rubrics, 41, 47, 66, 71, 95, 105, 117, 129, 135, 150, 159, 168, 181, 199, 211, 218, 236, 237, 252, 261
    tests, 25, 49, 73, 97, 119, 137, 161, 183, 201, 223, 239
Background information
    Adam's Rib, 103
    Animal Shelters in the Bible, 79
    Arctic Cod, 85
    Asthma, 115
    Balance Scale, 39
    Bicycle Helmet, Fitting a, 130
    Big Bang, 144
    Black Bear Mother and Cub Hibernating, 173
    Booster Seat, 131
    Bullhorn, 233
    Cactus, Barrel, 61
    Classification of People, 83
    Constellations, 152
    Dolphins, Bottlenose, 74
    Edible Plant Parts, 63
    Feathers, 81
    Fish, Largest, 84
    Gills and Lungs, 84
    Gospel, Explaining the, 122, A1
    Grams or Pounds, 39
    Hearing, 230
    Heart Rate, 115
    Hibernation, 173
    Honey, 123
    Image of God, 122
    Mammal Species, 83
    Meteorologist, 196–97
    Migration, 180
    Moon, Men on the, 155
    North Star, 153
    Oxygen, 81
    Phases of the Moon, 157
    Quarter Moons, 158
    Ribs, 103
    Scales, 81
    Science Notebook, 2
    Sharks and Dolphins, 84
    Skull, 112
    Sleet and Hail, 194
    Sound Waves, 229
    Spine, 112
    Spring Equinox, 174
    Sun
        at noon, 149
        distance, 147
        light, 147
    Telescopes, 152
    Temperature, Daily Changes, 189
    Vitelline Masked Weaver, 85
    Weather Balloon, 196
    Weather Reporting and Green Screens, 196
    Windsurfing, 190
    Worldview Terms, 21
    Zinnia, 61
biblical worldview, iv, 4, 19, A3
    Creation, 5, 13–15, 20–21, 53–54, 58–59, 64, 65, 70, 77, 103–4, 122–23, 141, 143, 205–9, 231, 254, 257, 261
    Curse, 21, 255
    dominion (rule over), 8, 16, 21–23, 40, 47, 79, 257
    evolution, 144
        Big Bang, 144
    Fall, 21, 59, 71, 93, 255
    Flood, 21, 198
    God's design, 60, 62, 64–65, 70, 78, 81, 84, 86, 94–95, 102–4, 117, 163, 166, 173, 230
    God's image, 8, 19–20, 22, 26, 103, 122–23, 254
    praise, 14, L25, 105, 110, 115, 121, 230–31, 250, 254, 261
    Redemption, 21, 256
    Scope and Sequence, A3
biomimicry, 94
cause and effect. *See under* reading skills
Creation. *See under* biblical worldview
Creation Corner. *See under* interest boxes
Curse. *See under* biblical worldview
design, God's. *See under* biblical worldview
diagrams
    brain, 115
    clouds, 193
    Earth Tilts, 166
    heart and lungs, 114
    Inside Your Body, 116
    Life Cycle of a Pumpkin, 69
    Outside Your Body, 109–10
    Parts of the Head, 106
    Path of the Moon, 156
    Path of the Sun, 148–49
    Phases of the Moon, 157–58
    plant parts, 61
    revolving earth, 153
    Robin's Life Cycle, 88–89
    Rotating Earth, 147
    Seasons, 153
    Shadows, 217
    Shape of the Moon, 157
    STEM Engineering Design Process, 18
    Water Cycle, 192
    See also reading skills
dominion. *See under* biblical worldview
enrichment
    additional activities by title
        Apple Decay, 120
        Bird Sounds, 81
        Build an Animal Shelter, 78
        Classifying Pretzels, 30
        Classroom Garden, 53
        Count Breaths, 114
        Different Leaves, 172
        Distinguishing Smells, 11
        Find Your Heart Rate, 114
        Germs and Hand Washing, 126
        Grow Leaves, 64
        Grow Roots from a Stem, 63
        How Big Is the Sun?, 146
        How Far Can Germs Spread?, 126
        How Much Light Passes Through?, 212
        Hunt for Flowers (online activity), 65
        I Spy with My Little Eye, 29
        Identify Main Needs, 78
        Identify Parts of a Plant, 61
        Identify Shelters, 78
        Identify Tastes, 12
        Light Travels in a Straight Line, 221
        Make a Paper Fan, 58
        Make a Source of Light Display, 208
        Make Faces, 12
        Make Shadows, 215
        Modeling the Phases of the Moon, 158
        Model the Cycle of the Seasons, 167
        Observe Birds with Webcams, 88
        Parents and Offspring, 90
        Partly Cloudy or Cloudy Day Picture, 193
        "Plant Parts" Song, 62
        Play "I Spy," 11
        Play "Simon Says," 113
        Practice Communicating, 36
        Reflecting Light, 155
        See Your Image, 22
        Senses Song, 13
        Show How Bones Protect the Body, 112

Photo Credits

Key: (t) top; (c) center;
(b) bottom; (l) left; (r) right;
(bg) background; (i) inset

Cover
front Toucan © iStock.com/edurivero; Rusty gears © iStock.com/GLYPHstock; Caterpillar © iStock.com/Captainflash; **back** © iStock.com/Mustafa Kocabasi

Unit Openers
1l Butterfly Hunter/Shutterstock.com; 1 (wing) Sukpaiboonwat/Shutterstock.com; 1 (magnifying glass) © iStock.com/kyoshino; 44–45 Milan Zygmunt/Shutterstock.com; 88–89 Puwadol Jaturawutthichai/Shutterstock.com; 124–125 Lambros Kazan/Shutterstock.com; 180–181 Thinkstock/Stockbyte/Thinkstock

Chapter 1
2 Janie Airey/Digital Vision/Getty Images; 4 Cathy Yeulet /123RF; 5 PhotoDisc, Inc.; 6 ALPA PROD/Shutterstock.com; 7 Monkey Business Images/Shutterstock.com; 8t Jiri Hera/Shutterstock.com; 8b photosbysuzi /Bigstock.com; 9 Mila May/Shutterstock.com; 10t PhotoDisc/Getty; 10bl Seregam /Shutterstock.com; 10br vitec/Shutterstock.com; 11l Getty Images/Thinkstock; 11r Getty Images/Hemera/Thinkstock; 13t Darren Baker/Shutterstock.com; 13c © iStock.com /BartCo; 13b © iStock.com/bluecinema; 14 © iStock.com/StefaNikolic; 15t © iStock.com /nicolas_; 15b Blue Jeans Images/Media Bakery 17 "Spinophorosaurus holotype" by Remes K, Ortega F, Fierro I, Joger U, Kosma R, et al. (2009) A New Basal Sauropod Dinosaur from the Middle Jurassic of Niger and the Early Evolution of Sauropoda. PLoS ONE 4(9): e6924.doi:10.1371/journal.pone.0006924 /Wikimedia Commons/CC By 2.5; 19 © iStock.com/wragg; 20 © iStock.com/vgajic; 21 © iStock.com/picture

Chapter 2
22–23 Chutima Chaochaiya/Shutterstock.com; 24l © iStock.com/3sbworld; 24c grafvision /Shutterstock.com; 24r–25l © iStock.com /emholk; 25c © iStock.com/AfricaImages; 25r Asia Images Group/Shutterstock.com; 26, 27b, 29, 35, 39, 40 both, 42 BJU Press; 27t © iStock.com/Ivantsov; 27ct, 33t, 37t © piai - Fotolia.com; 27cb Fedor Bobkov /Shutterstock.com; 28 © iStock.com /ayala_studio; 30 © iStock.com/FatCamera; 31 Elnur/Shutterstock.com; 32 Violart /Shutterstock.com; 33t © piai - Fotolia .com; 33b © iStock.com/RichVintage; 34, 37bc Raland/Shutterstock.com; 36, 37bl VladisChern/Shutterstock.com; 37br © iStock .com/kyoshino

Chapter 3
46–47 Walter Quirtmair/Moment/Getty Images; 50 Photographer Chris Archinet /Moment Select/GettyImages; 51 taka1022 /Shutterstock.com; 52tl anne willette /Shutterstock.com; 52tr Hurst Photo /Shutterstock; 52bl Moko Photography /Shutterstock.com; 52br © 2009 JupiterImages Corporation; 53tr Tamara Kulikova /Shutterstock.com; 53i © iStock.com/Kazakov; 53 © iStock.com/Difydave; 54 Anton Foltin /Shutterstock.com; 55 BJU Photo Services; 56 Zaitsava Olga/Shutterstock.com; 57t Monkey Business Images/Shutterstock.com; 57b © iStock.com/wingmar; 58tl domnitsky /Shutterstock.com; 58tr © iStock.com /joakimbkk; 58c Noah Strycker/Shutterstock .com; 58bl Blue Jeans Images/Media Bakery; 58br © iStock.com/undefined undefined; 59t harmpeti/Shutterstock.com; 59b © iStockphoto.com/IgorDutina; 60 BJU Press;

© iStock.com/ryasick; **222t** Westend61/Jaak Nilson/Media Bakery; **222b** Chuck Eckert /Alamy Stock Photo; **223t** Eddie J. Rodriquez /Shutterstock.com; **223c** fivespots /Shutterstock.com; **223b** H. Mark Weidman Photography/Alamy Stock Photo; **224** BJU Photo Services; **224i** © iStock.com/dszc; **225t** Carolina K. Smith MD/Shutterstock.com; **225ct** OgnjenO/Shutterstock.com; **225cb** Shannon Alexander/Shutterstock.com; **225b** © iStock .com/kevinjeon00; **229** © iStock.com/wragg; **231** Andrey_Popov/Shutterstock.com; **232** © iStock.com/Steve Debenport; **233t** Goran Bogicevic/Shutterstock.com; **233c** Syda Productions/Shutterstock.com; **233b** Romolo Tavani/Shutterstock.com; **234** robert_s /Shutterstock.com

Glossary
237t Apollo 14 Shepard/Nasa/Wikimedia Commons/Public Domain; **237c** BJU Press; **237b** Sebastian Kaulitzki/Shutterstock.com; **239t** michaeljung/Shutterstock.com; **239c,** **240t** Littlekidmoment/Shutterstock.com; **239b** Francescomoufotografo/Shutterstock.com; **240c** Anton Petrus/Shutterstock.com; **240b** Lambros Kazan/Shutterstock.com; **241t** © iStock.com/ kyoshino; **241b** u3d/Shutterstock.com; **243t** © iStock.com/Ivantsov; **243b, 247t** © piai - Fotolia.com; **244** iStockphoto/Thinkstock; **245t** © iStock.com/SolStock; **245b** © iStock.com /cunfek; **246** both BJU Photo Services; **247c** Elnur/Shutterstock.com; **247b** Imageman /Shutterstock.com; **248t** © iStock.com/ra-photos; **248b** Getty Images/Hemera/Thinkstock; **249t** Tetra Images/Media Bakery; **249b** PaulPaladin /Shutterstock.com; **250t** VladisChern /Shutterstock.com; **250b** BIGANDT.COM /Shutterstock.com